Legitimizing Science

Axel Jansen, PD Dr., has taught American history in the US and in Europe, currently at Eberhard-Karls-Universität Tübingen. *Andreas Franzmann,* PD Dr., teaches in the Department of Sociology at Johann-Wolfgang-Goethe Universität Frankfurt. *Peter Münte,* Dr., has been an assistant professor at the University of Bielefeld.

Axel Jansen, Andreas Franzmann, Peter Münte (eds.)

Legitimizing Science

National and Global Publics (1800–2010)

Campus Verlag
Frankfurt/New York

This volume has been printed with subsidies from:
Dr. Bodo Sponholz Stiftung für Wissen, Kunst und Wohlfahrt
Vereinigung der Freunde der Universität Tübingen (Universitätsbund) e.V.

Distribution throughout the world except Germany, Austria and Switzerland by
The University of Chicago Press
1427 East 60th Street
Chicago, IL 60637

Bibliographic Information published by the Deutsche Nationalbibliothek.
The Deutsche Nationalbibliothek lists this publication in the Deutsche Nationalbibliografie;
detailed bibliographic data are available in the Internet at http://dnb.d-nb.de

ISBN 978-3-593-50487-2 Print
ISBN 978-3-593-43262-5 E-Book (PDF)

Table of Contents

Section III: Legitimizing Fields of Investigation

Section IV: Global Science

Acknowledgments

This collection of essays emerged from a workshop organized by the editors at Eberhard-Karls-Universität Tübingen (Germany) in September 2013, except for Fabian Link's contribution on the Frankfurt School of Sociology, which we subsequently asked him for. The arrangement of papers on the nineteenth and twentieth centuries and their focus on the natural sciences as well as the social sciences reflect our overall approach, first, to retain an investigation of science within general sociology and history, and second, to retain a comprehensive view of curious investigation that is represented by the natural sciences, the social sciences, and the humanities together.

The workshop at Universität Tübingen on "Science and the Public in the Nation-State: Historic and Current Configurations in Global Perspective, 1800–2010" took place in the context of a research project by Andreas Franzmann and Axel Jansen on "Professionalization and Deprofessionalization in the Public Context of Science since 1970." The project is co-hosted by the Department of History at the University of California, Los Angeles (UCLA), and we would like to thank our colleagues in both institutions for their kind support. We thank the Volkswagen Foundation for sponsoring the research project that provides the intellectual backdrop for this book, and we also thank the sponsor of our conference and of the publication of this book, the Vereinigung der Freunde der Universität Tübingen (Universitätsbund) e.V., and our second sponsor for this publication, the Dr. Bodo Sponholz Stiftung für Wissen, Kunst und Wohlfahrt. Lars Weitbrecht provided organizational support at the conference and Ian Copestake of slovos.com helped proofread manuscripts. We thank both of them, and also Jürgen Hotz at the Campus Verlag for facilitating this book.

Section I:
Approaches

Legitimizing Science: Introductory Essay

Andreas Franzmann, Axel Jansen and Peter Münte

1. The Continuing Dependence of Science on a Plurality of Political Communities

The pursuit of science requires legitimacy that science itself cannot provide. The most obvious reason why such legitimacy is required today is that science costs a lot of money. At an accelerating pace during the nineteenth and twentieth centuries, scientists have had to raise funds to cover salaries and apparatus at institutions such as academies, universities or research institutes. But science has needed legitimacy, even at times when science was run by experimental scientists not employed to do research but pursue such interests on the side. Then as now, investigating nature by asking unfamiliar questions requires resources but also protection, freedom from political or religious constraints, the leisure to tackle fundamental problems without obvious practical value, and authorization through cultural and political affirmation. All of these matters point to the issue of legitimacy, and in the context of the modern nation-state such legitimacy relates to a political public and its endorsement.[1] At a time of increased

---

1 The relationship between science and public has caused a great amount of interest in the last two or three decades. See, for example, Steven Shapin, "Science and the Public," in *Companion to the History of Modern Science*, edited by Robert C. Olby et al. (London: Routledge, 1990) 990–1007. For the debate in Germany, see Peter Weingart, *Die Wissenschaft der Öffentlichkeit: Essays zum Verhältnis von Wissenschaft, Medien und Öffentlichkeit* (Weilerswist: Velbrück, 2005). This interest seems to arise from debates on the role of science in society. Concerning this connection, see Peter Münte's essay in this volume. In the fields of history, the relationship between science and the public has become an important topic in the context of attempts to reintegrate the history of science with general history. See, for example, Rüdiger vom Bruch, *Wissenschaft, Politik und öffentliche Meinung: Gelehrtenpolitik im Wilhelminischen Deutschland 1890–1914* (Husum: Matthiesen, 1980); Rüdiger vom Bruch, "Wissenschaft im Gehäuse: Vom Nutzen und Nachteil institutionengeschichtlicher Perspektive," *Berichte zur Wissenschaftsgeschichte* 23 (2000), 37–49.

global interdependencies, furthermore, this raises the issue of whether the legitimacy of science is shifting to a transnational and global plane.

The need for a legitimacy of science has been particularly evident in times of conflict. In the past, opponents of an experimental approach to testing truth claims have represented the church, cultural *Weltanschauungen*, or political ideologies. Conflicts have tended to unfold when the results of research questioned conventional explanations. Galileo, Kepler, Darwin, and Freud are prominent examples in the history of science.[2] At the beginning of the twenty-first century, debates on cloning and on stem cells are a reminder that science continues to be associated with provocations to world views and ethical convictions.[3] Such debates challenge politics to balance the demands arising from such beliefs with competing demands for scientific freedom and economic opportunities. While we have come to accept and demand from science technological innovation relevant for the economy and for society's other needs, science has remained a potential source of cultural, political, and economic instability. Hence this particular mode of truth-seeking continues to require the kind of protection, promotion, and authorization for which science has sought the political sovereign's patronage since early modern times. Science claims to work out a collectively binding understanding of the world. This presupposes a general acceptance of science as the source of such knowledge and the continuous integration of such knowledge in general education and political decision-making.[4]

From the Renaissance and into our own time, political, cultural, and economic elites have played a key role in shielding the experimental sciences from religious or cultural attack and in supporting and transferring authority to them. Such protection, promotion and authorization has been granted by elites in the emerging context of the modern state, but also through private philanthropy or foundations that have provided essential support. Their decision to support research often reflected a broader na-

---

2 Joseph Ben-David, "The Ethos of Science in the Context of Different Political Ideologies and Changing Perceptions of Science," in *Scientific Growth* (Berkeley: University of California Press, 1991), 533–59.

3 See Axel Jansen, "Stem Cell Debates in an Age of Fracture," in this volume.

4 As in most debates in the sociology and history of science, the focus here is on the kind of science that evolved into the empirical or natural sciences that were institutionalized in Europe from the seventeenth century. Fabian Link's contribution to this volume demonstrates, however, that similar questions may well be asked with respect to the social sciences, in general, and with respect to a critical theory of society, in particular.

tional commitment to the role of science in society.[5] By supporting research financially or by endorsing such work symbolically, they bestowed public affirmation and significance on the larger scientific enterprise. Today, the principles of this approach have become relevant in all areas of political leadership and administration that touch on scientific knowledge. The relationship between the state and science has not merely served to protect science but also to endorse its particular commitment to establishing truth-claims on behalf of a wider community. Such an endorsement of science has become an important element in national cultures and their self-perception. For scientists, public affirmation of their work has translated into cultural prestige and leverage.

The emerging legitimacy of science may be studied with particular effectiveness by focusing on a period when its social and political position remained unsettled. The founding of the Royal Society in seventeenth-century England provides a well-known case in point.[6] After the Puritan interregnum, a small group of natural philosophers including Robert Boyle was able to commit the returning king to provide patronage and his seal for the founding of a scientific organization. The king's protection and endorsement of the Royal Society implied that after its founding period in the 1660s, no one else could lay claim to discovering the laws of nature in the name of the king and of the nation he represented. But Charles II had to leave it to the Royal Society's active nucleus to define experimental philosophy because the king himself could not provide that definition. The Royal Society used this privilege to establish principles of scientific activity, among them the rule that claims to findings had to be established through experiments among witnesses, that experiments had to be recorded, and that results were to be transferred to the Society's records. While a general endorsement of such principles would not take place for decades or even centuries, important norms of modern science had been recognized by an official institution representing the king, norms that otherwise would not have had the standing that they came to have. Without official endorsement such principles would have remained subject to fundamental ques-

5 Joseph Ben-David, *The Scientist's Role in Society; a Comparative Study* (Englewood Cliffs, NJ: Prentice-Hall, 1971).

6 Michael Hunter, *Establishing the New Science: The Experience of the Early Royal Society* (Woodbridge: Boydell & Brewer, 1995); Peter Münte, *Die Autonomisierung der Erfahrungswissenschaften im Kontext frühneuzeitlicher Herrschaft: Fallrekonstruktive Analysen zur Gründung der Royal Society*, 2 vols. (Frankfurt: Humanities Online, 2004).

tions concerning their relevance, validity, and authorship. Science would not have been protected against philosophical and theological attacks on experimental methods, and demands that they be replaced by other methods such as philosophical introspection or revelation. Charles II had delegated the power to define science as a mode of truth-seeking through experiment-based philosophy, and the Royal Society assumed responsibility for this particular set of universalistic principles shared by those committing themselves to the scientific project.

While the Royal Society's founding context was distinctly British, it remains of significance well beyond this particular state. The Royal Society raised a standard of aspiration for experimental philosophers in other countries and they soon sought to emulate that model. The *Académie royale des sciences* established similar principles for France, effectively adopting the aspirations for scientific achievement and the responsibility for protecting and enhancing this particular mode of investigating nature. The Paris academy served this role even though the state kept it on a much shorter leash, paying researchers a salary and charging them with official state business.[7] The British and the French institutions have provided a template for other countries and their histories suggest that the institutionalization of experiment-based science took place by association with a political sovereign.[8]

For science to unfold, it had to be embedded in a particular community through political representatives who bestowed legitimacy on this particular mode of testing ideas. Such a community, of course, is always particular and not universal, because it is bound to a concrete country with its own territory and history. An essential tension exists, therefore, between the universalistic endeavor of science (a generalized methodology aiming at a universal validity of research results), on the one hand, and political communities, on the other.[9]

---

7 Roger Hahn, *The Anatomy of a Scientific Institution: The Paris Academy of Sciences, 1666–1803* (Berkeley: Univ. of California Press, 1971); Peter Münte, "Institutionalisierung der Erfahrungswissenschaften in unterschiedlichen Herrschaftskontexten. Zur Erschließung historischer Konstellationen anhand bildlicher Darstellungen," *Sozialer Sinn* 1 (2005): 3–44.

8 James E. McClellan, *Science Reorganized: Scientific Societies in the Eighteenth Century* (New York: Columbia Univ. Press, 1985).

9 Brigitte Schroeder-Gudehus, "Nationalism and Internationalism," *Companion to the History of Modern Science*, edited by Robert C. Olby et al. (London and New York: Routledge, 1990), 909–1007.

The rise of science in the wake of its empowerment by the political sovereign since the seventeenth century opens up two key questions. The first concerns the impact on science of significant changes in the legitimacy of political power. How has the role of science shifted during and after political revolutions? What has the role of science been as it carried over from a monarchic or aristocratic state into a democratic nation-state, and what has been the impact of such momentous transformations as the emergence of the public sphere and the rise of mass media in modern democracy? Different assumptions about the role of subjects or citizens within a state's political sphere, for example, surely must have had an effect on the role assumed by science. All of this, of course, points to the more general question of how the history of science relates to political history.

The other question concerns the national and global history of science as different states chose to empower it from the seventeenth century: How has science derived legitimacy from endorsement in some countries while being stifled in others, and how has the legitimacy of science evolved from an association with key supporters such as national political elites, intellectuals, occupations, and industries? Much like China, Brazil, and India in recent decades, France, Britain, Germany, and the United States in previous centuries have all created specific traditions of science funding, lobbying structures, and legitimizing discourses that have impacted public agendas, expectations, and controversies about science policy and the development of science disciplines. While each country's tradition is unique, global dynamics of science emerge on their basis. Among transnational effects of national patterns of science organization are shifts in centers of science, with researchers looking to particular countries or regions for the development and validation of important work.

The present volume provides an opportunity to explore the legitimacy of science historically by taking as a point of departure an assessment of present challenges and problems. Hence this collection of essays does not seek to identify and trace "origins" of modern experimental science—transformations that precede the nineteenth century. This book provides a platform for looking back from the early twenty-first century to identify, chart, and compare developments that have turned out to be important or representative in legitimizing science since 1800. If the authority of science has rested on its endorsement by the political sovereign, what has been the history of that relationship in the age of the modern nation-state?

In this introductory essay, we will proceed by first taking a step back to explain how we became interested in the science-politics nexus. We will then turn to a trend that has come to characterize the relationship between science and the public during the past two centuries: the growing emphasis on the utility of research. A presentation of select historical tokens to illustrate this point will then help prepare the ground for concluding questions on the role and integrity of science in a globalized world.

2. Legitimizing Science as a Profession

In recent years, the editors of this volume have been involved, with Ulrich Oevermann, in helping develop in the history and the sociology of science a revised concept of professionalization.[10] While sociologists of science have focused their investigation on institutions of knowledge production and the cultural formation of scientific knowledge, our interest in the vocation's political legitimacy relates to the pragmatic requirements, the prerequisites, and the specific demands arising from the essence of scientific activity: research.[11]

We begin by asking what goes on when empirical scientists try to make sense of uncharted realities. While this focus to us seems central in identifying the "unnatural nature of science," it has been absent from the recent

---

10 See Ulrich Oevermann, "Theoretische Skizze einer revidierten Theorie professionalisierten Handelns," in *Pädagogische Professionalität. Untersuchungen zum Typus pädagogischen Handelns*, edited by Arno Combe and Werner Helsper (Frankfurt: Suhrkamp, 1996), 70–182; Ulrich Oevermann, "Wissenschaft als Beruf: die Professionalisierung wissenschaftlichen Handelns und die gegenwärtige Universitätsentwicklung," in *Die Hochschule* 14, no. 1 (2005): 15–51; Peter Münte and Ulrich Oevermann, "Die Institutionalisierung der Erfahrungswissenschaften und die Professionalisierung der Forschungspraxis im 17. Jahrhundert: Eine Fallstudie zur Gründung der Royal Society," in *Wissen und soziale Konstruktion*, edited by Claus Zittel (Berlin: Akademie Verlag, 2002), 165–230; Münte, *Autonomisierung der Erfahrungswissenschaften*; Andreas Franzmann, *Die Disziplin der Neugierde: Zum Professionalisierten Habitus in den Erfahrungswissenschaften* (Bielefeld: Transcript, 2012); Axel Jansen, *Alexander Dallas Bache: Building the American Nation through Science and Education in the Nineteenth Century* (Frankfurt and New York: Campus, 2011).

11 For the main paradigms in the sociology of science, see Bettina Heintz, "Wissenschaft im Kontext: Neuere Entwicklungen in der Wissenschaftssoziologie," *Kölner Zeitschrift für Soziologie und Socialpsychologie* 45, no. 3 (1993): 528–52; Uwe Schimank, "Für eine Erneuerung der institutionalistischen Wissenschaftssoziologie," *Zeitschrift für Soziologie* 22, no. 1 (1995); Peter Weingart, *Wissenschaftssoziologie* (Bielefeld: Transcript, 2003).

"practical turn" towards the situational realities of science.[12] In our work, we have come to assume that scientists engaged in research are not involved in solving established puzzles with established tools but that they engage with their curiosity in trying to identify new questions so as to advance their field through resolving them. The demands of their work leads them to develop a particular habitus, a habitus that is shaped by and informs a self-sufficient investigative perspective on a reality that will never conform to evolving theories about it.[13]

This approach offers an alternative to the main paradigms in the sociology of science and an answer to a key question in the sociology of the professions. The classical sociology of the professions could not explain particularly well what distinguishes science and other professions from vocations that are not professionalized.[14] Any explanation that goes beyond an institutional description of vocations claiming professional status would need to show, after all, how such claims are justified (or unwarranted) by pragmatically serving specific needs and responsibilities.

Work on this question has come to conclude that professions are distinct from other vocations in that they engage, not in solving problems by

---

12 Wolpert's perspective is similar to ours. See Lewis Wolpert, *The Unnatural Nature of Science*, (Cambridge: Harvard Univ. Press, 1994). On the "practical turn" in the sociology of science, see Andrew Pickering. ed., *Science as Practice and Culture* (Chicago: Univ. of Chicago Press, 1992); Moritz Epple and Claus Zittel, eds., *Science as Cultural Practice*, vol. 1, *Cultures and Politics of Research from Early Modern Period to the Age of Extremes* (Berlin: Duncker & Humblot, 2010).

13 The term "habitus" is commonly associated with Pierre Bourdieu's work, but we use it to depict the specific attitudes and responses elicited by problem-solving in science. Compare, for example, Pierre Bourdieu and Loïc Wacquant, *An Invitation to Reflexive Sociology* (Chicago: Univ. of Chicago Press, 1992), to Andreas Franzmann, *Die Disziplin der Neugierde: Zum Professionalisierten Habitus in den Erfahrungswissenschaften* (Bielefeld: Transcript, 2012). In referring to a scientific mindset, Max Weber uses the concept of *Geistesaristokratie* ("intellectual aristocracy"). Max Weber, "Wissenschaft als Beruf," in Max Weber, *Gesammelte Aufsätze zur Wissenschaftslehre* (Tübingen: Mohr, 1988), 582–613, quotation on p. 587.

14 For the "classical" sociology of professions, see Alexander M. Carr-Saunders and Paul Alexander Wilson, *The Professions* (Oxford: Clarendon Press, 1933); Talcott Parsons, "The Professions and Social Structure," *Social Forces* 17 (1939), 457–67; Talcott Parsons, "Professions", in *International Encyclopedia of the Social Sciences*, 12 (1968), 536–47; Thomas Humphrey Marshall, "The Recent History of Professionalism in Relation to Social Structure and Social Policy," in *Canadian Journal of Economics and Political Science* 5 (1939): 325–40; Everett C. Hughes, "The Social Significance of Professionalization," in *Professionalization*, edited by Howard M. Vollmer and Donald L. Mills (Englewood Cliffs, NJ: Prentice-Hall, 1966), 62–70.

only using technical standards derived from the established knowledge in their field, but in coping with crises for which no solution is at hand. Professions deal with crises that cannot be reduced to well-defined problems, and they try to resolve them on behalf of others, such as a patient, a client, or (in the case of science) on behalf of humanity at large.[15] In the case of science, researchers deal with crises of explanation and validity, crises they identify in the explanatory power of their field's theory when confronting that theory with unexplained observations. And they do so as part of a community of investigators that has come to develop and share convictions on how to do science, and on how to identify sound answers to scientific questions.

The specific nature of the activity in which empirical scientists are engaged explains why an assessment of their work through an evaluation in a market or through an assessment by administrators would be inadequate. An evaluation will have to turn to autonomous collegiate cooperation and critique rather than outside control and standards. Professional autonomy has evolved on different levels: (1) As part of a professionalized habitus, it includes the individual researcher's internalized standards of critique and refinement; (2) Professional autonomy involves criticism in a universe of discourses through colleagues and collegial control elicited through procedures of peer-review and evaluation; (3) Professional autonomy is made possible through institutions such as academies, associations, university departments and research institutes, all of which provide the field with a platform for its ongoing work, with the jurisdiction required to enforce adherence to its standards among colleagues, and procedures to raise and distribute budgets and to codify rules and standards for scientific work.[16]

It is one thing to develop an interest in the particular mode of investigation that empirical science has come to stand for, but quite another to

---

15 For a comparison of science to other professions, see Ulrich Oevermann, "Theoretische Skizze einer revidierten Theorie professionalisierten Handelns," in *Pädagogische Professionalität. Untersuchungen zum Typus pädagogischen Handelns*, edited by Arno Combe and Werner Helsper (Frankfurt: Suhrkamp, 1996), 70–182.

16 While sociologists and historians have investigated the collegiate role of scientists as well as their institutional settings, an empirical investigation of the scientist's internalized habitus has remained a desideratum. Such a habitus was sometimes referred to rather philosophically as "professional ethics." In his recent study on this subject, Andreas Franzmann mobilized the close-reading approach of objective hermeneutics to interpret interviews with researchers, deducing from these interviews tacit assumptions informed by internalized routines and beliefs. See Franzmann, *Disziplin der Neugierde*.

claim to speak for it and to enforce professional standards with the authority of a wider community. This is where authority comes into play. The political sovereign provides empirical scientists with protection and sometimes with financial support, but also with the authority to deal with the profession's affairs. In early modern times, the court provided patronage for individual scientists, bestowing "social and cognitive legitimation" on such individualists as Galileo.[17] With the founding of institutions such as the Royal Society, the *Académies royale* and subsequent national academies in other counties, the practice of science received a continuous institutional foundation empowering not just one scientist, but the general logic of research represented by the academy. The king's endorsement entrusted scientists with organizing the profession so as to effectively safeguard on behalf of the sovereign the advancement of science.[18] With the advent of the democratic nation state, such institutional support and endorsement of science then took place on behalf of the people. The nation-state came to assume the role of client and supporter of science as it began to dedicate itself to the protection and support of the freedom of scientific inquiry and education.[19] In this sense, nation-states through their endorsement of scientific institutions such as academies, universities, scientific associations, or research institutes entered a "contract" with experimental science by accepting, in principle, that science would challenge and test ideas about how the world works even if science came up with new explanations that undermined established beliefs or world views.[20] This development re-

---

17 Mario Biagioli, *Galileo Courtier: The Practice of Science in the Culture of Absolutism* (Chicago: Univ. of Chicago Press, 1993), 354.

18 The resulting embeddedness of the profession as a community in a wider community is the central theme in William J. Goode, "Community within the Community: The Professions," *American Sociological Review* 22 (1957): 194–200.

19 For a recent presentation of this argument, see Alfons Bora and David Kaldewey, "Die Wissenschaftsfreiheit im Spiegel der Öffentlichkeit," *Freiheit der Wissenschaft: Beiträge zu ihrer Bedeutung, Normativität und Funktion*, edited by Friedemann Voigt (Berlin: De Gruyter, 2012), 9–36.

20 This view takes for granted that the institutionalization of science is a component of building a political community, and it differs from another approach in the sociology of science prominent in Germany, i.e. an approach informed by systems theory. For the latter, see Rudolf Stichweh's contribution to this book, "Transformations in the Interrelation between Science and Nation-States: The Theoretical Perspective of Functional Differentiation." See also Niklas Luhmann, *Die Wissenschaft der Gesellschaft* (Frankfurt: Suhrkamp, 1990); Rudolf Stichweh, "Differenzierung des Wissenschaftssystems," in *Differenzierung und Verselbständigung: Zur Entwicklung gesellschaftlicher Teilsysteme*, edited by Renate Mayntz et al. (Frankfurt and New York: Campus, 1988), 45–115; Rudolf

sulted in a system of institutionalized training at universities where students internalized the scientist's role and its logic of inquiry. Eventually, this mindset would be directed at a growing number of subjects outside the natural sciences even if its proper adjustment to an investigation of culture, society, politics, and economies remains disputed. In this volume, such a broadened conception of science (in line with a German conception of *Wissenschaft*) is reflected in contributions on the history of sociology and philosophy by Fabian Link and on the history of Islamic studies by Andreas Franzmann.

So this is how the autonomy of science as a profession played out and how it was institutionalized. But the legitimacy of science has always had to go well beyond this framework. Science has never been self-referential in establishing the foci of its work, and questions researchers have chosen to pursue have not been provided by curiosity or the state of research alone. The legitimacy of science in public and in politics has drawn on a variety of motives, including cultural and utilitarian promises and competitive struggles for funding within and among disciplines. From the inception of institutionalized research science in the seventeenth century, utilitarian promises have played an important role in bolstering research, among them prospects for developing useful technology in such areas as agriculture, navigation, and medicine.[21] But the significance of such utilitarian prospects grew stronger and became dominant as science turned into a successful enterprise. In countries supporting science, administrations, the military, and industries became dependent on technological applications

---

Stichweh, *Wissenschaft, Universität, Professionen: Soziologische Analysen* (Frankfurt: Suhrkamp, 1994; Peter Weingart, *Die Stunde der Wahrheit? Zum Verhältnis der Wissenschaft zu Politik, Wirtschaft und Medien in der Wissensgesellschaft* (Weilerswist: Velbrück, 2001). Approaches informed by systems theory commonly focus on an exchange of services or accomplishments by self-referential subsystems of society. We argue that the state's empowerment of science to cope with crises of explaining reality on behalf of a wider community represents a relationship structurally similar to that between a physician and a patient. A physician is "empowered" by his patient to cope with his/her health crisis. Unlike physicians, however, scientists cope with more general crises that are relevant for all humans, not just one patient. In a strict theoretical sense, therefore, the client of science is not concrete for it is neither a person nor any particular community. But this universalistic and abstract client is nevertheless represented by individual communities that are able and willing to dedicate themselves to the universalistic program of science. This structural similarity to the relationship in other professions such as medicine is what we mean when we consider a community to be a "client" of science.

21 See Merton's famous study on science in its formative period. Robert K. Merton, *Science, Technology and Society in Seventeenth-Century England* (New York: Harper & Row, 1970).

derived from investigating their underlying principles. When curiosity-driven research translated into spectacular technological solutions, furthermore, the success of science through technology has led to the demand that science should assume a more significant role in education. The growth of universities in many countries in the late twentieth century has had the effect of associating larger segments of the population with institutions dedicated to science (the *Massenuniversität* in Germany) while engaging a smaller percentage of university students in "real" research. Significant investments by nation-states in research and education have gone hand in hand with the growth of management structures, and this has also further changed the relationship of science to the public.

While these developments during the past two centuries may be understood within the context of individual states, they have taken place at a time of accelerating globalization since 1970. In our next and somewhat longer section we will focus on challenges to professionalized science in the context of technology-oriented states since 1800. We will close our introduction with a brief section on issues arising from globalization.

3. Legitimizing Science: The Challenge of Utility

3.1. Science and Technology

While technology is much older than science, science and technology have been associated ever since modern science was institutionalized in the seventeenth century.[22] Because of this link, matters related to technology-development have influenced the justification and support of curiosity-driven research.

Prior to World War II, science and technology had had a long interactive history in weapons technology, chemicals and pharmaceuticals.[23] Fran-

---

22 On the difference between science and technology that we have in mind here, see Wolpert's lucid observations in his *Unnatural Nature of Science*, 25–34.

23 Alex Roland, "Science, Technology, and War," in *The Modern Physical and Mathematical Sciences*, edited by Mary Jo Nye, vol. 5 of *The Cambridge History of Science* (Cambridge: Cambridge Univ. Press, 2002), 559–78; John P. Swann, "The Pharmaceutical Industries," *The Modern Biological and Earth Sciences*, edited by Peter J. Bowler and John V. Pickstone, vol. 6 of *The Cambridge History of Science* (Cambridge: Cambridge Univ. Press, 2009), 126–40.

cis Bacon considered the discovery of nature's secrets and the production of useful knowledge two sides of the same coin.[24] The founding of the Royal Society took place on the utilitarian assumption that science and technology were tightly intertwined.[25] During the eighteenth century, France had taken the lead in associating the interests of the state with the elite *Ecole d'Artillerie* or the *Corps des Mines* and the *Corps des Ponts et Chausées*.[26] Such developments carried on and expanded during the nineteenth century. But World War II provided a singular opportunity for science administrators to lay claim to authority well beyond the core functions of exploring nature. Physicists came to rely and depend on massive government funds legitimized by the Manhattan Project and national security. Their success in developing technology provided them with political leverage as they assumed influential roles in policy-making. Political scientist Donald K. Price argued that scientists constituted a "fifth estate" and the scientific community a model of democracy.[27]

Science seemed to provide the tools that made or broke a state's international influence and power in the contested terrain of the Cold War.[28] Following claims by scientists to cultural leadership in the US during the Cold War, and through a representation of science as a tool to solve all sorts of societal problems, the public came to associate science

---

24 Francis Bacon, *The New Organon*, edited by Lisa Jardine and Michael Silverthorne (Cambridge: Cambridge Univ. Press, 2000).

25 Thomas Sprat, *The History of the Royal Society of London for the Improving of Natural Knowledge* (1667), edited by Jackson I. Cope and Harold Whitmore Jones (St. Louis: Washington Univ. Press, 1966).

26 Charles Coulston Gillispie, *Science and Polity in France: The Revolutionary and the Napoleonic Years* (Princeton: Princeton Univ. Press, 2004); Terry Shinn, "Science, Tocqueville, and the State: The Organization of Knowledge in Modern France," in *The Politics of Western Science, 1640–1990*, edited by Margaret C. Jacob (Atlantic Highlands, NJ: Humanities Press, 1994), 47–80.

27 Paul Josephson, "Science, Ideology, and the State," *The Modern Physical and Mathematical Sciences*, edited by Mary Jo Nye, vol. 5 of *The Cambridge History of Science* (Cambridge: Cambridge Univ. Press, 2003), 590–91; Joseph Ben-David, "The Ethos of Science: The Last Half-Century," in *Scientific Growth* (Berkeley: Univ. of California Press, 1991 [1980]), 485–500, esp. 492. The key source for this observation, of course, is Vannevar Bush, *Science. The Endless Frontier: A Report to the President by Vannevar Bush, Director of the Office of Scientific Research and Development, July 1945* (Washington DC: United States Government Printing Office, 1945), https://www.nsf.gov/od/lpa/nsf50/vbush1945.htm. See also Don K. Price, *The Scientific Estate* (Cambridge: Belknap Press, 1965).

28 For the bomb's effect on American politics and a culture of fear, see Ira Katznelson, *Fear Itself: The New Deal and the Origins of Our Time* (New York: Liveright, 2013).

ever more closely with technology. References to "pure research" had begun to be replaced in the 1960s with terms such as "basic" or "fundamental" research, suggesting that science was merely a first step in developing technology. At the same time, sociologists supplied keywords such as "postindustrial" or "knowledge" society, setting the stage for what Ben-David a few years later called a "scientific utopia." Resources for knowledge came to be considered essential components of economic growth.[29] The close association of science and technology in many countries blurred an understanding of the distinct and limited capabilities of scientific research. It helped produce a technocratic ideology that reduced society to an apparatus.[30] The rise of scientism eventually prompted a reaction.

The context for science and for technology-development shifted dramatically in all Western countries during the sixties when the very idea of scientific progress met growing academic criticism, and the legitimacy for scientific work and for its institutions began to be reviewed by an increasingly discerning public.[31] Following periods shaped by world wars and political and social crises, national publics in Europe and in the United States established or reestablished a self-assured role vis-à-vis science that encouraged a critical view of promises associated with science. Ben-David has argued that this was the period when an overly optimistic assessment of science ("scientism") faced a critical reevaluation but also the rise of an "anti-scientific" movement.[32] A critique of science addressed scientists' "complicity" with the military-industrial complex, nuclear power, chemical

---

29 Ben-David, *Centers of Learning*, 174.

30 See, for example, Helmut Schelsky, *Der Mensch in der wissenschaftlichen Zivilisation* (Cologne: Westdeutscher Verlag, 1961).

31 Essential for the field of the philosophy of science: Thomas S. Kuhn, *The Structure of Scientific Revolutions* (Chicago and London: Univ. of Chicago Press, 1976 [1962]). Kuhn's work went along with a general shift in the sociology of science where criticism of Robert K. Merton's work began to set the tone. This shift is well-documented in two volumes: Peter Weingart, ed., *Wissenschaftssoziologie I: Wissenschaftliche Entwicklung als sozialer Prozeß* (Frankfurt: Athenäum, 1973) and *Wissenschaftssoziologie II: Determinanten wissenschaftlicher Entwicklung* (Frankfurt: Athenäum, 1974). Critical perspectives on science and technology where developed by others as well, including Herbert Marcuse, *One-Dimensional Man* (Beacon: Boston 1964) and Jürgen Habermas, *Technik und Wissenschaft als "Ideologie"* (Frankfurt: Suhrkamp, 1968).

32 Joseph Ben-David, "The Ethos of Science in the Context of Different Political Ideologies and Changing Perceptions of Science," in *Scientific Growth* (Berkeley: Univ. of California Press, 1991), 533–59.

disasters, and environmental pollution. It also aimed at the role of scientists in colonial affairs, in producing social inequality, and in developing psychological methods for assessing and dealing with minorities and deviant behavior by administrations and in schools. A shift towards a more critical public reception of science usually took place when issues arose from prominent fields of research that came to stand for the scientific project at large. Their resolution came to shape the subsequent public and academic discourse on science. In his contribution to this volume, Shiju Sam Varughese sketches such developments for India.[33]

In the US after 1945, the field of physics had become the "public face" of science. Physics represented technological achievements relevant for the military and consumers. The secrecy of nuclear facilities added to the field's aura but also shielded from public scrutiny work attributed to it. The sixties, however, witnessed the transformation of the public sphere in the transatlantic region that brought about a reassessment of the state's role and responsibilities towards its citizens as well as a reconsideration of science and technology in modern democracies. In the US, polls indicated that Americans, despite successes such as the 1969 moon landing, considered quality-of-life issues to be more relevant than the space race.[34]

The torch symbolizing science to the public was passed from physics to biology during a controversy about the safety of recombining (altering) the DNA of a living organism, a debate that was considered by some contemporaries as helping provide the critical public assessment that nuclear technology had not received.[35] The decade witnessed a "swing from the physical to the life sciences" as public critique and public hopes came to focus on biology.[36] This shift also led to a transfer of focus from federal to private funding. Physics during the Cold War had stood for federal support

---

33 Shiju Sam Varughese, "The State-Technoscience Duo in India: A Brief History of a Politico-Epistemological Contract," in this volume.

34 Daniel Kevles, *The Physicists: The History of a Scientific Community in Modern America*, Revised edition (Cambridge: Harvard Univ. Press, 1995), 398.

35 Joachim Radkau, "Hiroshima und Asilomar: Die Inszenierung des Diskurses über die Gentechnik vor dem Hintergrund der Kernenergie-Kontroverse," *Geschichte und Gesellschaft* 14, no. 3 (Jan. 1, 1988): 329–63.

36 Agar, *Science in the 20th Century and Beyond*, 508. For a statistical overview of US science spending, see, for example, James Edward McClellan and Harold Dorn, *Science and Technology in World History: An Introduction* (Baltimore: Johns Hopkins Univ. Press, 2006), 418. Also consider recent data on global private and public R&D funding by Scienceogram UK, http://scienceogram.org/blog/2013/05/science-technology-business-government-g20.

within the wider political atmosphere concerned with national security but biotech came to be associated with markets and opportunities.[37] The growth of biotech drew global attention and established a new competitive arena for scientific, technological, and economic leadership. Industry continued to rely on universities for basic research and the training of scientists but public commentators both inside and outside of academia (among them historians and sociologists) differed in their assessments of what some conceived of as a privatization of science or as the emergence of "technoscience". The shift towards biology and biotechnology from the 1970s provided new opportunities and challenges for legitimizing science.

At a time when the old dream of science as a source for technological solutions finally seemed to come into its own, therefore, a cluster of transformations set in: In many countries, a critical public increasingly reflected on the societal consequences of research practices and technologies; a reassessment of the state's role included a reevaluation of the support of science where the state's role had been strong; the intellectual framework that guided the debate came to use market-models even in the case of science organization; and an academic discourse on science increasingly focused on innovative modes of knowledge production, a top-down managerial approach to innovation, and on the regulation of science and technology. While this shift towards the utility of science was most pronounced in the sphere of science studies and in science management, it played out in education as well.

3.2. Science as a Basis for Modern Education

Before 1810, the modern empirical sciences were largely confined to institutions not in charge of education. In the seventeenth and eighteenth centuries, academies had empowered research science and it was from the nineteenth century that empirical science began to be implemented within institutions of higher education.

At that time, universities were affiliated with religious denominations in the United States or had become associated with emerging territorial powers in Europe. With the rise of nation-states after 1800, education began to be secularized in many countries as states sought to educate their citizens.

---

37 The broader context was a "rediscovery of the market" in political debate. Daniel T. Rodgers, *Age of Fracture* (Cambridge: Harvard Univ. Press, 2011), chap. 2.

After German universities began to expect relevant contributions to research from professors in addition to education, other countries such as England from the 1860s and France from the 1880s sought to emulate their success.[38]

In this volume Dieter Langewiesche charts the role of science in society that university presidents in German-speaking countries conveyed to the public in their annual addresses. Langewiesche points to the significance of formulating that role within the context of the state. In his essay on Alexander Dallas Bache, Axel Jansen explains how leading US scientists during the nineteenth century sought to implement experimental research in educational institutions. By proposing to include research science in university and school curricula, researchers questioned established educational contents and challenged patterns of elite formation. Developments that are frequently discussed under the distinct rubrics of a popularization of science and a history of education together advanced the legitimacy of science during the nineteenth century.

Such legitimacy came to be represented through different institutions in different countries. In German-speaking countries, universities assumed a key role. During the nineteenth century, leaders of American science had expected the founding of a national academy to provide public acknowledgment of the role of science in American society. While public universities in Germany associated research science with the state, most leading research universities in the US were private. The public role of research science was acknowledged through the profession's contribution to educating elites and through the endorsement of research at universities by private philanthropy (such as the Rockefeller and Carnegie foundations) during the 1920s and 1930s. Before World War II, however, research science remained one of several competing intellectual approaches offered to students at universities as traditions of a "liberal", humanistic, and Classics-based education remained strong. Andrew Jewett has reminded us that "the contextual understandings of science that emerged in the 1930s took their shape from a desire to augment science's cultural influence rather

---

38 Roger Geiger, *The American College in the Nineteenth Century* (Nashville: Vanderbilt Univ. Press, 2000); Roger L. Geiger, *To Advance Knowledge: The Growth of American Research Universities, 1900–1940* (New York and Oxford: Oxford Univ. Press, 1986); Joseph Ben-David, *Centers of Learning: Britain, France, Germany, United States* (New Brunswick, NJ: Transaction Publishers, 1992); Christophe Charles, "Grundlagen," *Vom 19. Jahrhundert zum Zweiten Weltkrieg 1800–1945*, vol. 3 of *Geschichte der Universität in Europa*, edited by Walter Rüegg (Munich: C.H. Beck, 2004), 43–76.

than to challenge it, as is typical today."[39] Before World War II, in other words, advocates of research science in the US continued to operate with a sense of mission.

Things changed with World War II, just as they did for the public perception of science through technology. The public role of science was transformed against the backdrop of the war and its sequel, the Cold War, as a sense of fear and national crisis continued to provide a rationale for nationwide initiatives to expand national integration through education. The expansion of higher education after 1950 introduced a new topic for legitimizing science. After 1960, more than 10 per cent of youths aged 18 to 21 were enrolled in colleges and universities in all developed countries, and that number has continued to grow.[40] The success and impact of university-based research in the US during World War II led countries elsewhere to try to emulate it, much like American universities had sought to emulate German successes during the nineteenth century. Many countries from the 1950s increased investment in research and in education. For the key period of university expansion, Ben-David observed in 1977 that every country

wanted to imitate the American model exactly when there emerged in that system new and not sufficiently recognized problems. The American system ... had grown into a position of leadership through an intricate division of labor between hierarchically arranged graduate schools and colleges and between university and industrial research. However, the imitators of the American system abroad were only faintly aware of this background of American university research. Their model was the American system's exceptional effort and success during the Second World War, and its post-Sputnik boom in research. [...] This created a mirage of a vast university system educating about half of the relevant age group at good institutions that conducted research at a respectable level.[41]

Universities in many countries had become centers of research science but they were now also viewed as a means to advance a country's economic standing and enhance social justice through facilitating occupational opportunities. Access to higher education and the need to expand higher

---

39 Andrew Jewett, *Science, Democracy, and the American University: From the Civil War to the Cold War* (Cambridge: Cambridge Univ. Press, 2012), 235; Laurence R. Veysey, *The Emergence of the American University* (Chicago: Univ. of Chicago Press, 1965).
40 Ben-David, *Centers of Learning*, 161.
41 Ibid., 121.

education became a key political issue in many countries because it was associated with economic success and social justice.

The growth of student populations has also had the effect of strengthening those areas within universities not dedicated to research but to general education. Many more students attended universities and not all of them were carved out to be research-minded. Not all students could enter the professions either, for which universities in the past had prepared them. The widening of subjects taught at universities in recent decades reflects the broader responsibility that institutions have taken on. While a much larger cohort of the population in many countries today will have had exposure to higher education and to institutions representing "science", such exposure has not always been associated with an immersion in the peculiar explorative outlook represented by research. While science and science-derived technology has come to shape all walks of life, even university students may not be susceptible to how it works. Max Weber's famous example of the commuter who rides the streetcar but hardly understands how it works reflects not only the rationalization of everyday life, but also the inability to follow its underlying developments.[42]

Such developments have gone hand in hand with a shift in the finances of higher education. In the US, the Higher Education Act of 1972 introduced student loans. Today, about seventy percent of American college students take on loans and the total volume of student loans exceeds one trillion dollars.[43] The effect of shifting the financial burden of higher education from the state to students and their families has been that science-based education has turned from a public good into a personal investment. Changes in numbers have surely led to changes in educational content. The larger the cohort in higher education, the more such education will have to accommodate students seeking to pursue careers in administration and business. In the US, such functions could perhaps be relegated to second or third tier institutions. In Germany, research universities implemented new degree programs in line with European policy ("Bologna"). For many students at German universities (and young researchers emerging from its degree programs), university education is no longer associated with curiosity-driven research.

---

42 According to Weber, such developments are part of an ongoing process of "intellectual rationalization." "Wissenschaft als Beruf," 593.

43 Andrew Delbanco, "Our Universities: The Outrageous Reality," *The New York Review of Books,* July 9, 2015, http://www.nybooks.com.

3.3. The Rise of Science Administration

The two areas we have discussed so far to sketch a trend towards utility in science—technology and education—are connected with a third area in which utility-orientation has played out: administration. The management of science expanded rapidly after 1945 as the US and other countries began to invest in science by building up large structures to accommodate research.[44]

The growth of such structures has challenged the professional autonomy of science because administrations could develop interests distinct from researchers in their organization. As a general trend, expanding science administrations have appreciated routinized and formal procedures, procedures that tend to develop a life and persistence of their own, instead of spontaneous adjustment to unpredictable research. In Max Weber's terminology, we have witnessed a shift in emphasis from "material" towards "formal rationality."[45]

Discourses informing science administrators since the 1970s have highlighted practical solutions, rates of innovation, and opportunities for translation to technology. As a result of administrative developments, even institutions controlled by researchers themselves have been inclined to adopt perspectives of planning and control because they are in line with

44 Alex Roland, "Science, Technology, and War," in *The Modern Physical and Mathematical Sciences*, edited by Mary Jo Nye, vol. 5 of *The Cambridge History of Science* (Cambridge: Cambridge Univ. Press, 2002), 564. John Agar considers as one prominent theme in the twentieth-century history of science the "extraordinary and unambiguous importance of the working world of warfare in shaping science." Jon Agar, *Science in the 20th Century and Beyond* (Cambridge: John Wiley & Sons, 2012), 6. Important effects of outside intervention probably consist in delaying or accelerating work in relevant research fields, in bloating areas of research that suddenly have proliferating budgets and opportunities or depressing others through diminishing budgets and expectations. For the growth of science administration, also see the introduction Gili S. Drori, John W. Meyer, Francisco O. Ramirez, Evan Schofer, "Introduction: Science as a World Institution," in *Science in the Modern World Polity: Institutionalization and Globalization* (Stanford: Stanford Univ. Press, 2003), 3–7.

45 Max Weber, *Wirtschaft und Gesellschaft* (Tübingen: Mohr, 1980 [1921]). See also Wolfgang Mommsen, "Personal Conduct and Societal Change," in *Max Weber, Rationality and Modernity*, edited by Sam Whimster and Scott Lash (Abingdon, UK and New York: Routledge, 2006 [1987]), 35–51.

science policy.[46] Such developments have evolved into a search for "new forms of science governance" that consider scientists, not peers in self-governed bodies, but employees of research companies infused with ideas of a "New Public Management."[47] NPM has played a significant role in Europe where it evolved in the context of public service reform, providing an agenda for implementing management ideals derived from the private sector. During the 1990s, NPM has prompted public university administrations to tap private sources for funding, accept target agreements between the state and university departments, and adopt modes of reporting and monitoring performance.[48] Science administration has established its own courses of study as well as its own degrees, associations, and journals.

Such transformations do not take place overnight and they are by no means complete. We are not suggesting that structures of science, in Europe or elsewhere, have been absorbed completely by utilitarian interests aimed at technology development or by discourses focusing on society's "great challenges" such as climate change, green energy, migration, or educational reform. The wider scientific culture has produced proponents as well as opponents in such debates, and in debates about a legitimate role of science in society.

From the perspective of a professionalization of science, therefore, the potential utility and practical value arising from research science has in-

---

46 To cite an example of a critical investigation of ties between industry and academia: Martin Kenney, *Biotechnology: The University-Industrial Complex* (New Haven and London: Yale Univ. Press, 1986).

47 Literature on a New Public Management includes Uwe Schimank, "Die akademische Profession und die Universitäten: 'New Public Management' und eine drohende Entprofessionalisierung," in *Organisation und Profession*, edited by Thomas Klatetzki and Veronika Tacke (Wiesbaden: VS Verlag, 2005), 143–64; Bettina Heintz, "Governance by Numbers: Zum Zusammenhang von Quantifizierung und Globalisierung am Beispiel der Hochschulpolitik", in *Governance von und durch Wissen*, edited by Gunnar Folke Schuppert and Andreas Voßkuhle (Baden-Baden: Nomos, 2008), 110–28; Andreas Knee and Dagmar Simon, "Peers and Politics: Wissenschaftsevaluationen in der Audit Society," in *Governance*, edited by Folke and Voßkuhle, 173–85; Richard Münch, *Akademischer Kapitalismus: Zur politischen Ökonomie der Hochschulreform* (Berlin: Suhrkamp, 2011); Andreas Franzmann und Peter Münte, "Von der Gelehrtenrepublik zum Dienstleistungsunternehmen. Ausschnitt aus einer Deutungsmusteranalyse zur Erschließung kollektiver Bewußtseinslagen bei Protagonisten der gegenwärtigen Universitätsreform," in *Zwischen Idee und Zweckorientierung: Vorbilder und Motive von Hochschulreformen seit 1945*, edited by Andreas Franzmann und Barbara Wolbring (Berlin: Akademie Verlag, 2007), 215–29.

48 For an overview see Walter J. M. Kickert, ed., *Public Management and Administrative Reform in Western Europe* (Cheltenham, UK and Northampton, MA: Edward Elgar, 1997).

creasingly challenged the profession in various countries to reassess its role and to reassert its autonomy. Reassessments have concerned the role of research in the light of demands for practical solutions and in view of demands for a "relevant" university education. The professional autonomy of research science has played out in specific national settings, but some developments since 1970 have evolved on a global scale. This raises the question of how globalization plays out for science as a profession.

4. Global Challenges to Science as a Profession

In recent decades, the professions and the nation-state have both come under scrutiny in academic discourses on modernity. Science has increasingly been considered to be a mere component of developing technology, a component whose economic productivity and sustainability were to be optimized. At the same time, however, transnational developments suggested that nation-states were quickly losing their political grip and relevance. The rise of military alliances, the globalization of markets and organizations, environmental issues such as climate change, the emergence of global communication networks, and the internet all seemed to point towards the growing relevance of social movements and modes of governance transcending traditional states.[49] In recent years, academic criticism that takes aim at traditional agents of modernization such as the professions or the state seems to have lost some steam. But real-world developments persist that continue to challenge the integrity of both professions and nation-states. In concluding this essay, we would like to identify three dimensions of globalization that touch on science as a profession: (1) the development of supranational structures for the promotion of science, (2) the increased international mobility of researchers, and (3) the emergence of global frameworks for comparing and evaluating science.

---

49 Overviews include Emily S. Rosenberg, ed., *A World Connecting: 1870–1945* (Cambridge: Harvard Univ. Press, 2012); Akira Iriye and Jürgen Osterhammel, eds., *Global Interdependence: The World after 1945* (Cambridge: Harvard Univ. Press, 2014). Insightful through its focus on a particular industry: Peter Borscheid, "Introduction," in *World Insurance: The Evolution of a Global Risk Network*, edited by Peter Borscheid and Niels Viggo Haueter (Oxford: Oxford Univ. Press, 2012), 1–34.

Supranational structures of science have grown rapidly since World War II but their professional implications are most severely felt in Europe.[50] They have been of particular concern here because the promotion of science has traditionally been in the hands of national states but is now being superseded by a supranational "European Research Area," a new platform for research funding whose effects on national traditions are not yet evident. Nina Witjes and Lisa Sigl describe how European states are building up infrastructure for internationalizing science and technology development and they propose that such growth has been significant enough to speak of a new field, that of the "Internationalization of Science, Technology, and Innovation (STI)."[51] On the level of European institutions, Arne Pilniok's contribution to this volume suggests a number of questions about the evolution of an organizational structure that anticipates replacing national ones.[52] The national setting has played a key role for the cohesion and integration of the research profession in the past. How will such cohesion evolve on the supranational European plane? How will Europe alter the public legitimacy of science? Perhaps an emerging bureaucratic superstructure continues to rely on national science systems for legitimacy and funding. Or the European research profession and community will evolve into an integrated counterpart to an emerging European public and its institutions, similar perhaps to integrational ambitions by nineteenth-century leaders of American science.

The second issue concerns the increased international mobility of researchers. Beyond the issue of a "brain drain" among students and postdocs, such mobility raises questions when leading researchers move to a different country. How does their decision to move abroad affect their commitment to the scientific profession, represented through national

---

50 For a chart depicting the rapid growth after 1950 of science organizations and structures such as national science policy organizations, inter-governmental science organizations, science ministries, nongovernmental science organizations, and international science education organizations, 1870–1995, see Drori et al., "Introduction", *Science in the Modern World Polity*, 3. See also Elizabeth Crawford, Terry Shinn, and Sverker Sörlin, "The Nationalization and Denationalization of the Sciences: An Introductory Essay," in *Denationalizing Science: The Contexts of International Scientific Practice*, edited by Crawford, Shinn, and Sörlin (Dordrecht: Kluwer, 1993), 1–42, esp. 1–3.

51 Nina Witjes and Lisa Sigl, "The Internationalization of Science, Technology & Innovation (STI): An Emerging Policy Field at the Intersection of Foreign Policy and Science Policy?," in this volume.

52 Arne Pilniok, "The Institutionalization of the European Research Area: The Emergence of Transnational Research Governance and its Consequences," in this volume.

associations and organizations? Would such mobility result in a weakening of national science, and by implication, of science as a profession?[53] In the past, national organizations and associations have played a key role in representing the professionalized discourse in a given field. The founding of international organizations such as, for example, the International Society for Stem Cell Research (ISSCR) suggests that this discourse is being lifted onto a new plane, but international organizations have no political counterpart. This raises the question of the role of these organizations. Are they a result of efforts within the scientific community to ensure that the field adheres to standards of good practice even where national organizations may not be able to enforce them? Such questions seem particularly relevant in countries like China, India, or Singapore because they aspire to world class research through significant investments and competitive international hiring. In emerging nations, such ambitions will likely result in efforts to stimulate processes of professionalization and create the cultural setting within which research can thrive and recruit new students and researchers. In order to create sufficient momentum, such efforts will ultimately have to be in line with a broader political and cultural commitment to freedom of speech and freedom of research. Shiju Varughese delineates the peculiar interests associated with developing science in the context of emerging public awareness of large-scale technology in India.[54]

A third dimension of globalization relates to the rise of science administrations discussed above. The management of science and its institutions is increasingly shaped by discourses that emerge from the social sciences, theoretical premises that structure and shape the evaluation and assessment of science around the globe. A comparison between the relative achievements of science in particular countries or by individual institutions, has, to be sure, been around for a long time. The essays by Dieter Langewiesche and Axel Jansen respectively, show that spokesmen of science sought to explain the relevance of scientific institutions within national settings, tak-

---

53 There have long been concerns about a "brain drain" in various countries. For an interesting case of how such concerns have been instrumentalized by ambitious German academics abroad, see Axel Jansen, "Patriotismus- und Eliteninszenierung im deutschen Hochschulreformdiskurs. Analyse des 'offenen Briefes' der Initiative Zukunft Wissenschaft vom September 2005," in *Zwischen Idee und Zweckorientierung: Vorbilder und Motive von Hochschulreformen seit 1945*, edited by Andreas Franzmann and Barbara Wolbring (Berlin: Akademie Verlag, 2007), 185–94.

54 Shiju Sam Varughese, "The State-Technoscience Duo in India," in this volume.

ing for granted that science provided an arena for national competition.[55] In academic circles, a discussion of the merits of particular universities and researchers is common when making decisions about one's own career or when advising students on theirs. Such assessments, however, have been eclipsed in science administrations and in public by scores and rankings established through measurements developed by the social sciences. Numbers, perhaps, are more readily accessible than differentiated judgments by scholars, and they provide a sense of unflinching objectivity.[56] As Tobias Werron points out in his contribution to this volume, global rankings have become an important reference for public science policy in nation-states.[57]

From the perspective of science as a profession, therefore, all three dimensions of the internationalization and globalization of science result in a question well beyond the scope of the sociology of science or the history of science: What future issues will demand a renewal of the public legitimacy of science? And within what political framework will such legitimacy evolve?

---

55 See Dieter Langewiesche, "State—Nation—University" and Axel Jansen, "Science in an Emerging Nation-State" in this volume.

56 Theodore M. Porter, *Trust in Numbers: The Pursuit of Objectivity in Science and Public Life* (Princeton: Princeton Univ. Press, 1995).

57 Tobias Werron, "Universalized Third Parties: The Role of 'Scientized' Observers in the Construction of Global Competition between Nation-States," in this volume.

Transformations in the Interrelation between Science and Nation-States: The Theoretical Perspective of Functional Differentiation

Rudolf Stichweh

1. Two Function Systems in Modern Society

Science and politics are two function systems in modern society with rather different histories and trajectories. In the case of science the most important period for its emergence as an autonomous function system is the so-called "second scientific revolution" of the eighteenth and nineteenth centuries (at its core 1750–1840), a concept which will be explained in the following.[1] Politics as nation-state politics arises significantly later. If one looks at the nation state as the primary and universal form of internal differentiation of the system of (world) politics, this only takes its present form in the process of the final decolonization of the world after 1960. Of course, major developments in the concept of the nation and its institutional realizations go back to the eighteenth century.

If science and politics are both function systems in society, one has to explain the concept of a function system. A function system is a macro system in society based on more elementary levels of system formation (interactions among participants in co-presence, formal organizations consisting of members), which includes the systems on these levels and builds on this basis a global communication system which is specified by its function, that is by a specific type of problem solving which constitutes the unique contribution of this function system to society. Only the function system of science offers systems of thought to society whose primary description is that their propositions are true. Only religion is a communication system which opens the mundane lifeworld of individuals to tran-

---

[1] Stephen G. Brush, *The History of Modern Science: A Guide to the Second Scientific Revolution, 1800-1950* (Ames, IA: Iowa State Press, 1988); Peter Watson, *The German Genius: Europe's Third Renaissance, the Second Scientific Revolution, and the Twentieth Century* (New York: Harper Perennial, 2011).

scendence. Only the political system incessantly produces collectively binding decisions in society. And society is the still more extensive macro system, encompassing all these functional perspectives and the problem solutions offered by them, which in its present-day condition clearly has to be analyzed as a world society, i.e. as a worldwide communication system. Besides science and the (world) political system there exist a significant number of other functional domains: law, the economy, sports, education, religion, the arts, etc. The theory of functional differentiation is thus the theory of these global communication systems.[2] This essay will focus on two of them—science and politics—and try to analyze and understand their shifting interrelations in the last 250 years. In doing this I privilege the perspective from science to politics as one of the formative environments of science, and in this chapter I do not analyze the opposite direction, from politics to the ways this political system is changed by influences from science.

2. The Universality of Science

Even if we go back to late medieval Europe and its learned institutions (monasteries, universities) the characterizations of the knowledge and science practiced there point to the claim of a universality of science. Universality first of all means spatial and temporal invariability. Scientific propositions if they are true at all will be true at every place in the world. Differences of place may matter for the relevance of propositions—a description of an animal or plant may be of limited relevance at a place where these organisms do not occur—, but these propositions are nonetheless true even at places where empirical observations testing the truth claims are not possible. Temporal invariability implies that no automatic temporal decay is built into truth claims. Of course, we know that a proposition may be refuted at some later point in time. But such a refutation can never

---

2 Rudolf Stichweh, "The History and Systematics of Functional Differentiation in Sociology," in *Bringing Sociology to International Relations. World Politics as Differentiation Theory*, edited by Mathias Albert, Barry Buzan, and Michael Zürn (Cambridge: Cambridge Univ. Press), 50–70; Rudolf Stichweh, "Comparing Systems Theory and Sociological Neoinstitutionalism: Explaining Functional Differentiation," in *From Globalization to World Society. Neo-Institutional and Systems-Theoretical Perspectives*, edited by Boris Holzer, Fatima Kastner, and Tobias Werron (New York and London: Routledge, 2015), 23–36.

claim that a truth claim is "outdated" in a simple temporal understanding. It has to show that it is untrue or alternatively has been superseded by a more general perspective. But even in this latter case it is not simply a question of change in time or of temporal decay.

Furthermore there is the social invariability of science. Scientific truths are true for everybody, even for those who do not yet know and will perhaps never know about them. And finally there is the thematic universality of science, in which science claims, and is justified in claiming, that there are no limits to aspects of the world on which it can do research. These multiple ways in which science has since medieval Europe claimed universality is clearly a contingent structure of expectations addressed to the social and physical world. But the improbable rise of this contingent structure of expectations is very relevant for the (success) story of the differentiation of science.

3. The Globality of Science

What is the difference between the medieval universality and the contemporary globality of science? It is obvious that they represent consecutive stages in the differentiation of science. Globality is to be understood as the spatial ubiquity of those beliefs and structures and institutions which stand for the strange idea of the universality of science. Since the second half of the twentieth century these beliefs and institutions can be observed everywhere in the world and their interconnectedness is ensured by a global, and intensifying, reciprocity of observations. In science what is done at one place is easily to be observed from any other place, and these observations are clearly reciprocal observations by which the global commonality of practices and beliefs is very much strengthened. From this results the astonishing phenomenon that the knowledge systems of scientific disciplines are singularities whereby in each scientific discipline there is one global knowledge system which is clearly diversified in itself but does not arise in the form of conflicting representations of the world, as is the case in other knowledge systems such as medicine and religion. In the latter two you have competing knowledge systems arising historically in different world regions, but which later become global systems with a global followership. In science there is no such thing as a plurality of versions of science deriving from different world regions, but instead what is observed is the globalization of only one tradition which has claimed universality from its very beginnings.

4. Social Inclusion into Science

Besides the (spatial) globality of science, and very much furthered by the ubiquity of certain beliefs, expectations and institutions, we observe something that can be seen in every function system, namely a tendency towards the social inclusion of potentially every person in the world into the possibilities of participation in science. But science is different from other function systems. Regarding the possibilities of access to the role of a professional scientist, universal inclusion is real and exists in relevant parts of the world in the contemporary situation. There are no longer any exclusions of social categories of people from participation as professional scientists. But looking at those who do not become scientists by profession, there is no complementary and universal client role defined for them in science. Scientific professionals do not care for all the non-scientific people as their scientific clients, and in this respect science is not a profession, as it does not have clients in the way the classical professions of medicine, law, education and the clergy have.[3] Instead there are much more indirect ways of participating in science, the most important among them probably being the universal inclusion into education (secondary education, higher education), and finding there, in schools and universities, curricula which have been scientized since the early nineteenth century. This indirect inclusion into science via education seems to be the major path. But compared to other function systems, this indirect way is very much a singularity which explains the astonishing autonomy of science in modern society.

There are other structures and semantics which have to be looked at. There is the normative idea of utility which was a dominant conceptual figure in Enlightenment discourse. And a few decades later, in the early nineteenth century, the idea of applied science arises, as a type of scientific research of its own. But neither concept belongs in the domain of social inclusion, but instead they formulate aspects of the responsiveness of sci-

---

3 Everett C. Huges, *The Sociological Eye. Selected Papers on Institutions & Race* (Chicago: Aldine Atherton, 1971).

ence towards societal needs which only indirectly touch on the question of the inclusion of individuals into the knowledge processes of science.

A further development regards the idea of the popularization of science. This already existed as a postulate in the nineteenth century, and inspired public lectures by prominent scientists and scientific associations for lay people in larger cities. But these were relatively small initiatives. After 1950 popularization became ever more important and it was complemented by the new expectation of "Public Understanding of Science" which is a relatively recent idea (post World War II), but which now is addressed to each individual professional scientist with the expectation that he/she contribute to it. This expectation becomes part of the project culture of modern science, with the implication that in many proposals for projects you will have to explain your contribution to a "Public Understanding of Science," which makes out of it a nearly universal concern. But nonetheless this remains a somewhat ceremonial addendum to normal scientific practice and as such does not impact upon everyday activities.

5. The University as a Global Scientific/Educational Organization

From its beginnings in the decades around 1200, the university was a surprisingly global and not a regional or local organization. It was clearly related to nearly all of the aspects of universality, globality and social inclusion discussed above. The name it used for itself was "Studium Generale" and this term also meant the spatial extension of human capital produced by the university, i.e. the degrees awarded by the university were supposed to be valid anywhere in Europe, and it confirmed the thematic universality of the university as a knowledge institution (the potential inclusion of any subject whatsoever into the curriculum of the university) as finally the social inclusion to which even the medieval university was amenable. From its beginnings the university was not an institution limited to the higher strata, and the fact that it was never the educational institution of the European nobility ensured a certain distance from social structures of political domination.

For nearly six hundred years the university was clearly a European and not a world institution. The first universities established on other continents were mainly related to the European colonial powers. From the six-teenth to the eighteenth centuries it was primarily in Latin America and later in North America that some universities and colleges were established and these were furthered either by the educational orders (especially the Jesuits[4] and sometimes the Dominicans) of the Catholic powers or by individual migrants who had studied at European universities (early univer-sities in the English colonies in North America). The other continents of Asia, Australia and Africa only followed in the nineteenth and twentieth centuries.[5] The real globalization of universities and the globalization of Western scientific and learned knowledge, as far as it depends on educa-tional and university institutions, is mainly a phenomenon of the last two hundred years.

6. Res Publica Literaria

One of the most remarkable inventions of early modern scholarship (from the sixteenth to eighteenth centuries) is the idea of a transnational cosmo-politan corporation of all people of learning and science, which we find in the major European languages as *res publica literaria, république des lettres, republic of letters* or *Gelehrtenrepublik*. Of course, the social system consisting of all European scholars and scientists is not really a corporation (organi-zation) and therefore the other contemporary term is very enlightening: *republic*. This is clearly a political term. A republic is a kind of political order which is distinct from other political orders. It is not a monarchy and not an aristocracy but a much more egalitarian political order—and it is auton-omous towards other principles of ordering. This is one of the earliest and most precise forms of claiming autonomy for scholarship and claiming an autonomous social system which is undoubtedly a transnational, pan-European social system. It is defined by its republican status and by its global distribution constituting a world, as is aptly claimed in a dissertation

---

4 Ronnie Po-chia Hsia, "Jesuit Foreign Missions. A Historiographical Essay," *Journal of Jesuit Studies* 1, no. 1 (2014):47–65.
5 There is one college in the Philippines established by Dominicans in 1611.

published in Leipzig in 1698: "The republic of letters is dispersed through the whole world and therefrom takes its name 'learned world'."[6]

7. The Paradox of Nationalization

The discontinuity which introduces a new form of the social organization of science is in one relevant respect identified in Friedrich Gottlieb Klopstock's "Die deutsche Gelehrtenrepublik" (1774). This was a new term and phenomenon, a "republic of science" with a national prefix and was intentionally declared by Klopstock to be a "secession from the Latin republic:" "you separate yourself from it and leave this union of many years, and you dare to do it with your language, to examine how it will be able to go through the sciences now articulated in it or be unsuccessful in it. We know, I answer, that we separate and we know what we dare to do."[7]

This statement by Klopstock is more programmatic than institutional. The scientific academies of Europe which were, besides the universities, the most important institutional supports of the internationalization of science do not yet take the path to national science in these years.[8] But in the domain of education and higher education one observes a development nearer to the postulates of Klopstock. Since the 1770s there has been in many European countries a literature on "national education" as a newly articulated imperative, and here "national education" means a reorganization of the curricula and other practices of universities to ensure that they serve the presumed interests of the respective nation. This literature on "national education" was closely coupled with the principled opposition to the transnational Catholic educational orders, especially the Jesuits, which

---

6 Johann Georg Pritius, *De republica literaria* (Leipzig, 1798); citation in Johann Erhard Kappen, "Versuch einiger Anmerkungen über Saavedra gelehrte Republic," in *Die gelehrte Republic*, edited by Diego de Saavedra Fajardo (Leipzig, 1748), 201–80.

7 Friedrich Gottlieb Klopstock, "Die deutsche Gelehrtenrepublik" (1774), in Friedrich Gottlieb Klopstock, *Werke und Briefe*, vol. VII.1, edited by Rose-Maria Hurlebusch (Berlin and New York: Walter de Gruyter, 1975) 129; citation on p. 33, Lutz Danneberg and Jörg Schönert, "Zur Transnationalität und Internationalisierung von Wissenschaft," in *Wie international ist die Literaturwissenschaft?*, edited by Lutz Danneberg and Friedrich Vollhardt (Stuttgart: J. B. Metzler, 1995), 7–85.

8 Rudolf Stichweh, "Wissenschaftliche Akademien aus soziologischer Perspektive. Organisierbarkeit und Organisationen im Wissenschaftssystems der Moderne," *Acta Historica Leopoldina* 64 (2014): 79–89

resulted in the abolition of the Jesuit order by the Pope in 1773 (revoked in 1814).

A further relevant context is the strong emphasis on the utility of science and higher education characteristic of the Enlightenment. "Utility" clearly is a norm for knowledge processes which favors local or national relevances as against global perspectives on knowledge.

8. The Rise of Modern Scientific Structures in the Context of Nationalized Communication (1770–1830)

Approximately between 1770 and 1830 we observe the emergence of many of the structural novelties which become definitive for modern science: The differentiation of scientific disciplines and the attendant rise of closely knit communities of disciplinary specialists. Then there is the fact that a growing number of disciplines exist in the environment of each single discipline. From this arises a "milieu interne" (Claude Bernard) of these other disciplines which becomes ever more relevant for specialists in other disciplines, who then compare theories, methods and learn from and compete and conflict with these other disciplines, and from this there emerges as a structural effect that the external environments of science relatively lose in importance. As is the case in other function systems, the rise of internal complexity/differentiation is the major reason for the autonomy of the function system of science.

These changes are accompanied by the rise of specialized publications in specialized journals around which disciplinary communities are formed. Then there is the imperative of the discovery of novelties enabled by the institutionalization of the "research imperative," research being the method for the incessant production of novelties which now defines science.[9] For research as an activity and for the discovery of novelties one needs theories and/or methods (experimental, mathematical methods). The act of publishing research-based novelties becomes highly standardized as a form of communication, and science adds as its most elementary and its most consequential reward the citation of publications, which besides being a re-

9 Steven R. Turner, "The Prussian Universities and the Research Imperative, 1806–1848" (PhD diss., Princeton University, 1973).

ward at the same time brings about the self-referential closure of science, defining boundaries and incessantly redefining boundaries (of problem situations, research programs, disciplines, and the system of science itself).[10]

What is remarkable is that these transformations are paradoxically coupled to a temporary restriction of communication to national scientific communities which are linguistically national scientific communities. Communication in science is for the first time in history communication in national languages. This does not exclude communication in foreign languages, but their mode of reception now often means that you read them in translation—and these translations are inescapably selective and interpretive. What are the causes of this paradoxical nationalization of scientific communication at the moment of the genesis of the modern forms and structures of scientific communication?

The most important cause of this paradoxical development can be seen in the relevance of social inclusion into the processes of production of science. Science before the modern period, the early modern science of the universities and academies, was very much a small-scale and elite phenomenon. This was changed by the switch from the international languages of early modern science (primarily French and Latin) to scientific communication in national languages. This switch to national languages and to publication in national language journals has two major implications. First there is a significantly higher number of participants who can be included into the production of science this way. These participants may be local and regional practitioners (apothecaries, doctors, teachers, lawyers, etc.) who would have been excluded by the international languages of science. They may also be university students, as the scientific revolution in German universities in the early nineteenth century saw a continuity of progression from participation in introductory courses to seminars, in which original research could already be carried out by relatively young students. This was facilitated by the use of national languages. The second major consequence of doing science in national languages is that the communicative distance from science to applications and technology becomes shorter in a national language. Therefore science which finds its modern form in national scientific communities is not only a socially more inclusive scientific system, it is at the same time a system of science which becomes

---

10 Rudolf Stichweh, "Die Autopoiesis der Wissenschaft," in Rudolf Stichweh, *Wissenschaft, Universität, Professionen. Soziologische Analysen*, 2nd ed. (Bielefeld: Transcript, 2013), 47–72.

more relevant for the environments which are near to it in the national sociopolitical system in which it is located.

9. Structural Changes of the System of Science (1830–2015)

Since 1830 we observe an ongoing process of ever new changes and transformations in the social system of science. There is first of all the incessant restructuring of universities. The reform period around 1800 was—at least in the German universities—dominated by the rise of the philosophical faculty and its disciplines as the new paradigm of scientificity. In the subsequent one hundred fifty years, there was only one change of comparable weight: The emergence of the technical university as a new type of higher educational institution for a new type of scientific discipline (technical sciences/engineering sciences). Afterwards it became very important for the system of the sciences that technical universities did not remain institutions separated from the classical universities, and so they began to include disciplines from the social sciences and the humanities. There are many countries today in which one or several technical universities are the most excellent institutions they have. Universities throughout the world did not focus on being very specialized institutions which only cultivated a few scientific subjects, instead the diverse university in which you can study many subjects or disciplines became the dominant type. This characteristic was very important for the system of science so that it did not become a system characterized by isolated segments but instead was a system defined by a very dynamic "milieu interne" of many scientific disciplines interacting with one another.

The rise of technical universities and engineering disciplines correlates with an emphasis on applied knowledge types: technology, applied science, research and development—and these have expanded and grown together with scientific knowledge at the other pole of the spectrum, namely fundamental, basic, pure scientific knowledge that is proud of its purity. This bipolarity that characterizes the space of scientific knowledge is another factor that acts against separation, as one can combine terms from the two sides of this distinction and speak of "fundamental development" as an activity type, a term one often hears in industry.

Besides a diversification of universities another important change is the pluralization of the organizational infrastructure of science.[11] Universities and academies of science constitute the two premodern organizational types in the system of science, both of which have experienced continuities over hundreds of years, as well as reforms and discontinuities. Since the late nineteenth century, research became an organizational reality of its own, leading to the rise of research institutes which are not places for scientific elites, as is the case in the academies of sciences, but are strictly organized along thematic lines of institutionalizing research on "important" (societal, scientific) problems in relatively large scale institutes that did not exist in universities. There is here a parallel to the rise of the research institute late in the nineteenth century, with the emergence of the research laboratory in industry serving the need for innovation in specific industrial enterprises and achieving this by cultivating the whole spectrum of applied research, from unexpected discoveries in fundamental research to the development of industrial prototypes and products.

A third change in the second half of the nineteenth century is the return of international science. German and then English renewed the availability of global scientific languages; there is international techno-scientific cooperation (telegraph cables, etc.) and cooperation in measurements which demand that observations are made at many places in the world (terrestrial magnetism, etc.); travel becomes easier and faster, and conferences and world exhibitions and other events are occasions for international travel by scientists; and there is the temporary or permanent migration of scientists and students to attractive places for conducting research.

Of the many transformations of twentieth and twenty-first century science, I will only single out one other momentous change: The rise of cooperation among scientists as the normal way of pursuing scientific research. Why did it happen? Behind cooperation can be seen the rise of complexity in research. Research in many disciplines became so complex and demanded so many different competences (methods, theories, mathematics, instruments, different places of observation) that it could no longer be pursued alone. Cooperative science thus involves ever increasing numbers of participants and as it often does not find them locally or nationally it becomes ever more international. The emergence and rapid growth (after 1950) of scientific cooperation adds an equally global network structure of

11 Stichweh, "Wissenschaftliche Akademien."

collaborations and publications to the worldwide community structure of disciplinary science. Thus besides the organizations of science and the disciplinary communities of the world system of science, the small world network[12] of collaborators became a third core social structure of the worldwide function system of science.[13]

10. The Structural Couplings of the Global System of Science to the Other Function Systems of Society

There is one last analytical perspective which has to be articulated here. A momentous change in the social system of science which characterizes its fate in the twentieth and twenty-first century is that its enormous internal complexity which is the basis of its autonomy is complemented by the complexity of its structural couplings with the function systems of its societal environments. The concept of structural coupling means a contiguity of two social systems which induces structures in both systems which are internal to the respective systems, but are related in their transformations over longer periods of the evolution of both systems.[14] The diversification and intensification of its structural couplings is clearly a watershed in the history of the social system of science—and to understand the growth and relevance of science in the early twenty-first century you have to understand and analyze this history of its structural couplings.

At this juncture I will only mention some important cases of the structural coupling of modern science, though an extensive analysis of the history and sociology of the function system of science in modern society is a fascinating task for future research that has not really been done until now.

---

12 David Easley and Jon Kleinberg, *Networks, Crowds, and Markets. Reasoning about a Highly Connected World* (Cambridge: Cambridge Univ. Press, 2010).

13 On "Eigenstructures" of society, see Rudolf Stichweh, "The Eigenstructures of World Society and the Regional Cultures of the World," in *Frontiers of Globalization Research: Theoretical and Methodological Approaches*, edited by Ino Rossi (New York: Springer, 2007), 133–49.

14 Niklas Luhmann, "Operational Closure and Structural Coupling: The Differentiation of the Legal System," *Cardozo Law Review* 13 (1991): 1419–41; Humberto Maturana Romesin, "Autopoiesis, Structural Coupling and Cognition: A History of these and other Notions in the Biology of Cognition," *Cybernetics & Human Knowing* 9, no. 3–4 (2002): 5–34.

The first case is the structural coupling of science and higher education which is the core event in the reform period of universities and sciences around 1800[15] and remains the most important structural coupling of science to date. It is accompanied by the structural coupling of science and the curricula of secondary schools which occurs a little bit later via the introduction of numerous scientific disciplines as school subjects into secondary schools. This is ongoing and has been intensified in our day, for example by the preference for MINT-education articulated in many contemporary school systems in the world.

Then there are two further couplings which became prominent approximately one hundred years later: The coupling of science and the economy via industrial research and industrial laboratories as a way for industry to conduct the research itself that was necessary for the transformation of science into the development and preparation of prototypes and products for the markets that industrial actors perceived as their own (ca. 1880). There is also the coupling of science and politics via the rise of "science policy" as a policy domain of its own (the word *Wissenschaftspolitik* seems to have been used since 1910) which is in a fascinating way coupled to the emergence of the institutions of self-government of science in a remarkable illustration of the way structural coupling works in function systems. A third parallel case that can be identified is the coupling of science and the emerging function system of health concerns for which university clinics are very important, and subsequent distinctions such as biomedical research vs. clinical research. Finally at the end of the nineteenth century we observe the structures of the modern patent system institutionalizing global comparisons of what is known in technical problem situations and coupling law, science and economic opportunities linked to valid patent claims.[16]

Religion obviously is a case apart. In the medieval situation of the rise of universities as quasi-monastic institutions, the interdependencies of science and religion are extremely strong. But in this case we observe a loss of structural couplings in history. When in the seventeenth and eighteenth centuries physicotheology is invented as a way of coupling new scientific research with a belief in God and the admiration of God's creation, this is

15 Rudolf Stichweh, *Zur Entstehung des modernen Systems wissenschaftlicher Disziplinen. Physik in Deutschland 1740-1890* (Frankfurt: Suhrkamp, 1984).

16 Christian Mersch, *Die Welt der Patente. Soziologische Perspektiven auf eine zentrale Institution der globalen Wissensgesellschaft* (Bielefeld: Transcript, 2014).

a tradition no longer existent in the late nineteenth century. Therefore the most important form of structural coupling in the present day can be seen in the astonishing belief that theology *is* science and in the application of historical and philological methods to religious texts.

There are numerous other cases that can only be suggested here: The coupling of science and ethics is an interesting case, as ethics did not differentiate as a function system in society; the coupling of science and the public sphere via popularization and the public understanding of science, which is not a coupling to another function system but is more about the rise of public roles in science which are an internal environment of the system of science; the couplings of science to concerns of nation-states not only via science policy but more generally by what is today often called a national innovation system, which includes political actors, economic actors and scientist-entrepreneurs.[17] I have not mentioned sports and the arts, but these are all cases deserving of thorough analysis.

Finally, there is a very important case to be noted.[18] In the twentieth century something emerges which we may call the responsiveness of science to society. The diagnosis of anthropogenic climate change is a good example for it. This is not a structural coupling to a specific function system, but a generalized capacity of modern science to identify macro problems which are problems not only of specific function systems but of the encompassing macro system of (world-)society. This tells us something about the autonomy and strength of modern science, which although it is an observer internal to society succeeds in some cases to observe society as if it were an observer from outside, and which can in this way identify and analyze and even work on problem solutions for crucial problems of the future of humanity.

---

17 Richard R. Nelson, *National Innovation Systems: A Comparative Analysis* (New York: Oxford Univ. Press, 1993).

18 Rudolf Stichweh, "Analysing Linkages between Science and Politics. Transformations of Functional Differentiation in Contemporary Society," in "Interfaces of Science and Policy and the Role of Foundations" (Stiftung Mercator, Essen, March 2015), 38–47.

Section II:
Science in Emerging Nation-States

State—Nation—University: The Nineteenth-Century "German University Model" as a Strategy for National Legitimacy in Germany, Austria, and Switzerland

Dieter Langewiesche

1. The German University Model as Self-Description in Speeches by University Presidents

This study is focused on self-descriptions of and within the German university model, and as such it looks at self-depictions by "the German university" rather than the real composition of universities within the German-speaking world. "The reality of the university is not the reality of its idea,"[1] but from the nineteenth century and into our own times, the development of the university was directed by providing an authoritative model for implementation.

Many have spoken and written about the idea of the German university, but the most important platform for staging such ideas during the nineteenth and well into the twentieth century were festivities that provided occasions for speeches by university presidents.[2] Each year such

---

Translated by Axel Jansen.

1 Rudolf Stichweh, *Wissenschaft, Universität, Professionen. Soziologische Analysen* (Frankfurt: Suhrkamp, 1994), 254.

2 My research has been part of a project pursued in tandem the Historical Commission of the Bavarian Academy of Sciences. See Dieter Langewiesche, "Rektoratsreden—ein Projekt in der Abteilung Sozialgeschichte," in Historische Kommission bei der Bayerischen Akademie der Wissenschaften, *Jahresbericht 2006* (Munich: Bayerische Akademie der Wissenschaften, 2007), 47–60; Langewiesche, "Rektoratsreden. Schlüsselquelle der Universitäts- und Bildungsgeschichte," *Akademie Aktuell. Zeitschrift der Bayerischen Akademie der Wissenschaften* 2 (2008): 68–72. On its website the Historical Commission provides a list and sometimes entire transcripts of presidential addresses given at German univer-

speeches helped mark occasions when a new president assumed office, or in the context of a foundation's festivities and similar events. In his speech, the president would address members of the university, lecturers, and students in particular, but presidents also had in mind those among the audience who represented the state or society at large: ministers or high state officials, in the monarchical era members of the dynasty, representatives of the city, churches, the military, and later also those representing particular societal interests, political parties, or parliaments. Press coverage was common. In Vienna, for example, the president of the University of Technology (*Technische Hochschule*) in 1912 greeted delegates of the government departments, particularly the department of culture and education, the ethnarchy of Lower Austria and additional civil and military agencies, presidents of other universities in Vienna and of corporate bodies and associations, as well as men of science and the arts, of industry and trade whom he did not mention by name. He honored their presence as a token of the influence achieved by other technical universities, an influence he ascribed to the impact of these figures on those institutions.[3] In 1921, the president of the University of Vienna could welcome the President of Austria[4] much as his predecessors before the end of the monarchies in Germany and Austria had frequently welcomed members of the dynasty among their audience.[5] In their speeches, presidents thus spoke to the university as well as those in power and whose political influence allowed them to shape public opinion in the state and society. Each year, speeches were given at all the universities in the German language area, and they

---

sities. It also provides access to the presidential addresses given at Swiss universities. http://www.historische-kommission-muenchen-editionen.de/rektoratsreden.

3 K. K. Technische Hochschule in Wien, *Bericht über die feierliche Inauguration des für das Studienjahr 1912/13 gewählten Rektors, o. ö. Professor der darstellenden Geometrie Dr. Emil Müller am 26. Oktober* (Wien: Verlag der K. K. Technischen Hochschule, 1912), 31–61. Müller presented on "Das Abbildungsprinzip."

4 Gustav Riehl, "Über Entwicklung und Forschungswege der neueren Dermatologie," *Die Feierliche Inauguration des Rektors der Wiener Universität für das Studienjahr 1921/22 am 25.10. 1921* (Wien: Universität Wien, 1921), 52–69.

5 Döllinger, for example, in 1866 welcomed two Bavarian princes and Brock in 1860 the Prussian prince regent and several princes. Johann Josef Ignaz von Döllinger, *Die Universitäten sonst und jetzt. Rectorats-Rede gehalten am 22. Dezember 1866* (München: J. W. Weiß, 1867); August Böckh, *Rede zur Jubelfeier der Königlichen Friedrich-Wilhelms-Universität zu Berlin gehalten in der St. Nikolai-Kirche am 15. Oktober 1860* (Berlin: Verlag von J. Guttentag, 1860). Böckh spoke on "The Contemporary Context and the Spirit in which the University was Founded."

each described the German university model as superior to others due to its efficiency and as an expression of distinct national characteristics.

Presidential addresses came in two particular varieties. The president could speak about research in his field and this allowed him to demonstrate the nature of university education, while he could also discuss questions pertaining to society and the state. In the case of the latter, he also claimed to be speaking on behalf of science (*Wissenschaft*). Though I will not go into this huge subject, I wish to throw light on these speeches with respect to the strategies employed to justify the German university model and to defend it in times of crisis.

The term "German university model" refers to what is commonly called the *"Humboldt'sche Universität"*—the German university as conceptualized by Wilhelm von Humboldt. This term has been in use since the second half of the twentieth century, and somewhat earlier in Berlin—specifically around 1910 and in the context of the centennial. But it was not in use earlier than that. The term *"Humboldt'sche Universität"* is thus of recent coinage, and though the nineteenth century produced this type of university it did not know this label. In the last decade, this point has been sufficiently established by scholars,[6] and I merely note this at the beginning

---

6 Sylvia Paletschek, "Verbreitet sich ein 'Humboldt'sches Modell' an den deutschen Universitäten im 19. Jahrhundert?" in *Humboldt International. Der Export des deutschen Universitätsmodells im 19. und 20. Jahrhundert,* edited by Rainer Christoph Schwinges (Basel: Schwabe & Co. Verlag, 2001), 75–104; Paletschek, "The Invention of Humboldt and the Impact of National Socialism: The German University Idea in the First Half of the Twentieth Century," in *Science in the Third Reich,* edited by Margit Szöllösi-Janze (Oxford: Berg, 2001), 37–58; Paletschek, "Die Erfindung der Humboldtschen Universität. Die Konstruktion der deutschen Universitätsidee in der ersten Hälfte des 20. Jahrhunderts," *Historische Anthropologie* 10 (2002): 183–205; Paletschek, "Zurück in die Zukunft? Universitätsreformen im 19. Jahrhundert," in *Das Humboldt-Labor. Experimentieren mit den Grenzen der klassischen Universität,* edited by Wolfgang Jäger (Freiburg im Breisgau: Universität Freiburg, 2007), 11–15. For my own studies on this subject, see n. 19. Heinz-Elmar Tenorth provides a fair consideration of the various positions relativizing the Humboldtian myth in his "'Mythos Humboldt'—eine Notiz zu Funktion und Geltung der großen Erzählung über die Tradition der deutschen Universität," in *Intuition und Institution. Kursbuch Horst Bredekamp,* edited by Carolin Behrmann, Stefan Trinks and Matthias Bruhn (Berlin: Akademie Verlag, 2012), 81–92; Tenorth, "The University of Berlin: A Foundation between Defeat and Crisis, Philosophy and Politics," *Bildungsgeschichte: International Journal for the Historiography of Education* 4 (2014): 11–28; Tenorth; "Wilhelm von Humboldts (1767–1835) Universitätskonzept und die Reform in Berlin—eine Tradition jenseits des Mythos," *Zeitschrift für Germanistik* 20, no. 1 (2010): 15–28.

so that no one wonders why I am not using the standard terminology. There are other reasons for this which I will discuss later.

Presidents of all universities within the German language area, without exception, had this university model in mind and they continuously inferred a specific type of university that had evolved in the nineteenth century. It took shape in the German states, was quickly adopted in Switzerland and, after the middle of the century, in the Habsburg Monarchy as well, before radiating out, with modifications, beyond the German language region.[7] In 1924, at the annual meeting of the *Europäischen Studentenhilfe des Christlichen Studentenweltbundes*, a university student organization, Carl Heinrich Becker lectured on what was perceived to be the peculiar nature of the university model and how it was tied to the nation.[8] Becker— who was a distinguished professor of Oriental Studies, and in 1921 as well as from 1925 to 1930 held positions as Prussian minister of science, art, and education—condensed the self-perception that university presidents had proclaimed in their speeches at universities in the German language area, and tried to convey this to the foreigners among his audience as something that was thoroughly German.

Without exception, all presidents described the connection between research and teaching as the core of the German university model. Becker called it the "principle of research that is responsible only to itself" that was tied to the "inner freedom to ask any question."[9] This corresponded to the freedom of education among students because one could devote oneself to science (*Wissenschaft*) "only in voluntary chastity and responsibility."[10] Becker thus reiterated standard assumptions. In their respective speeches, presidents came up with many ways to express them, and so did Becker. But in the lingering shadow of World War I, he modified this self-perception to an extent that even critics did not question it. In order to

---

7 See Walter Rüegg, "Themes," chap. 1, Christophe Charle, "Patterns," chap. 2, Edward Shils and John Roberts, "The Diffusion of European Models outside Europe," chap. 6, in *A History of the University in Europe*, vol. 3, edited by Walter Rüegg (Cambridge: Cambridge Univ. Press, 2004); Schwinges, ed., *Humboldt International*.

8 Carl Heinrich Becker, "Vom Wesen der deutschen Universität," in *Die Universitätsideale der Kulturvölker*, edited by Reinhold Schairer and Conrad Hoffmann, Jr. (Leipzig: Quelle & Meyer, 1925), 1–30; Carl Heinrich Becker, *Internationale Wissenschaft und nationale Bildung. Ausgewählte Schriften*, edited by Guido Müller (Cologne: Böhlau, 1997), 305–28. Cf. Guido Müller, *Weltpolitische Bildung und Akademische Reform. Carl Heinrich Beckers Wissenschafts- und Hochschulpolitik 1908–1930* (Cologne: Böhlau, 1991).

9 Becker, *Universität*, 308.

10 Ibid., 309.

align it with the demands of the postwar world as he saw it, he gave up a core element of the university model, its educational concept.

As part of the ritual of the presidential speech, participation in research and its instructional simulation were praised and continued to be praised as the most advanced mode of education based on the conviction that this kind of education shaped an academic habitus. Regardless of later occupation, a university education and participation in learning through research qualified students to engage with the unknown, to handle unsolved problems. The research university was thus an educational institution—because research and teaching based on research facilitated education.[11] That was the German university model's core as coherently presented in presidential addresses. "At least once in his life, at the end of his period at the university, the educated young man should be sufficiently advanced that his knowledge may correspond to the medium stage of scientific research"— proclaimed Rudolf Virchow in his presidential address in Berlin in 1887. Because the "aim of a university education" was "very high [...]: a general scientific and moral education and full knowledge of the discipline."[12]

Becker suggested that this type of education, produced an elite that the thriving nation required. But, after World War I Becker saw an emergence among his contemporaries of a "new humanitarian ideal" that replaced the "liberal individualism of science" with the "spirit of community" from which would arise the "true educated man of a new time." Hence there was no more "education through science" but the shaping of a "coherent personality" through new forms of community.[13] Becker expected the university to lead this development as well, because he considered the German university model to be at the vanguard of innovation.

Becker addressed two other aspects which require our attention in the context of strategies for national political legitimation that were implied by

---

11 For details, see Dieter Langewiesche, "Bildung in der Universität als Einüben einer Lebensform. Konzepte und Wirkungshoffnungen im 19. und 20. Jahrhundert," in *Meta-morphosen der Bildung. Historie—Empirie—Theorie*, edited by Edwin Keiner et al. (Bad Heil-brunn: Klinkhardt, 2011), 181–90; Dieter Langewiesche, "Die Universität als Bildungs-institution," in *Festschrift für Prof. Dr. Heinz-Elmar Tenorth aus Anlass der Verleihung des Er-win-Stein-Preises 2011* (Berlin: Berliner Wissenschaftsverlag, 2012), 29–41; Langewiesche, "Bürger bilden in der Universität," in *Bürger bilden*, edited by Otfried Höffe and Oliver Primavesi (Berlin: Walter de Gruyter, 2015), 155–84.

12 Rudolf Virchow, *Lehren und Forschen. Rede bei Antritt des Rektorats der Königlichen Friedrich-Wilhelms-Universität am 15. Oktober 1892* (Berlin: Verlag von August Hirschwald, 1892), 8.

13 Becker, *Universität*, 327–28.

the German university model. He homogenized the self-perception of universities by "de-federalizing" it, and avoided a problem directly or indirectly addressed in many presidential addresses, namely the tension between the universality of science, on the one hand, and the nationality of universities, on the other. Becker gave his speech at a time of a crisis that would not leave the German university model untouched, and as such it is especially valuable for selecting the foci of my analysis.

2. Investigative Foci

The first of these concerns language. The tone of Carl Heinrich Becker's speech "About the Nature of the German University" made it unmistakably clear that fundamental criticism was not permitted. He wished to reform universities, but he shielded the German university model through his use of a semi-religious but secular language that turned the model's preservation into a national duty: "One may speak of the essence of the German university only with reverent awe." This is how he began his speech, describing the university as a "kind of keeper of the holy grail," and at the same time as a "sacred national site for the entire German people." He continues I this vein, stating that "[t]he entire belief in the value of ideal entities finds its most immediate expression in the value bestowed on the university as an idea." To pursue research and to teach in such a university was thus a calling: "Anyone who devotes his life to it assumes the rank of a priest, but also shares in priestly duties and shares in responsibilities." Such a responsibility to the entire nation "provides the scholar with a preferential role. Every professor is a prebendary to the nation."[14] At a symposium in 1865, the physician Rudolf Virchow took a similar view when he postulated that the "character of German science has acquired much of the true moral earnestness that our people engage in all of their work and which is the real essence of the religious sentiment. I do

---

14 Ibid., 307–08.

not hesitate to say that for us science has become religion."[15] Science thus ennobled its practitioners.[16]

The university is seen as a place where the nation's scientific nobility is performed and where it shapes the future elite. As such this constitutes a spectrum of strategies of legitimation for national politics expressed through presidential addresses and in terms that appear throughout the nineteenth and twentieth centuries. Becker's sacred and inflated language may suggest that the defenders of the German university model were unprepared for the democratization, and particularly the republicanization, of the state, particularly in Germany and Austria in any event. Things were different in Switzerland, however, and thus this issue of openness towards the different political arrangements of the state is an aspect that I will focus on.

From his Prussian-German perspective, Becker pushed this issue of perspectival differences aside. But I shall place it in the center. Becker writes that the German university model was shaped by "the tight connection with the Prussian state [...] and with German Idealism" and that it had thus led "the entire higher intellectual life in Germany" in a direction that was different from other western European cultures."[17] The German university model, in this way, was a uniquely German solution (a "*Sonderweg*"). This point is made again and again in presidential addresses, but outside of the University of Berlin, however, speakers did not have Prussia in mind when they made this point. Instead, they either raise the standard of a national culture (even after the founding of the German national state, Austrian and even Swiss universities would be included), or they refer to the German national federation. From their inception,

---

15 Rudolf Virchow, *Ueber die nationale Entwickelung und Bedeutung der Naturwissenschaften. Rede gehalten in der zweiten allgemeinen Sitzung der Versammlung deutscher Naturforscher und Aerzte zu Hannover am 20. September 1865* (Berlin: Verlag von August Hirschwald, 1865), 17–18.

16 This may frequently be found in presidential addresses, particularly when expressing a view of the university's role in establishing new elites. There were frequent references to the "organization of a new aristocracy" by the university, sometimes with the Christian qualification that "piety is the soul of education." D. Philipp Bachmann, *Über das Interesse der christlichen Sittenlehre an dem allgemeinen Begriff der Bildung. Rede beim Antritt des Prorektorats der Königlich Bayerischen Friedrich-Alexanders-Universität Erlangen am 4. November 1910 gehalten von D. Philipp Bachmann. K. Ordentl. Professor der systematischen Theologie und der neutestamentlichen Exegese* (Erlangen: K. B. Hof- und Universitätsdruckerei von Junge & Sohn, 1910), 6, 13.

17 Becker, *Universität*, 313. He is referring to philosopher and educator Eduard Spranger, *Wandlungen im Wesen der Universität seit 100 Jahren* (Leipzig: E. Wiegandt, 1913).

German universities have belonged to different states. From the nineteenth century onwards, therefore, the strategy to legitimize and secure the German university model had to embrace both a loyalty to the particular state and its ruler and a commitment to the German nation beyond individual states.

3. Legitimizing the German University Model through Openness to the Idea of the German Nation

I begin by asking what type of nation nineteenth-century presidents invoked in their German university model, keeping in mind that even in Berlin, no one at the time referred to this model as "Humboldtian". What we find is the entire spectrum that shaped the political debate about the German nation outside the universities. This comes into focus only when we avoid equating "national" and "nation-state", and when we avoid taking for granted the view that the national state constituted a natural endpoint for German national history.[18] Prior to the creation of the German national state in 1871, contemporaries could mean several things when speaking of German national unity. The spectrum extended from unity as a state to a federation in the tradition of the Holy Roman Empire of the German Nation. Presidential addresses reflect this diversity and so I will provide a sketch of national concepts as they were designed in Berlin, Jena, and Munich,[19] and investigate the implications of politically antagonistic conceptions of a German nation for strategies to legitimize the German university model.

Presidential addresses given in Berlin were dominated by the Prussian-German image of a development, preordained by history, towards the

---

18 Dieter Langewiesche and Georg Schmidt, eds., *Föderative Nation. Deutschlandkonzepte von der Reformation bis zum Ersten Weltkrieg* (Munich: C. H. Beck, 2000).

19 For details (and on the case of Munich), see Dieter Langewiesche, "Humboldt als Leitbild? Die deutsche Universität in den Berliner Rektoratsreden seit dem 19. Jahrhundert," *Jahrbuch für Universitätsgeschichte* 14 (2011): 15–37; Langewiesche, "Die 'Humboldtsche Universität' als nationaler Mythos. Zum Selbstbild der deutschen Universitäten im Kaiserreich und in der Weimarer Republik," *Historische Zeitschrift* 290 (2010): 53–91; Langewiesche, "Selbstbilder der deutschen Universität in Rektoratsreden. Jena—spätes 19. Jahrhundert bis 1948," in *Jena. Ein nationaler Erinnerungsort?*, edited by Jürgen John and Justus H. Ulbricht (Cologne: Böhlau, 2007), 219–43.

creation of the German national state, without Austria, and with the Prussian king as German emperor. Within this image, university presidents in Berlin assigned a key role to the founding of their university. The Prussian military and the founding of the University of Berlin constituted "an intellectual armory to accompany the martial," as the university president explained in 1859.[20] In this image the founding of the University of Berlin is turned into a pillar of the German national state. During the Franco-Prussian War in 1870, the president of the University of Berlin even extolled his university as an "intellectual *Leibregiment* [household-bodyguard regiment] of the House of Hohenzollern."[21] The founding of that institution subsequently became part of a German national myth in which the fight against Napoleon constituted a prelude to the national state of 1871. "The founding of our Friedrich-Wilhelm-University among the rubble of the Reich was a German deed and a step to the emperor's throne," explained the president of the University of Berlin in 1882.[22] No one in Germany since 1871 could lay claim to a stronger form of legitimacy in national politics. Subsequent presidents of the University of Berlin renewed their claim each year by celebrating, over and over, their university's epic narrative.

The leadership that presidents of the University of Berlin claimed for their institution provided national political legitimacy that was not based on references to higher scientific achievements in research and education or by pointing to a specific policy success, such as heralding Berlin with the authorship of a Humboldtian university model. In his 1875 speech, the university president Theodor Mommsen emphasized the University of

---

20 August Boeckh, "Festrede gehalten auf der Universität zu Berlin am 15. October 1859," in Boeckh, *Gesammelte kleine Schriften*, edited by Ferdinand Ascherson (Leipzig: B. G. Teubner, 1866), 3.19–32, 28. Boeckh addresses Prussia's contribution to German science.

21 "Der deutsche Krieg. In der Aula der Berliner Universität am 3. August 1870 gehaltene Rektoratsrede," in *Reden von Emil du Bois-Reymond in zwei Bänden*, edited by Estelle du Bois-Reymond (Leipzig: Veit, 1912), 1:393–420, 418.

22 Ernst Curtius, *Rede zur Gedächtnisfeier König Friedrich Wilhelms III in der Aula der Friedrich-Wilhelms-Universität am 3. August 1882 gehalten von Ernst Curtius, z.Z. Rector der Universität* (Berlin: Buchdruckerei der Königlichen Akademie der Wissenschaften, G. Vogt, 1882), 12. Plans for a university were drawn up in response to the humiliating defeat in the Battle of Jena. "It was a decision made by the king, considered among the most noble men, carefully organized and quickly acted upon, a work on which all friends of the fatherland raised themselves, at which men such as Scharnhorst and Schleiermacher shook hands as equals, an epoch of patriotic history in the most humble forms." Ibid., 6.

Berlin's strategy for claiming an elevated status among its peers by pointing to those faculty and students who had been killed in action during the Franco-Prussian War of 1870, and by connecting that conflict to the anti-Napoleonic wars and the founding of his institution:

Asking for the best university in Germany is not less legitimate than asking for the greatest German scholar; we know nothing of it and we don't ask for it. But we do know and it is not a question that in the greatness of its foundation, in its wonderful initial blessing, no other institution of higher education in Germany may compare itself to us. Most German universities have been called into being through careful consideration or the blessing of luck; some have been created through an emperor's whim or by chance; our institution has sprung from the fight for life and death of a people and from the genius of the highest national danger.[23]

In Jena, national history was recalled in a different way, but this history was also used to legitimize the university. Here presidents pointed to the university's research achievements and tied them to national politics. "Our university has never been of mere local significance," explained jurist Richard Loening in his 1897 presidential address: "in the past three centuries there has not been one important advancement in Germany's intellectual life, not one important turning point in scientific knowledge [...] to which our Academia Salana has not contributed in a most significant way, in which she has not been a decisive factor."[24]

The political province was stylized as the German nation's intellectual center and as a key location of its history. As Bismarck had remarked when he visited the city, there was "[n]o Sedan without Jena,"[25] and its university presidents readily had acknowledged this confirmation of the significance of their city and university in national politics. In the nineteenth century and beyond, university presidents in Jena developed a key motif in their speeches, namely that it stood as a bolster to national politics, an intellec-

---

23 Theodor Mommsen, *Reden und Aufsätze* (Berlin: Weidmann, 1905; repr. 2001), 17–31, 17. First published 1875 in Berlin.

24 Richard Loening, *Über ältere Rechts- und Kulturzustände an der Fürstlich Sächsischen Gesammt-Universität zu Jena. Rede gehalten bei der akademischen Preisvertheilung am 19. Juni 1897 in der Kollegienkirche zu Jena* (Jena: Universitäts-Buchdrucker G. Neuenhahn, 1897), 3. Loening taught law.

25 Friedrich Nippold, *Der Kurfürst-Konfessor Johann Friedrich. Rede gehalten zu seinem Säkular-Jubiläum am 30. Juni 1903, nebst Universitätsbericht über das Jahr 1902/03 von Julius Pierstorff, o. ö. Professor der Staatswissenschaften, d. Z. Prorektor* (Jena: Universitäts-Buchdrucker G. Neuenhahn, 1903), 4, 3. In 1870, a decisive battle against France took place near Sedan. Nippold taught history of religion.

tual and scientific world capital, and of their university's cultural and political significance for the German nation. This line of argument was retained during the Third Reich and can be seen reiterated on the university's website today.[26]

The University of Munich is my third example. University presidents in Munich, of course, did not lay claim to Prussian-German historical mythology to legitimize the German university model. Wilhelm Heinrich Riehl assumed the office of university president in 1883, and he provided the most sophisticated interpretation of the interplay between the university and politics. The title of his address was "The University as a Home." (He used the German term *Heimat*.[27]) A cultural historian and a founder of folklore studies (*Volkskunde*) as an academic discipline at German universities, Riehl situated the German university in a *Heimat* of several layers. First: Munich—not a municipal university but the city shaped the university. Germany at the time could not ignore this because the rank of a university (as indicated by its hiring success) depended on its size. Hence the universities in large cities such as Berlin and Munich or Vienna led the way. They attracted the most students and the professors could generate the highest income as they lectured to the largest fee-paying audiences. While other universities provided an entrée or advancement for academic careers, Berlin and Munich were ultimate destinations.[28]

A second dimension of *Heimat* was the state or *Land*. According to Riehl, all German universities were shaped by political developments rather than developments in fields of research: "Decisive innovations in the spirit and form of our universities have emerged from the nature of the modern state." But he explained that the state needed the university: "In our time, it is not redundant to recall that the rank of states is related to the independent support of science [*Wissenschaft*]." As such there was no relevant state without a university: "States that do not even have a university are petty states indeed." Hence, as the state creates and finances the university, the university establishes the state's rank.

---

26 Friedrich-Schiller-Universität Jena, "Geschichte der Unviersität Jena," http://www.uni-jena.de/Geschichte.html.

27 Wilhelm Heinrich Riehl, *Die Heimat der Universität. Rede an die Studierenden beim Antritt des Rektorates der Ludwig-Maximilians-Universität gehalten am 1. Dezember 1883* (München: Kgl. Hof- und Universitätsdruckerei von Dr. C. Wolf & Sohn, 1883). For details, see Langewiesche, "Humboldt als Leitbild?"

28 Marita Baumgarten, *Professoren und Universitäten im 19. Jahrhundert. Zur Sozialgeschichte deutscher Geistes- und Naturwissenschaftler* (Göttingen: Vandenhoeck & Ruprecht, 1997).

A third dimension of *Heimat* was Germany itself. Riehl considered the German national state as a successor of the Old Empire. In this he deviated from his colleague in Bonn who considered that "the old Germany of rubbles and of statelessness" had been given a new form by Prussia through the founding of the universities in Berlin, Breslau, and finally Bonn.[29] Riehl had a positive view of the Empire, and that is why he could call all universities in Germany *"Reichs-Universitäten"* (universities of the empire). He thus described the claim made by all of them for scientific equality:

It is wrong to insist on uniformity in these matters, to create a standard charter for all. This would compromise the spirit of the 'German university'. This is because it was the individual life (which does not preclude unification) that gave rise to the inner strength of our institutions of higher education. There is no 'first university" in Germany, no normal or central university, no central-empire-university. At all times, we have had shifting predominant, outstanding universities, depending on the influence of great masters and their schools, but never has one dominated all others and in all disciplines.

Riehl considered the *Reichsuniversität* (the university of the empire) to be a fourth type of *Heimat*. Like many of his contemporaries, he was aware of the German cultural nation in addition to the national state. The German university model, therefore, extended well beyond the borders of the national state. In his words, the "northernmost German university is not Königsberg, but Dorpat, the southernmost not Freiburg but Bern; German universities in Austria, in Switzerland, and in Russia lie within our larger national academic *Heimat* just like Berlin or Munich."

This was no imperial transgression of German state borders but part and parcel of a tradition of thinking in terms of a nation as culture, a nation not instigated by statehood. In this sense, too, Riehl's 1883 presidential address proves to be a highly political alternative to the Prussian-German mythology proposed in similar speeches in Berlin—a mythology

---

29 Aloys Schulte, *Die Schlacht bei Leipzig. Mit einem Schlachtenplan. Rede, gehalten bei Übernahme des Rektorates der Rheinischen Friedrich-Wilhelms-Universität zu Bonn am 18. Oktober 1913* (Bonn: Marcus & Weber, 1913), 21. Charles E. McClelland, *State, Society, and University in Germany, 1700–1914* (Cambridge: Cambridge Univ. Press, 1980), considers the founding of the University of Bonn to be of equal importance for the Prussianization of the Rhineland as the stationing of troops there (ibid., 146). To him, the German university model is the Prussian model, its victory complete with the founding of the German Reich (ibid., 152). McClelland's analysis has been disproven by recent scholarship (see n. 6 and 19 above).

that ossified during the second half of the nineteenth century even if it was accepted only in Berlin. Riehl countered by proposing his empire-based, federal historical model that contained no leading university, neither on political terms nor in terms of scientific achievement. In tune with his colleagues in other non-Prussian states, he praised federal pluralism, a pluralism he wished to preserve.

Such a federal variation of national political legitimacy for the German university model was aimed at the perception, deeply rooted among Germans, that they were citizens of a number of states. This consciousness was preserved in the national state and provided the basis for German federalism. Legitimizing the German university model in this way—i.e. on the basis of the individual state, on federal terms, as well as nationally— had a much stronger appeal than alternate models that considered the University of Berlin to be the model German university. In this sense, it made sense for everyone not to infer the Humboldt model. The claim that all German universities were equals and each dedicated to the same idea had political implications as it called on the state to provide it with a material basis. By declaring the uniform German university model to be the basis for the German nation's position in the world and for the rank of individual states within Germany, university presidents demanded from political leaders among their audience that such ideas be institutionalized.[30]

Such demands were accepted remarkably well. In a 1902 book that has become a standard, Friedrich Paulsen has written that German universities in the second half of the nineteenth century no longer provided "the real center of national life," as German unity now rested on stronger pillars such as parliament and the economy, but universities remained significant

---

30 The publication by a Swiss-born theologian indicates just how effective such arguments were. He had studied in Germany, moved on to England and was then appointed to a chair in theology in Mercersburg, Pennsylvania: Philip Schaff, *Germany. Its Universities, Theology and Religion* (Philadelphia: Lindsay & Blakiston, 1857): "The Universities are the pride and glory of Germany. They exert more influence than similar institutions in any other country. They are the centres of the higher intellectual and literary life of the nation, and the laboratories of new systems of thought and theories of action" (ibid., 27). The university assumes "the first rank of all similar institutions in Germany not only, but in the world" (ibid., 63). When asked to recommend a university in Germany to Americans, he would advise that they "devote the winter in Berlin, and to divide the summer between Bonn and Heidelberg." If another year was available, he would recommend spending "the cold season at Halle, Leipzig, Munich, or Vienna, and the warm season at Göttingen, Tübingen and Zürich" (ibid., 62).

And they also profited from competition among the German states.[31] When the national state restricted the powers of its member states, university politics emerged as a field in which the states could continue to compete. When the King of Saxony, as *Rektor magnificentissimus* in Leipzig (an honorary title he had received in 1875) attended lectures by university professors, this also served to reinforce the institution against competition from Berlin.[32] After World War I, cabinet members and high-ranking state officials continued this tradition.[33]

In this way, the homogenous German university model was designed to facilitate competition. Riehl will have had this in mind when he stated the following in his Munich address: "By creating a privileged leading university in Germany, German universities will seize being the first universities in Europe." In Austria, there were complaints that the country lacked a "culturally beneficial competition among dynasties and governments." Such competition had resulted in "universities in the German province" (a phrase that would never have crossed the lips of a German university president) to be much better endowed than their Austrian counterparts.[34]

---

31 Friedrich Paulsen, *Die deutschen Universitäten und das Universitätsstudium* (Berlin: A. Asher & Co., 1902), 9–10, 93. Along these lines, the philosopher Hans-Georg Gadamer explained the quality of German universities in the empire by suggesting that "the great Prussia with its many universities and the monumental Berlin founding called into action small Saxony and small Baden as competitors who sought to keep up with their own universities." "Heidelberg als geistige Lebensform: Ein Gespräch mit dem Philosophen Dieter Henrich," *Frankfurter Allgemeine Zeitung*, Oct. 15, 2014, N4.

32 Jens Blecher, "Hoch geehrt und viel getadelt. Die Leipziger Universitätsrektoren und ihr Amt bis 1933," in *Die Leipziger Rektoratsreden 1871–1933*, edited by Franz Häuser (Berlin: De Gruyter, 2009), 7–34, 30–32.

33 Ibid., 32–33.

34 Richard Wettstein Ritter von Westerheim, "Forschung und Lehre," in *Die Feierliche Inauguration des Rektors der Wiener Universität für das Studienjahr 1913/14 am 20.10. 1913* (Wien: K. K. Universität, 1913), 47–78, 62–63. Concerning competition in the German university system, see Baumgarten, *Professoren und Universitäten* on hiring practices; for a different approach see Joseph Ben-David, "Scientific Productivity and Academic Organization in Nineteenth Century Medicine," *American Sociological Review* 25 (1960): 828–43.

4. The German University Model: Forms of Government, Concepts of Freedom

Another aspect to consider is the German university model and the form of political government. As long as monarchies existed in Germany, no university president admitted to any doubts concerning his university's loyalty to its ruler and to monarchy as an institution. German universities had commonly been founded by the dynasty in their respective German state. In their speeches, university presidents commemorated the founding of the university, and sometimes the transfer of the presidency coincided with festivities to honor the university's founding. The proximity to the prince thus bestowed the university with legitimacy and university professors reminded their audience of this proximity again and again. In Berlin, for example, the university was called a "monument to Hohenzollern,"[35] while in Dorpat the founding and support of the university by Czar Alexander I was praised,[36] and in Würzburg one felt that the Wittelsbach family had endowed the university so well that the university "did not need to avoid comparison with any other."[37] Even in Rostock (during the nineteenth century the least attractive university for professors and students[38]) the president praised the ruler as a "prototype of a real German prince" who endeavored to support his university to the utmost.[39]

---

35 *Rede zur Gedächtnisfeier* (see n. 22 above), 4.

36 Leo Mayer, *Festrede zur Jahresfeier der Universität Dorpat am 12. December 1871 gehalten von Leo Meyer, ord. Prof. der deutschen und vergleichenden Sprachkunde, nebst den Mittheilungen über die Preisaufgaben sowie dem Universitäts-Jahresbericht für das Jahr 1871* (Dorpat: C. Mattiesen, 1871). Alexander Brückner and Carl Weihrauch, *Festreden zur Feier des hundertjährigen Geburtsfestes Seiner Hochseligen Majestät Alexander I. am Stiftungstage der Universität Dorpat am 12. December 1877* (Dorpat: C. Mattiesen, 1878). Brückner taught the history of Russia and he presented on the life of Alexander I. and the founding of Dorpat University. Weihrauch was professor physical geography and meteorology and he presented on the uses of meteorology.

37 Wilhelm Wien, *Die neuere Entwicklung unserer Universitäten und ihre Stellung im deutschen Geistesleben. Rede für den Festakt in der neuen Universität am 29. Juni 1914 zur Feier der hundertjährigen Zugehörigkeit Würzburgs zu Bayern* (Würzburg: Stürtz, 1915), 3. Wien was court counselor and professor of physics.

38 Baumgarten, *Professoren und Universitäten* 257–58.

39 Franz Bernhöft, *Das neunzehnte Jahrhundert als Vorläufer einer neuen Bildungsstufe. Rede gehalten zur Universitäts-Feier am 28. Februar 1900* (Rostock: Univ.-Buchdr. von Adlers Erben, 1900), 3. Bernhöft was a lawyer.

Presidents put their university under the protection of the monarch who in several states was the university's formal head, but it was a conscious proximity that was celebrated. There was also the obedient panegyric[40], but more common was the proud claim that the university was the central place for science and that the state's power and future rested on its success. To convey this much was the dominant role of addresses in the various disciplines, particularly in the natural sciences.

When the chemist August Wilhelm Hofmann gave his presidential address in Berlin in 1881, he entitled it "A century of chemical research under the patronage of the Hohenzollern," but the actors of progress were chemists. As a result of their research Hofmann reported that the sugar industry really had "arisen in the lap of science." [41] The university as "one of the most organic elements of civilized states," as one medical scientist had written in 1855,[42] was an outlook that university presidents repeatedly presented to their audience. And because German universities, due to their specific form, were "central places for German education" (according to the president of the University of Erlangen in 1870), the "great deeds of our nation" on the battlefield were not a success of armies alone, but "to no lesser degree the natural result of the superiority of our education, order, and moral state of affairs." [43] The new buildings erected on university campuses in Germany ("grand palaces, glorious temples" according to the Hofmann in 1881[44]) represented the university's position in the state.

---

40 See, for example, Hugo Wilhelm Paul Kleinert, *Beziehungen Friedrichs des Großen zur Stiftung der Universität Berlin. Rede zur Gedächtnisfeier des Stifters der Berliner Universität König Friedrich Wilhelms III in der Aula der Universität am 3. August 1886* (Berlin: Königliche Akademie der Wissenschaften, G. Vogt, 1886).

41 August Wilhelm Hofmann, *Ein Jahrhundert chemischer Forschung unter dem Schirme der Hohenzollern. Rede zur Gedächtnisfeier des Stifters der Kgl. Friedrich-Wilhelms-Universität zu Berlin am 3. August 1881 in der Aula der Universität* (Berlin: Königliche Akademie der Wissenschaften, G. Vogt, 1881), 5.

42 Christian Gottfried Ehrenberg, *Antrittsrede bei der Übernahme des Rectorats der Universität zu Berlin am 15. October 1855 in der Aula der Universität gehalten von Dr. C. G. Ehrenberg, Professor der Medizin* (Berlin: Königliche Akademie der Wissenschaften, 1856). Ehrenberg discusses the "age of experiment" to which the natural sciences had given rise.

43 Karl von Hegel, *Die deutsche Sache und die deutschen Hochschulen. Rede beim Antritt des Prorectorats der Königlich Bayerischen Friedrich-Alexanders-Universität Erlangen am 4. November 1870 gehalten von Dr. C. Hegel, ordentlicher Professor der Geschichte, d. Z. Prorector* (Erlangen: Eduard Besold, 1870), 3–4.

44 Hofmann, *Ein Jahrhundert chemischer Forschung*, 55. The buildings at the Leipzig University are well documented in Senatskommission zur Erforschung der Leipziger Universitäts-

When the two Humboldt-monuments were erected in Berlin a few years later, Emil du Bois-Reymond pointed out that the university building, a "palace of science," was located across from the "palace of the dynasty." In combination with the Alexander- and Wilhelm-von-Humboldt memorials, the two palaces had long become "emblems of the Hohenzollern capital city." [45] In 1911, Max Lenz in his presidential address delivered an even stronger opinion on "freedom and power in light of the development of our university." He concluded by saying, that "if anything has become evident as a result of this century of cognition, it is the power of knowledge."[46]

University and prince stood at eye level with one another.[47] University presidents conveyed this impression in their politics of science and in their choice of symbols. Towards the establishment of the first republic, universities did not assume this attitude in Germany or in Austria. This comes as no surprise and presidential addresses also attest to it. Graz University in 1932 provides an example of the helpless earnestness emanating from such addresses after the end of the monarchy. The physicist Benndorf meditated on socialism and nationalism, about the "herd instincts" amongst the masses, about the necessary national indifference of the Catholic church, about the support of science as one of the "most noble tasks of the state devoted to culture," about elite formation. He also spoke of the "final

---

und Wissenschaftsgeschichte, *Geschichte der Universität Leipzig 1409–2009*, vol. 5, edited by Michaela Marek and Thomas Topfstedt (Leipzig: Leipziger Universitätsverlag, 2009).

45 Emil du Bois-Reymond, "Die Humboldt-Denkmäler vor der Berliner Universität. In der Aula der Berliner Universität am 3. August 1883 gehaltene Rektoratsrede," in *Reden von Emil Du Bois-Reymond. Erste Folge. Litteratur, Philosophie, Zeitgeschichte* (Leipzig: Veit, 1886), 480–517, 514. The speech was first published in 1883. The publication in the *Deutsche Rundschau* XXXVII (1883): 71 ff. reached a larger educated audience.

46 Max Lenz, *Freiheit und Macht im Lichte der Entwickelung unserer Universität. Rede zum Antritt des Rektorates der Königlichen Friedrich-Wilhelms-Universität in Berlin* (Berlin: Universitäts-Buchdruckerei von Gustav Schade, 1911), 22.

47 Ulrich van Wilamowitz-Moellendorff showed how finely granulated such ideas could be presented: Wilamowitz-Moellendorf, *Neujahr 1900. Rede zur Feier des Jahrhundertwechsels gehalten in der Aula der Königlichen Friedrich-Wilhelms-Universität am 13. Januar 1900* (Berlin: Weidmannsche Buchhandlung, 1900). He had the nineteenth century commence with the French Revolution and conclude with the first Kaiser of the German national state, the "father of the fatherland" (ibid., 8–9). But he then had his "review of the century's successes reach its apex with the prizes of science" (ibid., 19). The educated classes anticipated the new empire but it required military deeds to make it happen. Everyone played a role and such harmony remained untouched by anyone's demand for preeminence.

failure of the mechanisms of the private-capitalist economic order" and the necessary "transformation to a new economic order" (but not the Russian), before ending by hoping for a "great seasoned leader at the helm."[48] For the monarchical period the speeches infer a kind of freedom that is reflected in the university's freedom in research and education. For this is also where one felt superior to other countries, France in particular.

As we have seen, this particularly German type of freedom could be put to use on political terms. Positive tributes to the Revolution of 1848, for example, may be found in many presidential addresses.[49] Or when Rudolf Virchow, in 1865, called German reasoning the "engagement of the mind without authority," sought to ground it in the natural sciences, and demanded that these sciences should serve as a model for all reasoning. But his political reference remained muted. He positioned the sciences against "the feudal character where everyone remains in his fortress and wishes to be an independent German baron," and he elevated the German tax payer, not the prince, into a position of financial supporter of last resort for the natural sciences.[50] Emil du Bois-Reymond connected this German freedom, institutionalized by the German university model as the freedom of research and education, to the lecture fees paid by students. These fees turned the private *docent*, the unsalaried but fully qualified university lecturer, into an engine of innovation at German universities, and they would support the professors "to withstand a pressure from above in political and religious things." He even called the honorarium by students for lecturers "an element of academic freedom as such." Due to the lecture fees, all candidates for a professorship could prove themselves "year after year in the silence before the incorruptible jury of dues-paying German students." In the "fight for academic existence in which, according to steadfast natural laws, only he will prevail in the field who is inwardly

---

48 Hans Benndorf, *Die Aufgaben der Universität und ihre Bedeutung für Volk und Staat. Rede, gehalten bei der Inauguration als Rector magnificus der Karl-Franzens-Universität in Graz am 14. November 1932* (Graz: Leuschner & Lubensky, 1932).

49 See Bachmann, *Professoren und Universitäten*, 3.

50 Virchow, *Lehren und Forschen*, 29, 21. After an extended stay at Harvard and Columbia universities, chemist and philosopher Wilhelm Ostwald thought "monarchical Germany" had "the most democratic university organization" while democratic America had organized its universities in a "strictly monarchical" order. The American university president Ostwald considered an "absolute elected monarch for life." Wilhelm Ostwald, "Deutsche und amerikanische Universitäten," in Ostwald, *Die Forderung des Tages* (Leipzig: Akademische Verlagsanstalt, 1910), 538–49, 546. Ostwald's text dates to 1906.

truthful."[51] The physiologist, one of the best-known German professors at the time, frequently interpreted historical developments using biological analogies.

In Germany and Austria, the German university model was tied to the monarchy before World War I, and after the war there was regret that this form of government was gone. The dual loyalty to nation and monarchical state within that nation severely complicated a reorientation after the demise of all monarchies. German and Austrian universities were thus unprepared for parliament and republic and found it difficult to integrate with them. They could, however, continue to appeal to the nation as a source of legitimacy above the state, a source that was intimately connected to the German university model. But there was also an antithesis to the unloved republic. The case of Switzerland during the nineteenth century showed that this connection to the monarchy was not part of the German university model. It all depended on the model's political and social context.[52] In republican Switzerland, the educational responsibility bestowed on universities according to the German university model (education through participation in research or instruction on the basis of research) was interpreted in a different way from monarchical Germany and Austria.

At Swiss universities, university presidents translated this model into a concept that aimed at the formation of political elites on the basis of popular sovereignty and democracy. This would have brought universities in Germany and Austria into an unceasing conflict with their respective states. Their guiding principles for dealing with the political sovereign before 1918 kept educational institutions at arm's length from societal developments that aimed to strengthen democracy and parliament and weaken the role of monarchs. In a monarchical state, the distance to democracy and parliament could not be cancelled by appealing to the nation. In Switzerland, however, no conflict existed between national state and canton. In a nutshell, the form of the German presidential address related the university to two different levels of legitimacy: to the nation and to the individual state within that nation. University presidents demanded immediacy of relation to both: to the monarch and to an ideal power for

---

51 "Über Universitätseinrichtungen. In der Aula der Berliner Universität am 15. Oktober 1869 gehaltene Rektoratsrede," in *Reden von Emil du Bois-Reymond in zwei Bänden*, 2nd ed., edited by Estelle du Bois-Reymond, (Leipzig: Veit, 1912), 1:356–69, 366–67.

52 For Swiss universities, additional details in Langewiesche, "Bürger bilden in der Universität."

which the university claimed to be the predestined interpreter. In Germany, the nation could be played off against people and popular sovereignty, and many university presidents did so. Not in Switzerland, however. The Austro-Hungarian monarchy faced the most difficult situation because of the many nations there claiming autonomy. University presidents in Vienna thus refrained from engaging in national politics.[53] The idea of placing the university as a "stronghold of German science" against the political *Zeitgeist* did not enter presidential addresses before the onset of the republic.[54]

5. Universality of Science: Universities in the Nation

A final aspect to consider is that university presidents did not mention this tension. When they praised the German university model, they spoke about it decisively, but without much of a theoretical foundation and only in passing. They became somewhat more detailed and elaborate (even though they continued to undertheorize it and were not very explicit) when they presented developments in their discipline or problems in their research fields. To address such research problems in front of a large auditorium including many laypeople was by no means considered impertinent. An academic presentation was intended to showcase university education, to discuss unanswered questions in science and to learn that in the never ending research process mistakes may be just as revealing as the confirmation of research hypotheses. This conception of education (getting to know the limits of knowledge) claimed universality in its own right. It was dedicated to an idea of the university that connected the production of knowledge and the transfer of knowledge, knowledge that remained inde-

53 This assessment arises from a consideration of speeches delivered since 1900. In 1910, the new university president Edmund Bernatzik had spoken "On National Registers" ("Über nationale Matriken") to advocate a stronger acceptance of national rights. Bernatzik, *Die feierliche Inauguration des Rektors der Wiener Universität für das Studienjahr 1910/1911 am 20. Oktober 1910* (Vienna: K. K. Universität, 1910), 57–108. For Austrian presidential addresses, see Karl-Franzens-Universität Graz, Universitätsarchiv, "Inaugurationsreden und Rektoratsreden als universitäts- und wissenschaftshistorische Quellen und ihre Bedeutung der Universitäts-, Wissenschafts- und Kulturgeschichte," http://archiv.uni-graz.at/de/forschung/laufende-forschungsprojekte/inaugurationsreden.
54 Riehl, "Über Entwicklung und Forschungswege der neueren Dermatologie," 55.

pendent of its use in a particular time and place.[55] Its organizational setting within an individual state or its ideal location in a nation did not detract from its universality.

And there is another criterion that Rudolf Stichweh mentions in his considerations on the "universality of the European university" in the Middle Ages and in early modern times, a criterion that was met by the German university model (and the same holds true for most universities as they evolved in the German language area during the nineteenth century), namely the universality of research and education in the sense that all disciplines are represented at the university. The model of the "German university" and its idea of education sought to live up to this. In their speeches, university presidents persistently referred to the universality of all sciences within the university (according to du Bois-Reymond the "universality of teaching" was comprised of "all of human knowledge") in order to demarcate the German system and French science organized in special schools.[56]

The universality of knowledge in specific disciplines was amply demonstrated in academic speeches given by university presidents in which they focused on developments in their research fields. There were no territorial or national boundaries. According to Hofmann in 1881, many young researchers in his field of chemistry had gone to France, England, and Sweden in order to study chemistry as an experimental science. He then traced their careers as they became professors in Germany and reported on the innovations that they had contributed in their field.[57] In a similar vein, the president of the University of Rostock in 1902 told the history of his study, mathematics, as a European success story with roots in France and, to a lesser extent, Italy. His field approached "the ideal of a world language, not being tied to nations and political boundaries." But he ended his presentation with this confession: "The system of exact science has no nationality but we, the caretakers of science who wish to introduce German youth

55 Rudolf Stichweh, *Der frühmoderne Staat und die europäische Universität. Zur Interaktion von Politik und Erziehungssystem im Prozess ihrer Ausdifferenzierung* (Frankfurt: Suhrkamp 1991), 15–23, 23.

56 du Bois-Reymond, 359. See also Hermann von Helmholtz, *Über die akademische Freiheit der deutschen Universität. Rede beim Antritt des Rectorats an der Friedrich-Wilhelms-Universität zu Berlin am 15. October 1877 gehalten von Helmholtz* (Berlin: August Hirschwald, 1878).

57 Hofmann, *Rede zur Gedächtnisfeier.*

into its lively workshop, are committed to the fatherland and we take pride in serving it, a fatherland whose honor is our honor."[58]

Not all academic speeches end on such an emphatically national note but it was common to connect a commitment to the universality of science with its territorial or national use. It was also common to erect hierarchies of achievements in science. These mostly referred to the respective field or a state, and less frequently to a university. You will thus find this for Berlin more frequently than for other German universities, but this depended on the speaker's personality. In 1883, du Bois-Reymond considered the stars of small universities to have expired while "the sum of forces united at the University of Berlin and at the academy" were growing and made "Berlin the intellectual capital of Germany."[59] But this assessment was an exception. Others may have thought so, but as a university president you would refrain from saying it.

But it was also rare that someone—such as Ulrich von Wilamowitz-Moellendorff—would say that it was secondary in what areas science was advancing: "Every truth turns into common sense once it is stated. What does it matter who said it first? The source trickles into the creek, the creek into the stream, and the water of all streams flows into the eternal ocean. Thus in science. How may an individual or a people claim, this drop, this wave derives from me?" [60] As a strategy to legitimize the German university model and for the costly demands of all presidents to realize this model in every German university in its entirety, such an irenic attitude was of little use. The German Kaiser was more realistic at the University of Berlin's centennial celebrations when he demanded that they should "retain their Prussian-German character. Science was the common good of the entire cultural world and its achievements do not stop at any boundary-post. But still—just like any nation needs to preserve its particularity if it wished to sustain its independent existence and its value for the whole—

58 Otto Staude, *Die Hauptepochen der Entwicklung der neueren Mathematik. Rede gehalten in der Aula der Universität Rostock am 28. Februar 1902 von Prof. Dr. Otto Staude, als derzeitigem Rektor,* in *Jahresbericht der Deutschen Mathematiker-Vereinigung,* vol. 11 (Leipzig: B. G. Teubner, 1902), 280–92.

59 du Bois-Reymond, "Die Humboldt-Denkmäler," 503–04.

60 Wilamowitz-Moellendorff, *Neujahr 1900,* 29.

the Alma mater Berolinensis remained aware that it was a German university." [61]

One may characterize as a token of its time this insistence on the university's service to the nation and to the state that paid its bills. This is certainly the case. But one should also be aware that an idea of the nation was dominant at the time that considered the nation to be an eternally valid form of human organization. The German historian Wilhelm Giesebrecht in 1859 had written metaphorically that the nation was as though "God's thoughts were embodied" in it.[62] Many university presidents shared this perspective and they associated it with the university and with science. A protestant theologian in his presidential address at the University of Berlin in 1864 postulated that only if arranged in "nations and national communities of knowledge" could "humanity embrace the world, that which is worthy of knowledge."[63] He was convinced that Christendom had "cleaned the people's natural grounding [den Naturgrund der Völker]" and had united anything "truly human" with nationality. A labor of universalization that affected the university, because the "increasing penetration" of the national and the universal created "the strong living movement of our universities." He thus demanded of them scientific progress in a national framework but in a "universal spirit."[64]

The protestant theologian, by implication, also raised the question of universality and religion because he saw the national impulse in the German nation emerge from the University of Wittenberg.[65] Others were less cautious when they called Martin Luther the "greatest university teacher […] in the history of the world."[66] German universities are "not all or not

61 Erich Schmidt, *Jahrhundertfeier der Königlichen Friedrich-Wilhelms-Universität zu Berlin 10.–12. Oktober 1910. Bericht im Auftrag des Akademischen Senats von dem Prorektor Erich Schmidt* (Berlin: Universitäts-Buchdruckerei von Gustav Schade, 1911), 38.

62 Wilheml von Giesebrecht, "Die Entwicklung der modernen deutschen Geschichtswissenschaft," *Historische Zeitschrift* 1 (1859): 8.

63 Isaak August Dorner, *Über das Wesen und die Idee der Universität. Rede gehalten am 15. October 1864 von I. A. Dorner, d. Z. Rector der Königl. Friedrich-Wilhelms-Universität zu Berlin* (Berlin: Königliche Akademie der Wissenschaften, 1864), 10. Dorner developed the idea of the university, the production of knowledge and its universality but "within communities of knowledge."

64 Ibid. 8–9.

65 Ibid. 4.

66 Adolf Kirchhoff, *Rede des Antritts des Rectorats gehalten in der Aula der Königlichen Friedrich-Wilhelms-Universität am 15. October 1883* (Berlin: Königliche Akademie der Wissenschaften, G. Vogt, 1883), 4. Kirchhoff spoke on Luther, grammar schools and the classics,

exclusively protestant" but they "are rooted [...] in the spirit of Protestantism, and the more they move away from it, the less they are German." This attitude, expressed in 1859 by Böckh in his speech in Berlin,[67] was widespread at German universities. Catholicism and science were considered by them to be mostly irreconcilable. Even though they committed themselves to the German university model, Austrian universities were not entirely included in this perspective. This was Böckh's judgment[68], and Virchow raged against the "decay" (*Versumpfung*) of the University of Vienna because the Habsburg dynasty had managed to "subdue the protestant spirit."[69] In 1911 Max Lenz characterized plans for a university in Berlin that had been developed by the philosopher Fichte a century earlier as a "papacy of savants," and in doing so proffered the strongest condemnation that could possibly emerge from the university milieu in protestant Germany.[70]

The implementation of the medieval university by the pope had been part of its universality.[71] In the age of nation and national state, however, the commitment of a professor to the Catholic Church was considered a burden for respecting the universality of science in research and education.

During the nineteenth century, the German university model's standard to represent within each university the universality of science as *Universitas* came under pressure. It could not be overlooked that not a single full professor, as the representative of his discipline, could truly encompass his field. Some university presidents tiptoed around this issue by asserting that there existed "in truth, only one science" and that all faculties shared in it. It was thus a body of many limbs, an image that might have occurred to the botanist speaking about such questions.[72] But most professors dealing with this matter sought through concepts of education to preserve the universality of science as *Universitas* in the university, a concept that was part of the core of the German university model. Each field and area of

---

contemporary university studies, and the depletion of university education by focusing on exams.

67 Boeckh, *Festrede*, 25.

68 Ibid., 25.

69 Virchow, *Ueber die nationale Entwickelung*, 12.

70 Lenz, *Freiheit und Macht*, 6.

71 Stichweh, *Wissenschaft, Universität, Professionen*, 19.

72 Alexander Braun, *Ansprache bei der Eröffnung des Semesters am 15. October 1865 in der Aula der Königlichen Friedrich-Wilhelms-Universität* (Berlin: Königliche Akademie der Wissenschaften, 1865), 7–8. Braun discussed the unity of academic disciplines.

specialization within each field educated those who, as students, familiarized themselves with it—through research or research-based education. This concept did not require that the universality of science was tied to any single person surveying an entire field or moving about the various fields represented at a university. The educational concept associated with the German university model sought to convey the principle of research during a course of studies, and to do so permanently and for life. This was to be achieved regardless of field, with the only proviso being that university education was based on a given field's methodological principles.

6. Continuities

In conclusion, I offer a few limited comments on lines of continuity and transformation up to today. In doing so, I do not wish to address the question of whether "Bologna" spells the end for the German university model or has initiated its decline, but continue instead to follow the traces of presidential addresses. In Switzerland this tradition continues on, unlike in Germany and Austria where universities during the 1960s became large-scale organizations that were sharply criticized, and where rituals were abandoned as historical ballast. Such rituals have subsequently been revived, albeit in a different format.[73]

In a tradition that remained unique in Germany, the ritual of the presidential address was continued without interruption only in Cologne. The address served as an opportunity to discuss what the university is and should be. In 1975, for example, the educator Clemens Menze spoke about an issue that had always been of concern to university presidents: "The philosophical idea of the university and its crisis in the age of science." But unlike his predecessor, Menze had parted with the idea of the university as an educational institution. Instead, he took for granted a "perpetual conflict between science and education." The university of the future, the president told his audience, would need to focus on its main tasks: "the production of science and the transmission of science."[74]

---

73 I have developed these ideas on the basis of observations in Tübingen. Langewiesche, "Die Universität als Bildungsinstitution."

74 Clemens Menze, *Die philosophische Idee der Universität und ihre Krise im Zeitalter der Wissenschaften. Rektoratsrede* (Krefeld: Scherpe, 1975), 25, 28. Volume includes "An-

All of his predecessors had considered this to be "education," research and the transmission of research through teaching, and in this way they had legitimized the university in the state and to the entire nation. But this traditional belief in the educational value of research had not survived the compromising political role assumed by academics during National Socialism. Research does not produce values and ideals, but this had been misconceived in the German university's educational concept. Even after 1945, German university presidents had not adopted into their concept of university education Max Weber's sociological insight that no values arise from research. The disregard of this educational model was likely prompted by a loss of consensus in society about what education should consist of, particularly academic education. The various reforms that have been bestowed on the universities by politicians did not aim at "education" but at an increase in student numbers and the efficiency of degree programs. These issues were raised in Cologne at the traditional locus for such discussion, when a new president assumed office. In 1987, the incoming president opened the new academic year by presenting an annual report. This had become routine: the university as a business enterprise. But then a member of the scientific council in his presentation discussed "Developments in the University System—Opportunities and Problems."[75]

Politicans now formulated their expectations towards universities in a different way as universities had to adjust their legitimation strategies. This can be summed up by saying that they no longer told elites in the state and in society about the universities' achievements on behalf of the nation—formerly the key function of presidential addresses. Others now developed the criteria for evaluating German universities. Among them was the German Science Council (*Wissenschaftsrat*) which facilitates cooperation between science and politics and to which universities do not send representatives. Researchers are proposed by six science organizations including the Conference of University Presidents (*Hochschulrektorenkonferenz*), and they are then invited to join. The university is thus but one actor among many, with individual universities having no voice in this advisory board organized by the federal government and the individual German states.

---

sprache des scheidenden Rektors Prof. Dr. Wolfgang Isselhard anläßlich der Rektoratsübergabe."

75 Friedhelm Neidhardt, "Entwicklungen im Hochschulsystem—Chancen und Probleme," in *Reden anläßlich der Eröffnung des akademischen Jahres 1987/88 am 15. Oktober 1987 in der Universität zu Köln* (Köln: Universität, 1987).

The university's role in society, in other words, has fundamentally changed. The university president could no longer claim specific expertise to explain the world on scientific grounds. Instead a number of organizations and their experts have assumed that role and it is for this reason why individuals who work outside the university are invited on festive days to discuss societal issues. It is on such occasions that university and society seem to trade places in the auditorium.

The difficulties associated with this transformation continued to echo through a speech given at the inauguration of Cologne University's president in 1989 given by the assistant state secretary for education in North Rhine-Westphalia. At the time, he stood in for the secretary. Due to changes in Cologne and Bonn, he said, they had witnessed the disappearance of "the last universities controlled by tenured faculty [*Ordinarienuniversitäten*] from the German academic landscape." This completed their transformation into institutions controlled by the various groups within the universities (*Gruppenuniversität*). The universities, he continued, would be exposed to the "plurality of basic beliefs" which would replace the "formerly uniform spiritual world-view and conviction." Thus he described his perception of the old German university model and its transformation.[76] In his presentation on "Hierarchy in a Pluralistic World" he told the university how the ministry conceived of the future role of North Rhine-Westphalia's universities while the university president had just given a traditional speech in the area of his expertise: "On the Rearrangement of Greek Tragedies by Corneille and Racine: Ideas of Humanity and Notions of Fate." This speech proved futile in terms of both educational policy and strategies for legitimation. In the "old" university, such a speech would have had the function of presenting to the audience the meaning of academic education and why the university considered such education to be its most important contribution to society: To convey to students, how research works so that they may interiorize research for all aspects of their lives. What was absent were any thoughts on what would provide a basis for integrating the new *Gruppenuniversität* and what this implied for society, as the ministry merely notified the university of what was no longer desired. The university answered with a speech that was mere expertise and contained no educational program. Few will have understood why, on the

---

76 "Ansprachen anläßlich der Rektoratsübergabe am 11. Mai 1989," in Bernhard König, *Zur Umgestaltung griechischer Tragödien durch Corneille und Racine: Menschenbild und Schicksalsauffassung. Rektoratsrede* (Köln: Universität, 1989), quotations on pp. 38, 34, 39.

occasion of inaugurating a new university president, such a speech was given at all.

But there are continuities which are also apparent from presidential addresses. In 2003 the University of Düsseldorf, for the first time in 20 years, appointed a new president, and by this time the ritual had changed little. (The same holds true for the state of North Rhine-Westphalia.) The president spoke to more than 600 guests of "The 'Idea of the University' in Our Time." Among his audience were high-level state officials, politicians, business representatives, the general consuls of three European states and Japan. His predecessor had been bid farewell in a separate celebration in the city's theater. Cabinet ministers of both federal and state government spoke on this occasion, as did the mayor of Düsseldorf, the chairman of the Central Council of Jews in Germany, Wolf Biermann (who at the University of Düsseldorf a decade earlier had taught as the Heine Guest Professor), and the university president himself.[77]

This university president had read up on the tradition of presidential addresses, unlike the president of the University of Graz in 1993, a historian—of contemporary history—who thought his inaugural lecture was without precedent.[78] In Düsseldorf, autonomy was discussed along with the meaning of university education, global challenges, and concrete plans for the university.

Even in the "Best Practices Workshop" organized by the German Research Foundation (DFG) in 2014 as part of its program on "Research in Germany,"[79] continuities with the German university model can be identified, a model that had developed during the nineteenth century in a long decentralized federal process and which many continue to regard as a standard today. Under the heading of "International Research Marketing" much was discussed that university presidents had also debated many times. How do you successfully compete for the best talent and how do you keep it? What are the options for small universities? How do you attract attention? How do you "network"? At the time, the German university model was the strongest argument that university presidents could

---

77 *Magazin der Heinrich-Heine-Universität Düsseldorf* 4 (2003).

78 Helmut Konrad, *Inauguration des Rektors Dr. Helmut Konrad. Ordentlicher Professor für Allgemeine Zeitgeschichte am 12. November 1993 in der Aula der Karl-Franzens-Universität Graz* (Graz: Kienreich, 1994). Konrad spoke on twentieth-century Austrian history and on the new Austrian law regulating higher education, which he criticized. He thus engaged with themes that were common in presidential addresses.

79 "Internationales Forschungsmarketing," *Forschung. Das Magazin der DFG* 2 (2014).

mobilize when they intended to show why their institutions were worth the state's money and when they demanded larger budgets.[80]

---

80 Cf. Langewiesche, "Wissenschaftsmanagement im Selbstbild der deutschen Universität seit dem 19. Jahrhundert," in *Zwischen Stadt, Staat und Nation. Bürgertum in Deutschland. Hans-Werner Hahn zum 65. Geburtstag,* edited by Stefan Gerber, Werner Greiling, Tobias Kaiser and Klaus Ries (Göttingen: Vandenhoeck & Ruprecht, 2014) 2:759–69.

The Symbolic Formation of Science in its Historic Situation: Rudolf Virchow on Science and the Nation

Peter Münte

1. Science, Society, and the Formation of Societal Alliances

The relationship between science and society has become an increasingly important research focus in the sociology of science.[1] The simple reason for this seems to be that in recent decades this relationship has undergone dramatic changes which have provoked thought about the altered position of science in society. One important aspect of this development appears to be that the standing of science in society has been weakened in the face of a public that is considered to be critical of science with regard to the unintended consequences of its technological applications, an often alleged loss of belief in the objectivity of scientific expertise, and the constant observation of science by the mass media.[2] Against this backdrop, scholars of science and technology studies have felt they should look for a new basis

---

1 Peter Weingart, for example, in his recent introduction to the sociology of science focuses on an increasingly close "coupling" of science and other sub-systems of society (processes of politicization, economization, and medialization). Cf. Peter Weingart, *Wissenschaftssoziologie* (Bielefeld: Transkript, 2003); see also Peter Weingart, *Die Stunde der Wahrheit? Zum Verhältnis der Wissenschaft zu Politik, Wirtschaft und Medien in der Wissensgesellschaft* (Weilerswist: Velbrück, 2001).

2 For a prominent view of the relationship between science and public, see Peter Weingart, *Die Wissenschaft der Öffentlichkeit: Essays zum Verhältnis von Wissenschaft, Medien und Öffentlichkeit* (Weilerswist: Velbrück, 2005); Steven Shapin, "Science and the Public," in *Companion to the History of Modern Science*, edited by Robert C. Olby et al. (London: Routledge, 1990), 990–1007. For the problem of trust in science and the dialogue between science and public, see Brain Wynne, "Uncertainty and Environmental Learning: Reconceiving Science and Policy in the Preventive Paradigm," *Global Environmental Change* 2 (1992): 111–27; Brain Wynne, "Public Engagement as a Means of Restoring Public Trust in Science: Hitting the Notes, but Missing the Music?," *Community Genetics* 9 (2006): 211–20; for the problem of medialization and visibility, see Simone Rödder, *Wahrhaft sichtbar: Humangenomforscher in der Öffentlichkeit* (Baden-Baden: Nomos, 2009); Martina Franzen, *Breaking news: Wissenschaftliche Zeitschriften im Kampf um Aufmerksamkeit* (Baden-Baden: Nomos, 2011).

upon which to relate science to society as well as for securing the authority of its knowledge.[3] Another aspect is the emergence of increasingly expansive regimes of regulation that emphasize technological innovation, social utility, and sustainability, and foster external and countable success in science.[4] As a consequence of this development, sociologists have shown a growing interest in the autonomy required for the proper conduct of research and for the societal conditions that facilitate a support of a science that can be described as autonomous.[5]

These historiographic developments provide an opportunity to highlight a sociological problem that can be considered to be of fundamental importance. What sociologists of science sometimes refer to as the changing relationship between science and society implies complex processes of order formation, and the question is how these processes can be conceived theoretically and investigated empirically. The following study of such questions is based on particular assumptions. The changing relationship

3 On a new contract of science and society, see David H. Guston and Kenneth Keniston, "Introduction: The Social Contract for Science," in *The Fragile Contract: University Science and the Federal Government*, edited by David H. Guston and Kenneth Keniston (Cambridge: MIT Press, 1994); David H. Guston, *Between Politics and Science: Assuring the Integrity and Productivity of Research* (Cambridge: Cambridge Univ. Press, 2000).

4 For a general discussion of science policies connected to concepts of steering and governance, see Alfons Bora, "Wissenschaft und Politik: von der Steuerung über Governance zu Regulierung," in *Handbuch Wissenschaftssoziologie*, edited by Sabine Massen et al. (Wiesbaden: Springer, 2012), 341–53; Bettina Heintz, "Governance by Numbers: Zum Zusammenhang von Quantifizierung und Globalisierung am Beispiel der Hochschulpolitik," in *Governance von und durch Wissen*, edited by Gunnar Folke Schuppert and Andreas Voßkuhle (Baden-Baden: Nomos, 2008), 110–28; Andreas Knee and Dagmar Simon, "Peers and Politics: Wissenschaftsevaluationen in der Audit Society," in *Governance von und durch Wissen*, edited by Gunnar Folke Schuppert and Andreas Voßkuhle (Baden-Baden: Nomos, 2008), 173–85; Uwe Schimank, "'New Public Management' and the Academic Profession: Reflections on the German Situation," *Minerva* 43 (2005): 361–76.

5 For a discussion of the autonomy of science, see Martina Franzen et al., eds., *Autonomie revisited: Beiträge zu einem umstrittenen Grundbegriff in Wissenschaft, Kunst und Politik* (Weinheim: Beltz Juventa, 2014); David Kaldewey, *Wahrheit und Nützlichkeit: Selbstbeschreibungen der Wissenschaft zwischen Autonomie und gesellschaftlicher Relevanz* (Bielefeld: Transcript, 2013). For a discussion of how the concept of autonomy in science relates to a discourse on the order of modernity, see Peter Münte, "Die Autonomie der Wissenschaft im Ordnungsdiskurs der Moderne: Ein Versuch über den Formenwandel der modernen Wissenschaft," in *Autonomie revisited: Beiträge zu einem umstrittenen Grundbegriff in Wissenschaft, Kunst und Politik*, edited by Martina Franzen et al., (Weinheim: Beltz Juventa, 2014), 143–65.

between science and society is not properly understood if they are considered to be the processes of order formation within societal spheres such as science or politics. In studies of science that adhere to his model the focus is on how the internal structure of the system of science develops according to its own imperatives in a changing societal environment, and how politics, according to its specific imperatives, tries to "steer" science.[6] The changing relationship between science and society is also not appropriately grasped if the primary object of research consists in the changing relationship between actors from different spheres of society and their interest in expanding their reputation and institutional power. In such studies the focus is on how actors in science are more or less successful in finding allies, especially from other spheres of society, that fund and support them, because on the basis of such alliances scientists are able to pursue research programs and enforce knowledge claims. In return, they provide their allies with the knowledge required for a consolidation of institutional power.[7] In one perspective the relationship between science and society is analyzed as a process of interpenetration between self-referential and self-organizing sub-systems of society, while in the other it is seen as a process that is driven by the pursuit of interests, by the execution of power in the differ-

---

6 An approach that is developed in Niklas Luhmann, *Soziale Systeme: Grundriß einer allgemeinen Theorie* (Frankfurt: Suhrkamp, 1984). For its application to science, see Niklas Luhmann, *Die Wissenschaft der Gesellschaft* (Frankfurt: Suhrkamp, 1990); Rudolf Stichweh, "Differenzierung des Wissenschaftssystems," in *Differenzierung und Verselbständigung: Zur Entwicklung gesellschaftlicher Teilsysteme*, edited by Renate Mayntz et al. (Frankfurt and New York: Campus, 1988), 45–115. For the problem of political management control over science from a systems theory perspective, see Rudolf Stichweh, "Differenzierung von Wissenschaft und Politik: Wissenschaftspolitik im 19. und 20. Jahrhundert," in Rudolf Stichweh, *Wissenschaft, Universität, Professionen: Soziologische Analysen* (Frankfurt: Suhrkamp, 1994), 156–73; Marc Mölders, "'Geplante Forschung': Bedeutung und Aktualität differenzierungstheoretischer Wissenschafts- und Technikforschung," in *Schlüsselwerke der Science and Technology Studies*, edited by Diana Lengersdorf and Matthias Wieser (Wiesbaden: Springer, 2014), 111–21.

7 See Latour's and Callon's actor-network theory. Bruno Latour, *Science in Action: How to Follow Scientists and Engineers through Society* (Cambridge: Harvard Univ. Press, 1987). See also Fabian Link, "Shifting Alliances, Epistemic Transformations: Horkheimer, Pollock, and Adorno and the Democratization of West Germany," in this volume. A power-approach that focuses on how modern science has relied on supporters has been well-established in the history of science from the 1970s. Roger Hahn, *Anatomy of a Scientific Institution: Paris Academy of Sciences, 1666–1803* (Berkeley: Univ. of California Press, 1971); James R. Jacob, "Restoration, Reformation and the Origins of the Royal Society," *History of Science* 13 (1975): 155–76; Mario Biagioli, *Galileo Courtier: The Practice of Science in the Culture of Absolutism* (Chicago: Univ. of Chicago Press, 1993).

ent fields of society, and by the building of coalitions useful to these ends. Both approaches likewise ignore a basic fact of social life: Cooperation requires community-building and the ongoing negotiation of order through language.[8]

In contrast, the basic assumption behind what follows is that the changing relationship between science and society implies shifts in the societal meaning of science itself, and that the processes of symbolic formation that generate these shifts are an important subject of empirical research.[9] According to the view developed here, in discourses a shapeable societal meaning of science is negotiated. Nevertheless, such changes are negotiated not just with respect to institutional power; they also establish a general idea of societal cooperation that ascribes a specific function in society to science.[10] The societal meaning of science is also negotiated against the background of a given historic situation with its particular problems that challenge a given societal order.[11] For a discourse to generate an idea of order that is able to graduate to a concept that becomes binding for all, this discourse has to take place in public. A legitimate order of society and a legitimized understanding of science are then established as collectively binding ideas.[12] In the discourse in which the order of society is negotiated, actors that represent different spheres, such as science and politics, are also involved. The interplay of the contributions of these actors, apart from reflecting their own interests in power, can lead to a

---

8 A sociological approach that is useful but somewhat marginal in contemporary German sociology emphasizes the importance of community in contrast to society: Ulrich Oevermann, "The Difference between Community and Society and Its Consequences," in *Developing Identities in Europe: Citizenship Education and Higher Education. Proceedings of the Second Conference of the Children's Identity and Citizenship in Europe*, edited by Alistair Ross (London: CiCe, 2000), 37–61.

9 The notion of a "symbolic formation" draws on Hans-Georg Soeffner, *Symbolische Formung: Eine Soziologie des Symbols und des Rituals* (Weilerswist: Velbrück, 2010).

10 The mainstream of discourse analysis is, following Foucault, closely tied to the often critical analysis of power relations. See, for example, Johannes Angermüller et al., eds., *Diskursforschung: Ein interdisziplinäres Handbuch* (Bielefeld: Transkript, 2014). In contrast, the question raised here is how the ethos of activities at the center of societal spheres itself is negotiated.

11 The idea of a crisis challenging the order of society was introduced by Foucault when he sought to explain his concept of a "dispositive." Michel Foucault, *Dispositive der Macht: über Sexualität, Wissen und Wahrheit* (Berlin: Merve, 1978), 120.

12 The term "legitimation" can be used to refer to quite different problems of justification. It may relate to the justification of actions and of claims or (as is the focus here) to a particular order of society.

general idea of societal cooperation in which the societally relevant achievements of the different spheres of society are adapted to each other in a way that should allow them to cope with the pressing problems of the given historic situation. In the long run, such general ideas of order can lead to a transformation of institutions—institutions that facilitate and meet pressing problems of society. To understand the changing relationship of science and society thus implies grasping a complex nexus of order formation. In this nexus, processes of symbolic formation are omnipresent, and when shifts in the societal meaning of science and the related formation of societal alliances are the focus of our interest, then we ought to study such processes of symbolic formation.

For an analysis of such processes of symbolic formation, it is useful to focus on selected documents that promise to reveal details of such transformations, such as letters of dedication, speeches, addresses, polemic papers, academic treatises etc.[13] In taking on some of the questions outlined above, I will focus in this chapter on a speech by Rudolf Virchow, a speech that has received much attention in historical research on science in the context of German nation-building.[14] The analysis consists of three

13 Examples include Peter Münte and Ulrich Oevermann, "Die Institutionalisierung der Erfahrungswissenschaften und die Professionalisierung der Forschungspraxis im 17. Jahrhundert: Eine Fallstudie zur Gründung der 'Royal Society'," in *Wissen und soziale Konstruktion*, edited by Claus Zittel (Berlin: Akademie Verlag, 2002), 165–230; Peter Münte, "Institutionalisierung der Erfahrungswissenschaften in unterschiedlichen Herrschaftskontexten: Zur Erschließung historischer Konstellationen anhand bildlicher Darstellungen," *Sozialer Sinn* 6 (2005): 3–43; Peter Münte, "Strukturelle Motive der Beziehung von Wissenschaft und Herrschaft: Zur wissenschaftssoziologischen Bedeutung der Analyse von Widmungsbriefen am Beispiel der Widmung Christiaan Huygens' an Leopold de' Medici in Systema Saturnium," in *Die Kunst der Mächtigen und die Macht der Kunst: Neue Studien zur Kulturpatronage*, edited by Ulrich Oevermann et al. (Berlin: Akademie Verlag, 2007), 151–77; Andreas Franzmann, "Die Krise Frankreichs von 1870 und ihre Ausdeutung durch den Wissenschaftler Louis Pasteur: Eine Deutungsmusteranalyse," in *Krise und Institutionen*, edited by Carsten Kretschmann and Henning Pahl (Berlin: Akademie Verlag, 2004), 117–56; Andreas Franzmann, "Der 'gebildete Laic' als Adressat des Forschers: Sequentielle Analyse von Titel und Vorrede zur ersten Ausgabe von Justus von Liebigs 'Chemischen Briefen' von 1844," in *Wissenspopularisierung: Konzepte der Wissensverbreitung im Wandel*, edited by Carsten Kretschmann (Berlin: Akademie Verlag, 2003), 232–55; Axel Jansen, *Alexander Dallas Bache: Building the American Nation through Science and Education in the Nineteenth Century* (Frankfurt and New York: Campus, 2011).
14 Rudolf Virchow, *Über die nationale Entwicklung und Bedeutung der Naturwissenschaften: Rede gehalten in der zweiten allgemeinen Sitzung der Versammlung deutscher Naturforscher und Ärzte zu Hannover am 20. September 1865* (Berlin: August Hirschwald, 1865). References to and in-

parts. First, I will consider the speech from the point of view of symbolic formation. I will highlight strands of the speech in which Virchow opens up a discourse on social reality and its order, seeking to confirm or reorganize it. Second, I will contextualize the speech in order to unfold aspects of the historic situation that provide motives for Virchow's symbolic formation of order. At the center of the discussion will be a focus on motives that derive from the ethos of modern science as it was institutionalized in the end of the seventeenth century. Third, I will discuss the question of how the structure of science itself might be affected by nation-building as a process of modernization characteristic of the nineteenth century. In this connection the notion of a nationalization of science will be discussed that has attracted some attention in the recent history of science. In my conclusion, this will provide me with an opportunity to underline major differences between the nineteenth and the twentieth centuries with respect to their constellations of modernization.

---

terpretations of this speech in the historical research: Jutta Kolkenbrock-Netz, "Wissenschaft als nationaler Mythos: Anmerkungen zur Haeckel-Virchow-Kontroverse auf der 50. Jahresversammlung deutscher Naturforscher und Ärzte in München (1877)", in: *Nationale Mythen und Symbole in der zweiten Hälfte des 19. Jahrhunderts*, edited by Jürgen Link and Wulf Wülfing (Stuttgart: Klett-Cotta, 1991), 212–36; Constantin Goschler, "Deutsche Naturwissenschaft und naturwissenschaftliche Deutsche: Rudolf Virchow und die 'deutsche Wissenschaft'", in *Wissenschaft und Nation in der europäischen Geschichte*, edited by Ralph Jessen and Jakob Vogel (Frankfurt am Main: Campus, 2002), esp. 110; Yvonne Steif, *Wenn Wissenschaftler feiern: Die Versammlungen deutscher Naturforscher und Ärzte 1822 bis 1913* (Stuttgart: Wissenschaftliche Verlagsgesellschaft, 2003), 148–53; Tilman Matthias Schröder, *Naturwissenschaften und Protestantismus im Deutschen Kaiserreich: Die Versammlung der Gesellschaft Deutscher Naturforscher und Ärzte und ihre Bedeutung für die Evangelische Theologie* (Stuttgart: Franz Steiner, 2008), 90–92. See also Dieter Langewiesche, "State–Nation–University. The Nineteenth-Century 'German University Model' as a Strategy for National Legitimacy in Germany, Austria, and Switzerland," in this volume. The Interpretation of Kolkenbrock-Netz is quite detailed (216–25), but is restricted to an investigation of how concepts such as "science," "nation," "religion," "reason" etc. are linked to each other and how through narration a particular semantic of such discursive elements is formed. The intention here is to establish a sociological frame of analysis.

2. Rudolf Virchow's Speech as an Activity of Symbolic Formation

The document selected here for analysis is a speech by Rudolf Virchow (1821–1902), a German physician, anthropologist, and politician who became famous as the founder of cellular pathology and as a member of the liberal party, and in this role a leading opponent of Otto von Bismarck, Minister President of Prussia from 1862 and Chancellor of the German Empire from 1871. Virchow also was a protagonist of a Protestant ascendency in the Empire and he coined the term *Kulturkampf* (a "culture war" between the Catholic church and the Prussian state).[15] The speech in question was delivered at an 1865 meeting of the *Versammlung deutscher Naturforscher und Ärzte* (the Assembly of German Natural Scientists and Physicians) and its publication was subsequently based on a stenographic transcript. The *Versammlung deutscher Naturforscher und Ärzte* had been funded 1822 by the German naturalist and biologist Lorenz Oken with the aim of establishing continuous exchange between German natural science researchers and physicians.[16] Virchow's speech entitled *"Über die nationale Entwicklung und Bedeutung der Naturwissenschaften"* ("On the National Development and Significance of the Natural Sciences") essentially deals with the question of how the formation of the modern natural sciences correlates with nation-building. Virchow considers both the question of how nation-building is a prerequisite for the formation of the modern natural sciences and how the natural sciences on their part contribute to nation-building.

Regarding the relations between science and the nation-state, Virchow's speech at first glance is at best indirectly relevant. It seems not to be about the societal standing of science in relation to other spheres of society but instead connects the modern natural sciences to a process of community building. But we should nevertheless consider Virchow's speech as part of an effort to help form a societal alliance, namely that between science and a social movement. Movements are usually characterized by the creation of

15 Constantin Goschler, *Rudolf Virchow: Mediziner, Anthropologe, Politiker* (Cologne: Böhlau, 2002).

16 For the history of the *Versammlung*, see a collection of historical documents in Max Pfannenstiel, *Kleines Quellenbuch zur Geschichte der Gesellschaft Deutscher Naturforscher und Ärzte: Gedächtnisschrift für die hunderste Tagung der Gesellschaft* (Berlin: Springer, 1958). For a history of the *Versammlung* as a forum for cultural and political debate, see Steif, *Wenn Wissenschaftler feiern.* See also Schröder, *Naturwissenschaften und Protestantismus im Deutschen Kaiserreich.*

a public in which a collective awareness of major contemporary challenges is deployed and in which a certain ideal of living together in society is professed. In the case of Virchow, the central topic of the movement is nation-building. We may assume, therefore, that he would establish the societal significance of science in alliance with the German national movement. As a distinctive trait of the nineteenth century, furthermore, movements that evolve around the issue of nation-building also play an important role in the transformation of the political sphere.

In relation to the problem of symbolic formation, the speech may be viewed from three perspectives: as an *interaction* embedded in contexts in which an idea of order is unfolded; as a *presentation* of such an idea that is achieved step by step; and also as an achieved *pattern of interpretation* that may evolve in subsequent discourse and shape the minds of those involved in it. Features that made Virchow's speech part of a concrete interaction at the time will have been removed during the publication process, but this leaves us free to investigate the two other levels. In this section I will choose important strands of Virchow's speech and trace their unfolding. In the sections that then follow, the established pattern of interpretation will be discussed in a wider perspective.

The speech consists of at least four parts of different length. Virchow begins by bringing to mind the Assembly meeting's long history and points to their national significance through their contribution to a homogenization both in scientific as well as in civil life.[17] Virchow suggests that the meetings allow a community to form that has common aims and shares a specific work ethos. In this respect science should be a model for other branches of what Virchow calls *"deutsches Leben"* ("German life"). Thus in this first part of the speech Virchow recalls the Assembly's history and gives it a special meaning by assigning it a role in the larger process of nation-building. The community of scientists Virchow considered a model of the nation as a whole.

In this first and rather short part of his speech, Virchow focuses on a rather complex historic sequence and presents an interpretation of it by relating the association's history to another historical development: the forming of the German nation. By doing so, he attributes a meaning beyond the pursuit of specific scientific goals, and he also establishes a relationship between community-formation and work. He connects such

---

17 Virchow, *Über die nationale Entwicklung und Bedeutung der Naturwissenschaften*, 5–6.

community formation to the idea of working together, placing both into the context of modernization driven by a Protestant work ethic. It is illuminating how in this introductory passage Virchow provides complex historic processes with a suggestive meaning using sparse linguistic means. The symbolic formation of order takes place here through Virchow's interpretation of "essential" developments that have led to the present state of living together, i.e. by inferring a myth.

Taking into account that these historic processes could certainly have been interpreted differently, and that other aspects could have been given priority, the question arises to what extent an established habitus finds expression that is characterized by presenting these relations in this manner. This habitus seems to result from merging Protestantism with national consciousness and is expressed here by the symbolic forming of complex and opaque historic processes.

Virchow then turns to what he is presenting as the actual topic of his speech, namely his theme of the great developments in the science of his age.[18] It is not clear at the beginning what these great developments actually are. He refers to the names of great scientists inscribed on the walls of the room in which the meeting takes place and asks to what extent the increase in the number of German names is indicative of the fact that German scientists are recognizing their duty in the history of culture. Here we find a specific meaning attached to being a nation of culture, a nation with a cultural mission. Thus Virchow gives a special meaning to the contributions of German scientists to the progress of science and human culture. He does not consider them a record of former achievements that should then lead to national pride, but as indicators of fulfilling a duty. In this respect he deems Germany to be a special nation, distinguished by its sense of duty to be successful in conducting research. Here we also find a collectivist version of an ethic of work, as it is not individuals whom Virchow considers to be gifted with special talents to be developed through an appropriate calling, but rather it is the nation that he considers as providing the talents so that Germany can be a nation of science.

Virchow critically remarks that in selecting the names for the wall more Germans could have been incorporated, but he also concedes that in the present situation it is not yet possible to cover German assembly halls over and over with the names of German scientists. With his remark he projects

---

18 Ibid., 6–7.

a future of even greater success for German science. At this point in his speech, Virchow raises the question of whether science has a cosmopolitan or a national orientation. In what follows he then argues for the second position, and the main part of his speech is comprised of an argument for demonstrating that science is national in character.

In doing so, he again focuses on a historic process—the development of science itself—and he links it to his other topic of nation-building. This enables him to connect science with the idea of nations competing in the scientific arena. He locates such competition within a specific frame of reference, the progress of mankind, and he connects it with the idea of ethical commitment. As Virchow understands it, such competition is not primarily about gaining an advantage over other nations. Instead, nations compete to prove themselves in their service to mankind. While at the beginning of his speech, Virchow interpreted the *Versammlung* as a scientific organization that contributes significantly to building the German nation, in this subsequent passage he charts the self-conception of the evolving German nation in relation to competition with other nations dedicated to the pursuit of science. Again, Virchow describes complex and opaque historic processes in close relation to the formation of a collective identity.

It is precisely this point that prompts a need for clarification. That the identity of scientists should essentially feed off their being part of a particular nation that competes with other nations is a point that Virchow treats as being in need of a rationale. He indicates a conflict of different ideas of order regarding which he intends to take a stand. Having delineated this conflict, Virchow has thus prepared the ground for his speech's much more comprehensive part.[19] What kind of rationale does Virchow demand?

Virchow moves into the next section of his speech by introducing a thesis about the origin of science in modern understanding. His basic claim is that modern science emerged when circles of self-reliant men in different European nations began to think about nature in a manner related to the peculiar character of their own nation. Virchow's reference to the national background of scientific investigation seems to provide an antithesis to the common universalistic view of science. But how does Virchow actually develop his story? Virchow considers science in the modern sense as beginning in Germany during the Reformation, and provides a simple reason

---

19 Ibid., 7–27.

for this. Before the Reformation, the center of scientific activity had been located south of the Alps. German countries had made few contributions to research and instead had merely reproduced what had been done elsewhere. The Reformation then brought about a German intellectual life in its own right. Virchow's argument, of course, is not really about modern science but about centers of intellectual productivity. But Virchow also claims that the Reformation brought about a transformation of the structure of thought itself. Roman thinking, according to him, was authoritarian, whereas the Protestant spirit led to a critique of intellectual authority and a confident investigation of nature. As a result, Virchow suggests that after the Reformation, Italian science trailed behind.

He then moves on to work out his argument and emphasizes the importance of a national scientific life for the progress of science and points to the importance of national languages of science—languages that be come standard currency among researchers. According to Virchow, the nationalization of science produced a new dynamic of scientific growth. Science did not just contribute to the wealth of the people and the state by producing useful knowledge; it also gained influence over the way ordinary people thought. Such developments, Virchow believes, have been fruitful for science itself. When he suggests that there should be a "German science," therefore, he proposes that science should be productive in shaping the minds of ordinary people that may then become productive in science. In this respect, his concept of German science simply aimed at the democratization of knowledge through general education. Such a democratization of knowledge Virchow considers a productive resource and a propellant for scientific progress at large.

In his speech, Virchow arrives at a discussion of his own age. He refers to the former predominance of French science and in doing so opens up another emancipation narrative. Virchow considers his own generation's achievements in contributing to the various branches of science and points out that the character of German science has been affected by the character of the German nation—the work ethos Virchow considers peculiar to Germany. He proceeds to state that to the German nation, science has become a religion and that it has contributed greatly to its moral emancipation:

I do not hesitate to say that science to us has become a religion, and that through this true religious faithfulness of natural science researchers have also readied themselves for the loyalty to the law. I mean, it is no small thing for the significance, which the natural sciences have for the national essence that we may say of it: Natural science has contributed considerably to the moral liberation of our people.[20]

Virchow's assertion that science has become a religion implies that in his age science constitutes a key intellectual power that transforms the mind of the people so as to unify them intellectually.[21]

Subsequent to this key passage, Virchow moves on to stress how powerful the scientific method has become. Using the example of meetings by political economists in Germany, he suggests that political economy was the application of the scientific method to national life. German political economy, he suggests, in its conceptualization of society and the state differs from older Roman and French ways of thinking. What differentiates German political economy, according to Virchow, is that it constitutes a specific German understanding of the political sphere, an understanding derived from intensive studies of the nature of the nation. He concludes that this independent way of thinking, thinking without authority, should become the basis for shaping the whole nation. At this point of the argument, Virchow' story assumes manifestly particularistic traits, which once again relates to his overall plot of unfolding a story of national intellectual emancipation from supranational intellectual powers, a plot that restricts his narration. To the extent that Rome, France, and Germany are the only actors in his play, with Rome and France as supranational intellectual powers and Germany a nation, Virchow equates German science with true scientific thinking and productivity. Moving along, Virchow describes in more detail how science cannot be reduced to a source of useful knowledge. He had already pointed to the role of science in education. The scientific method, he suggests, should be considered the true method of the human mind. People everywhere should learn to think according to this method, and that, Virchow believes, implies "thinking naturally" and with an open mind. This would result in a similarity of thought, and, thus,

---

20 Ibid., 17–18. German original: "Ich scheue mich nicht zu sagen, es ist die Wissenschaft für uns eine Religion geworden, und in dieser wahrhaft religiösen Treue der Naturwissenschaft haben sich auch die Naturforscher mehr und mehr gerüstet für die Treue gegen das Gesetz. Ich meine, es ist kein kleines Ding für die Bedeutung, welche die Naturwissenschaft hat für das nationale Wesen, daß wir von ihr aussagen können, sie hat ein großes Stück mitgearbeitet an der sittlichen Befreiung unseres Volkes."
21 Cf. Kolkenbrock-Netz, "Wissenschaft als nationaler Mythos," 224.

science would be a driving-force for unifying the German nation. Virchow's comments clarify his idea that science has become a religion.

Virchow considers science to be the basis for a common intellectual life, a role once played by religion. That is why he insists that science could not be reduced to producing useful knowledge. While its role for knowledge-production was supported by German states, Virchow considers the role of science to be much more important science in revolutionizing the human mind using the scientific method, a role that requires science be enmeshed in the nation through a close association of scientists with their society. Virchow considers such a revolutionary role for science, a role leading to national unity, to be much more powerful than a political repression of free intellectual life in the German states.

Virchow moves on to recall the origins of modern science. Once again he suggests that the ability to conduct science in the modern sense is derived from a connectedness with life in the nation. While he appears to rehash his old theme, he now provides a slightly different version of the emergence of modern science. If German science had remained under Roman and French influence, he now suggests, German science would not have flourished. A new spirit of science only emerged when Germany was exposed to a better method of science developed in northern Protestant states such as the Netherlands and England. This spirit prevailed in Germany against external forces and developed into the original spirit of German independence. The crucial point here is that the historic constellation is enriched with further actors, other nations. Nevertheless, the structure of the story remains the same.

In the first version, the starting point was the Reformation, but now the starting point is the emergence of the so-called scientific method in Protestant nations to the north. Virchow's narrative is reflected in the older sociology of science that considered modern science to have emerged from Puritanism, one among numerous religious versions of Protestantism.[22] While Virchow seems to change his game, however, he merges back into his narrative of national emancipation. In the Protestant-north version of his story, Virchow concedes the universalistic character of modern science. Nevertheless, the story he tells is not suggesting that the scientific method should be an important part of nation-building in every country as it is in

---

22 Robert K. Merton, *Science, Technology and Society in Seventeenth Century England* (New York: Harper and Row, 1970); I. Bernhard Cohen, ed., *Puritanism and the Rise of Modern Science: The Merton Thesis* (New Brunswick and London: Rutgers Univ. Press, 1990).

Germany. The resulting story is that this mode of thought, after its cultivation in Germany, was threatened by powerful external forces that tried to suppress it. In the resulting fight for national independence the new method of science again turns into a particular German way of thinking connected to the special qualities of the German nation.

We can summarize Virchow's speech thus far by observing that he relates stories of origin that locate a point in history when something new emerged that had a crucial effect on the self-conception of those who were part of it. Virchow, in other words, constructs a myth of origin.[23] His myth revolves around an explanatory problem that would later be at the center of classic sociology of science, the question of how innovative science emerged. Virchow provides the simple answer that the rise of science correlates with nation-building. To justify this correlation he drafts two stories that differ in their structure, but lead to the same result. The first story is marked by the fact that nation-building in Germany is traced back to the Reformation, and the momentum of innovation in modern science is equated with the anti-authoritarian orientation attributed to Protestantism. This story can be considered flawed not only in so far as it shifts the beginnings of nation-building to the Reformation period and connects the emergence of modern science with Lutheranism in particular, but also in that it amalgamates nation-building in Europe primarily with the opposition of Protestantism and Catholicism and thus passes over the French Revolution (and lays the foundation of an understanding of the German nation that is grounded in Protestant domination).

Against this background the structure of the second story, which is different from the first with regard to its point of origin, is of particular interest. In this story the emergence of the modern natural sciences in the form of an improved scientific method constitutes the starting point, which is linked with the Protestant ethic. But in this second story it is linked with those manifestations of this ethic which are not located in Germany but in England and the Netherlands. Emanating from this point is the question how this science has detached itself from its local origins and spread across Europe and thereby flourished in different contexts of development in various ways. However, it is exactly this question that is not pursued in the story unfolded here. Moreover, the adoption of a better scientific method in Germany forms the starting point for a further story of emancipation.

---

23 See also Kolkenbrock-Netz, "Wissenschaft als nationaler Mythos," 218.

Emphasized here is the idea that this method was said to have been suppressed by hostile powers and adhered to in an act of national self-assertion.

If one asks what could have motivated this story in its particular selectivity, a specifically tinted emancipatory motive seems to play a prominent role. It is about the liberation from external powers. On the one hand, this means an intellectual liberation from foreign ways of thinking imposed from outside. On the other hand it is about national liberation from external powers. So the kind of nationalized Protestantism diagnosed in the beginning interestingly joins with decidedly emancipatory motives in the speech discussed here. The myth is also usable for creating a nationally minded scientific community which strives to distinguish itself through outstanding research achievements. In this respect it is interesting that the motive for doing science established here cannot adequately be described as being in competition with other science nations. The motive is instead a fight for self-assertion against powerful enemies, which in turn can motivate one's own aspiration towards dominance. So the structure of the myth offers a momentum of its own. The interesting follow-up question is, to what extent can such a momentum be reconstructed in research itself?[24]

After completing this argument, the fourth and final part of Virchow's speech is devoted to propositions aimed at achieving a closer understanding with the population.[25] I refrain from considering Virchow's concluding remarks in detail because they add little to our overall understanding of his speech.

---

24 In this respect, the reconstruction of Virchow's speech as a national myth of science can be related to a central problem in the classic sociology of science: the growth of science, which implies the question: Why are there differences in both the rate of growth and national contributions to its growth? For a discussion of the growth of science and its explanation on the basis of role-patterns and the structure of science-systems, see Robert K. Merton: "The Normative Structure of Science," in Robert K. Merton, *The Sociology of Science: Theoretical and Empirical Investigations* (Chicago: Univ. of Chicago Press, 1973), 267–78; Joseph Ben-David, *The Scientist's Role in Society: A Comparative Study* (Englewood Cliffs, NJ: Prentice-Hall, 1971); Joseph Ben-David, *Scientific Growth* (Berkeley: Univ. of California Press, 1991).

25 Virchow, *Über die nationale Entwicklung und Bedeutung der Naturwissenschaften*, 27–31.

3. Motives for Giving Meaning to Science

A consideration of Virchow's speech as an activity of symbolically assigning order raises further and more fundamental questions: Why does Virchow contribute to the formation of the relationship between science and nation at all, and what makes him suggest that specific role of science in nation-building? Furthermore, which motives matter with respect to a sociological understanding of the role of science in society? Such motives might be different in character. One motive for an engagement with the symbolic formation of order is that the negotiated order might fail to cope with pressing problems of life, that it is endangered by alternative ideas of order, or that it does not yet have a high esteem or is met with distrust. Against this background, the analysis of what is done by making order a subject of discourse becomes complex for the simple reason that it implies profound sociological and historical analyzes not just of the given societal order, but of the particular historical situation to be coped with. There has to be an idea of the motives for making an aspect of order a subject of discourse that results from a perceived need for consolidating or securing the given order or to adapt it to the necessities of the time. Thus, to place the analysis of the speech as an ongoing activity of symbolic formation in its context does not mean gathering up all the historical facts that might be relevant with respect to the document in question; it means forming an idea of a constellation which requires to be dealt with by symbolic formation. Therefore, to get an idea of the "deeper" sociological meaning of the speech, I will make conjectures about Virchow's motives that relate to problems of the societal order and I will formulate such conjectures on the basis of sociological theory as well as historical research.

I would like to draw attention to two facets of Virchow's presentation: First, Virchow does not simply act here as a scientist, but as the protagonist of a distinct understanding of science, of modern experiment-based science. As a protagonist of this specific understanding of science, Virchow was committed to developments he and his peers had advanced in German medicine. Second, their understanding of science could become relevant only if it had the support of powerful allies. And in the context of an emerging nation other allies can be considered relevant then in an estate-based society. In order to develop this thought a little further, I would like to infer the sociology of modern science and amalgamate it with historical studies of nineteenth-century Germany.

In order to situate Virchow's speech within the sociology of science, the question can be raised, what might prompt a scientist committed to the ethos of modern science, to relate his work to the process of nation-building? An analytical approach for asking such questions is, of course, different from the kind of document analysis used in the historical sciences, but I consider it necessary to relate the analysis to a sociological theory of society when a sociological understanding of Virchow's speech as an activity should be established in which society itself is formed. One possibility for defining science as a distinct sphere of society is to consider science as a complex of activities bound to a broad infrastructure used in the storage and systematization of knowledge.[26] The emergence of this complex implies the formation of specific social relationships, the relationship between scientifically educated experts and lay persons, whose knowledge is devalued by science and who therefore become dependent on experts' knowledge if they want to conduct their life according to the standard of rationality of their time.

Such a broad sociological understanding of science implies that the scientists' activity may affiliate itself with different conceptions of what matters in this activity and how it should be carried out, and this can imply the ascription of different functions to science in society and a varying internal order of science that depends on changing general ideas of how society as a whole should work. In this way, for instance, a learned person can be established in science characterized by his/her profound knowledge of a body of authoritative texts, and by his/her cultivating of it through commentary and critique. With its emphasis on experiment and observation, the modern understanding of science distances itself from such an understanding of more general activities of science. With the institutionalization of the modern empirical sciences in the seventeenth century a form of scholarliness was criticized, which drew upon ancient authorities and the cultivation of their theories in a scholastic atmosphere of debate, in favor of an understanding of science committed to empiricism and human industry.[27] A sociology of the different understandings of science has never

---

26 Such an approach can be contrasted with the more common approaches of functional differentiation that consider the orientation to values, basic problems of cooperation or codes as the basis for the constitution of societal order. For on overview of the different theories of differentiation, see Uwe Schimank, *Theorien gesellschaftlicher Differenzierung* (Opladen: Leske und Budrich, 2000).

27 The most prominent example of this criticism is Francis Bacon, *The New Organon* (Cambridge: Cambridge Univ. Press, 2000).

progressed beyond the rudiments, and in the history of science the development of systems of thought, scientific organizations and recently also the practices of "doing" science is mostly given priority. The history of the competing understandings of science is taken into account in particular where discourses about science are analyzed, in the attempts to define the norms and values of science or in the efforts to understand science as a vocation.[28] Nevertheless, the lived norms of science, its ethos, as expressed in the everyday reasoning and action of scientists, and the continual shift of this ethos from generation to generation, are not the subject of systematically conducted research.

With respect to Virchow's speech, an understanding of science that is specific to modern science is of particular interest. The ethos of modern science as it evolved in the course of the institutionalization of modern empirical science seems to be markedly complex. The following three groups of features can be considered formative to this understanding. First of all, the methodical description of the empirical manifestations of reality and the subsequent methodical construction of theoretical explanations should be mentioned. "Methodical" here means the controlled conduct and recording of a series of observations and experiments as well as theory constructed on the basis of defined mathematic and geometric means of thinking.[29] In doing so the explanation resorts to formulating mechanisms, which are meaningless for the normative orientation of human life. Thus science becomes an important motor of the disenchantment of the world.[30]

Second, the understanding of science in question merges with the idea of progress characteristic of modernity. This leads to the idea that ongoing research results in increasingly extensive and accurate data collection as

---

28 For norms of science, see Robert K. Merton, "The Normative Structure of Science," in Robert K. Merton, *The Sociology of Science: Theoretical and Empirical Investigations* (Chicago: Univ. of Chicago Press, 1973), 267–78. On science as a vocation, see Max Weber, *Wissenschaft als Beruf* (Berlin: Dunker und Humblot, 1984); Max Weber, "Der Sinn der 'Wertfreiheit' der soziologischen und ökonomischen Wissenschaften," in Max Weber, *Gesammelte Aufsätze zur Wissenschaftslehre* (Tübingen: Mohr, 1988); Ulrich Oevermann, "Wissenschaft als Beruf: die Professionalisierung wissenschaftlichen Handelns und die gegenwärtige Universitätsentwicklung," in *Die Hochschule* 14, no. 1 (2005): 15–51; Steven Shapin, *The Scientifc Life: a Moral History of a Late Modern Vocation* (Chicago: Univ. of Chicago Press, 2008).

29 This aspect was discussed by Max Weber in *Gesammelte Aufsätze zur Religionssoziologie I*, (Tübingen: Mohr, 1988), 141–42, n. 5.

30 Cf. Weber, *Wissenschaft als Beruf.*

well as in increasingly powerful theories, and that in the progress of science the already established research approaches are at least in principle to be abandoned in favor of better ones. Thus, besides the moment of accumulation a revolutionary quality is immanent in knowledge development.[31] At the same time the question arises, how on the basis of suitable institutions can a progressive research process be established along these lines? This can be accomplished by the inter-generational organization of research by powerful and persistent academies that are associated with the state and devoted to the idea of progress, but also according to such genuine modern ideas as creating all-inclusive communicative structures, opening up science to every talent, or establishing a well-organized and regulated system of scientific organizations.[32]

A third complex of characteristics results from the amalgamation of science and *Bildung*, to use the German term here. Due to this amalgamation, science is pursued freed largely from the pressure of coping with the necessities of life. Such a relief has far-reaching consequences for the structure formation in science itself. A suitably relieved science turns to ever more fundamental questions, opens up to the manifold manifestations of reality in their solely aesthetically experienceable diversity and elaborates the trains of thought formulated in this regard raising them to outstanding expressions of human reasoning.[33] The societal existence of such a science

---

31 This specific normative framework of doing science leads to debates now considered classic: Karl R. Popper, *Logik der Forschung* (Tubingen: Siebeck, 1989); Thomas S. Kuhn, *The Structure of Scientific Revolutions* (Chicago: Univ. of Chicago Press, 1976); Imre Lakatos and Alan Musgrave, eds., *Criticism and the Growth of Knowledge* (London: Cambridge Univ. Press, 1970).

32 On the organization of science by national academies, see James E. McClellan, *Science Reorganized: Scientific Societies in the Eighteenth Century* (New York: Columbia Univ. Press, 1985). The norms explicated by Merton ("The Normative Structure of Science") are meant to help the "institutional goal of science," i.e. "the extension of certified knowledge" (p. 270). On the relevance of institutional settings for the progress of science, see Ben-David, *Scientist's Role in Society*, and Ben-David, *Scientific Growth*. For a renewal of an institutionalist approach, see Uwe Schimank, "Für eine Erneuerung der institutionalistischen Wissenschaftssoziologie," in *Zeitschrift für Soziologie* 22, no. 1 (1995): 42–57.

33 For that aspect of science, see Ulrich Oevermann, "Theoretische Skizze einer revidierten Theorie professionalisierten Handelns," in *Pädagogische Professionalität. Untersuchungen zum Typus pädagogischen Handelns*, edited by Arno Combe and Werner Helsper (Frankfurt: Suhrkamp, 1996), 70–182; Peter Münte, *Die Autonomisierung der Erfahrungswissenschaften im Kontext frühneuzeitlicher Herrschaft: Fallrekonstruktive Analysen zur Gründung der Royal Society*, 2 vols. (Frankfurt: Humanities Online, 2004); Andreas Franzmann, *Disziplin der Neugierde:*

is always precarious, because it finds itself irrevocably in tension with the real basis of human life, the struggle for survival and competition over scarce resources. Furthermore, such a science seems to require specific forms of societal control of its societally relevant achievements. It has to be relieved from external control itself. Complementarily, forms of self-control arise, which are commonly described in sociology using the terms profession and professionalization.[34]

The outlined theory of an ethos crucial for modern science brings up the question of how a science committed to this ethos can be established in society. Science as one societal sphere amongst others is in various ways dependent on the achievements rendered in those other spheres. Apparently, politics is of particular importance there because it regulates and supports science. A science characterized by the ethos outlined depends heavily on politics. For, inasmuch as it searches for a fundamental understanding of reality by empirical and theoretical means, it has need of protection against competing authorities that themselves claim to provide an overall understanding of reality. Similarly, as it generates knowledge that is shared in a wider community, it has to be acknowledged and constituted as an instance of binding judgements that matter in education and in collective decision making. And to the extent that it serves as a source of *Bildung* in terms of the free development of the human mind and senses, the required freedom from the constraints of life must be granted to it.[35] Thus, a science that is committed to such an understanding of science requires a corresponding "science policy." It is important to note, that it is not just a mutual utilization of the achievements of different sub-systems of society takes place here, but a process of community building, in which the representatives of such different spheres as science and politics by being involved in the formation of the order of society act on behalf of a wider community (which does of course imply that they are also involved in the pursuit of interests and the execution of power).

Against this backdrop the development of the relations between modern science and politics can be studied from a specific perspective. The

Zum professionalisierten Habitus in den Erfahrungswissenschaften (Bielefeld: Transcript Verlag, 2012).

34 For the professionalisation of science, see Peter Münte, *Die Autonomisierung der Erfahrungswissenschaften*; Andreas Franzmann, *Disziplin der Neugierde*.

35 Franzmann, Jansen, and Münte, "Legitimizing Science: Introductory Essay," in this volume.

questions asked of such a history would be how the described ethos could be institutionalized in a system of science that depends on politics and its regulation, and how societal alliances between science and politics are formed that preserve this ethos. But a further question has to be asked in this connection: how are these alliances formed in the ongoing discourse on the order of modernity itself, and to what extent is it possible to integrate the ethos described into a general idea of how modern society should work? The crisis of a science that is committed to the ethos described above in the present age, seems not least to be due to the fact that the order of the present modernity excludes at least some of the features of this ethos.[36]

The starting point of such a history might be the patronage of science, which is distinctive for an estate-based society characterized by personal dependencies.[37] The institutionalization of modern sciences takes place in the hollow pattern of patronage as is also characteristic of Renaissance science. Nevertheless, with the foundation of such institutions as the Royal Society and the *Académie royale des sciences* this pattern assumes a generalized and impersonal shape and hence is overcome at the same time.[38] The result is the type of scientific academy distinctive of the eighteenth century: A self-organizing science that puts itself under the protection of a sovereign and partakes of his authority. On the downside, the still dynastically legitimizing sovereignty itself becomes the vehicle of a modernization initiated by the modern sciences. Accordingly, the question arises what happens when dynastic sovereignty disintegrates in the course of progressing modernization itself. How can science under the banner of nation-building establish new alliances of modernization which preserve the basic features of the ethos described? This is exactly the question posed by the document discussed here.

The specific features of the formation of the German nation and the German nation-state themselves are of importance for a sociological analysis of the speech, a field that is well-established in historical research.[39] It is

---

36 Such a view is developed in Münte, "Die Autonomie der Wissenschaft im Ordnungsdiskurs der Moderne."

37 Prominent in this respect is the study of Mario Biagioli, *Galileo Courtier.*

38 Münte, *Autonomisierung der Erfahrungswissenschaften.*

39 See Dieter Langewiesche, *Nation, Nationalismus, Nationalstaat in Deutschland und Europa,* (Munich: Beck, 2000); Dieter Langewiesche and Georg Schmidt, eds., *Föderative Nation: Deutschlandkonzepte von der Reformation bis zum Ersten Weltkrieg* (Munich: Oldenbourg, 2000); Dieter Langewiesche, *Reich, Nation, Föderation: Deutschland und Europa* (Munich:

important to remember the constellation that evolved in the wake of the French Revolution and the Napoleonic Wars, which brought forth a plethora of bigger or smaller German states as well as a distinct tension with the developments in France, which ranged from ambivalence to bitter enmity due to the experience of submission. In this context the idea of a federal nation is distinctive, as it allows dual nationality: having the citizenship of a German country that is under the rule of a German prince, as well as being part of a German nation which is not corresponding to a particular state and which is defined first of all by law and culture. For that reason it is not clear from the outset to what extent this constellation is heading for a German nation-state at all. Being part of this nation demands a commitment to a relatively diffuse national community, and finds its expression in "national activities" such as gymnastics, singing or, as in the case to be studied here, in an assembly of German natural scientists and physicians. This also implies, of course, that the evolving forms of unity can be disconnected from the real political issues and do not by any means contain participation in a political public, which deals with the central political decisions. Thus, the idea of a culturally unified nation can stand in firm contrast with that of a political nation. To understand German nation-building, the failure of the German Revolution and its suppression by the German princes, who strive to secure their privileges, is important. Against this background, in Germany nation-building takes place through a series of wars under the command of the hegemonic power of Prussia, which are not converted into territorial gains but in the realization of the lesser German solution excluding Austria. This leads to a state that is substantially reliant on a military power that defies control. As a consequence a state based on military power as well as a German culture considered exceptional become the fundament of the identity formation of the German educated classes. The speech analyzed here is situated in the context of a reinvigorated national movement prior to the foundation of the empire under Prussian leadership.

In view of the outlined background, a historical situation stands out that has comparatively sharp contours and in which two urgent problems can be identified that might be of importance for Virchow's speech as a social activity of symbolic formation. On the one hand the always pressing question of how a specific understanding of modern science, as repre-

---

Beck: 2008). On Germany before 1866, see Thomas Nipperdey, *Deutsche Geschichte 1800–1866: Bürgerwelt und starker Staat* (Munich: Beck, 1998).

sented by Virchow and his peers, and on whose basis they can distinguish themselves as modernizers in science and society, can be secured in historically viable alliances. On the other hand there is the question of what shape the German nation should take in the overall process of nation-building going on in Europe at this time. Thus, there is a quite simple explanation for the speech as an activity of ongoing symbolic formation: In the speech, answers to the two questions are intertwined in a manner that allows the formation of a powerful alliance of modernization which implies enforcing a particular understanding of modern science.

Such an explanation refers to quite general motives which of course do not determine the speech as an activity of symbolic formation and the resulting symbolic structure. Thus, a question raised is what can be considered as distinctive to the speech in this regard. What is most notable seems to be that the idea of science is not adapted to the requirements of nation-building. On the contrary, the emerging nation is defined in terms of that understanding of science that is represented by Virchow and his peers and should become the basic principle of the whole nation. Thus, the function of science in society first of all is to establish an overall scientific world view shared by all people as a common basis of their intellectual life. This function cannot be reduced to what is often called the "popularization of science." It implies that the societal function ascribed to science is community building with respect to the constitution of a shared view of the world. This of course raises a further question: Why does Virchow feel in such a powerful position to be able to articulate that claim in the way he does?

Here it might be significant that science itself, and not politics, is representing the German nation. The constellation of nation-building outlined above can explain an outstanding feature of Virchow's speech: the distance from the political sphere. An interesting fact here is the absolute absence of politics, both as a sphere of society with whose representatives the institutional framework in which science is conducted has to be negotiated, and as a sphere in which an institutional framework of binding decision-making is emerging, and in which decisions can be made that can then count as decisions made by a particular nation. Science is presented as part of a national movement that transcends political institutions and creates a community above the political sphere. Nevertheless, in the development projected in the speech, science depends on politics, because a system of education based on science would have to be built.

4. The Transformation of Science in the Context of Nation-Building

The starting point of the preceding reflections was that the societal meaning of science is subject to an ongoing negotiation in discourses on the cooperative order of modern society, and that alliances, which span different spheres of society, are of special importance in that respect. It was also claimed that a document such as the speech analyzed here should first of all be considered an activity that aims at the symbolic formation of order and that an understanding of such a formation implies explaining its motives, especially those that result from problems in the order of societal cooperation itself. Sociological and historical reflections are required to uncover these motives. In this regard it can be hypothesized that an ethos of science that initially tries to gain insights into the fundamental structure of reality calls for alliances that shield them against demands for utilization. It can also be assumed that the succeeding constellations in the history of modern society might provide an opportunity in varying degrees for the formation of such alliances. Against this background, the speech seems to indicate an alliance between science and a social movement, which nevertheless has implications for the relationship between science and politics.

In the analysis, priority was given to the reconstruction of a process of symbolic formation. Characteristic of this process was how a complex social reality and its inherent transformations were captured in an identity-establishing myth. In this myth science is presented as an important vehicle of nation-building, and in reverse the nation formed is interpreted in terms of science, namely as a cultural nation that proves itself among other nations of science serving the good of mankind through outstanding scientific results. Thus, there is a quite simple conclusion that can be drawn from the analysis. Supposing that this myth is so powerful that it is able to leave an imprint on the identity of the members of the emerging nation then it has tangible implications for the support of science and also for the science policy of this nation. Science would then have to be promoted not least as the source of pure cultural achievements.

Thus, Virchow's speech seems to be an example of a specific constellation in the history of modernity that allows the institutionalization of an understanding of science that gives priority to seeking fundamental insights into reality by linking it to the process of nation-building. But this linkage seems to also have implications for the understanding of science itself, which is shaped as a national endeavor. These implications are stressed in the historical research on the nationalization of science, where a somehow "nationalized science" is often contrasted with a former "cosmopolitan science."[40] Of special interest in this context is the question of the tensions between the international and national orientations of scientists.[41] Thus, in the historical research on the nationalization of science it is the adaption of science to the requirements of nation-building that seems to be the focus, while, especially in the older history of science, the process of nationalization is even considered to be a deviation from the very norms of science. As we have seen this conflict of order is central in Virchow's speech itself.

Nevertheless, from the point of view of a sociology of modern society, things seem to be more complex than such distinctions as universalism vs. particularism or internationalism vs. nationalism suggest. A more complex understanding of the process of the nationalization of science can be established by considering this process as part of an overall move towards modernization. At first glance the myth of German science seems to be a deviation from the norm of universalism which is thought to be constitutive of modern science.[42] But on closer examination the opposite turns out to be true. The concept of German science implies an argument for the

---

40 Ralph Jessen and Jakob Vogel, eds., *Wissenschaft und Nation in der europäischen Geschichte* (Frankfurt and New York: Campus, 2002); Carol E. Harrison and Ann Johnson, eds., *National Identity: The Role of Science and Technology*, special issue, *Osiris* 24 (2009); Lorraine Daston, "Nationalism and Scientific Neutrality under Napoleon," in *Solomon's House Revisited: The Organization and Institutionalization of Science*, edited by Tore Frängsmyr (Canton, MA: Science History Publications, 1990), 95–199; Elisabeth Crawford, "The Nationalization and Denationalization of the Sciences: An Introductory Essay," in *Denationalizing Science: The Contexts of International Scientific Practice*, edited by Elisabeth Crawford et al. (Dordrecht: Kluwer, 1992), 1–42; Jutta Kolkenbrock-Netz, "Wissenschaft als nationaler Mythos;" Joachim Fischer "'Nationale Wissenschaften' in den europäischen Naturwissenschaften," in *Nationale Grenzen und internationaler Austausch: Studien zum Kultur- und Wissenschaftstransfer in Europa*, edited by Lothar Jordan and Bernd Kortländer (Tübingen: Niemeyer, 1995), 334–43; Ludmilla Jordanova, "Science and Nationhood: Cultures of Imagined Communities," in *Imagining Nations*, edited by Geoffrey Cubitt (Manchester: Manchester Univ. Press, 1998), 192–211; Angela Schwarz, *Der Schlüssel zur modernen Welt: Wissenschaftspopularisierung in Großbritannien und Deutschland im Übergang zur Moderne (ca. 1870–1914)* (Stuttgart: Steiner, 1999), 332–55 (chap. "Wissenschaft und Nation"). For an older view, see Brigitte Schroeder-Gudehus, "Nationalism and Internationalism," in *Companion to the History of Modern Science*, edited by Robert C. Olby et al. (London and New York: Routledge, 1990), 909–1007.

41 Brigitte Schroeder-Gudehus, "Nationalism and Internationalism."

42 Such a deviation is assumed by Merton in his "The Normative Structure of Science," 270–71.

democratization of knowledge and in this respect the people are included in the emerging system of science through scientific education. Here the question emerges of what the norm of universalism means exactly with regard to science. It can refer to the fact of building a worldwide structure of communication and a more or less homogeneous overall community of scientists. But it can also refer to a process of intellectual inclusion of the people through scientific education. In this second respect Virchow can be considered an important advocate of universalism.

This raises the interesting question whether science and the nation-state, as is often argued, can be associated with one or the other side in the distinction between universalism and particularism at all. This is debatable, especially if one proceeds on the assumption that the norm of universalism is itself formed in a discourse of modernization, which begins to disintegrate all conceivable particularism under the banner of progressing modernization. Thus the norm of universalism in this understanding implies the formation of ever broader communities.[43] Community building in the sense of establishing solidarity ties and shared knowledge allows for cooperation between random members of this community.[44] On that basis ever broader forms of collaboration can emerge that are thereby in functional terms more and more differentiated. The bigger the crowds of people gathering in ever broader communities, the more specific the forms of cooperation that can emerge, and thus an all the more differentiated inter-

---

43 There is no consensus on what "universalism" actually means. Daston refers to an "international and indeed supranational entity" in contrast to "intellectual nationalism." Daston, "Nationalism and Scientific Neutrality under Napoleon," 95. Merton distinguishes between two types. First, truth claims "are to be subjected to *preestablished impersonal criteria*" (emphasis in the original). But this is not so much an explication of a norm of science as an implication of what it means to claim that a proposition is true. Merton has in mind developments in contemporary Germany and Russia. But even in political ideologies of such states truth depends on reality as even racist claims claim to represent reality. Second, careers have to "be open to talents." This norm is commonly associated with modernity in general. It is difficult to, see what the first type has to do with the second. Cf. Merton, "The Normative Structure of Science," 270–71. For a conception of universalism as an ethos of science subject to historic change, see Gabriele Metzler, "Nationalismus und Internationalismus in der Physik des 20. Jahrhunderts: Das deutsche Beispiel," in *Wissenschaft und Nation in der europäischen Geschichte*, edited by Ralph Jessen and Jakob Vogel (Frankfurt and New York: Campus, 2002), 285–309, esp. 288–89.

44 On evolutionary advantages through community-building, see Michael Tomasello, *A Natural History of Human Thinking* (Cambridge: Harvard Univ. Press, 2014).

play of highly specialized activities results.[45] The norm of universalism understood in this way can have effects on both science and politics. But it leads to entirely different structure formations in both spheres.

The norm of universalism implies in science that science is seen as an infinite process of augmenting knowledge based on a coordinated world-wide discourse involving all qualified scientists. The practical problem is to establish such a discourse, as it would seem to require the organization of scientific discourse and work by powerful scientific organizations and by scientific publication systems. But more is required. There must also be a particular attitude of cooperation in science. Theories, for example, should no longer be considered as a truth discovered by the insights of a great philosopher, but as the basis on which the work of the scientific community can proceed by being transformed into hypotheses that are to be discussed on the basis of empirical findings. For this there has to be a universally accepted and accessible stock of empirical data, but more importantly there must be an overall homogenization of the scientists themselves. All relevant people have to engage in science on the basis of a similar understanding of science and a shared knowledge base. If such a homogenization takes place, the result is objective knowledge. All relevant people in the end would thus agree in their judgements, despite working in highly dynamic fields of research, and this would be achieved without the leadership of an intellectual authority, but on the basis of the authority of facts themselves. As is well known, scientists do in fact dissent in their judgements, and this problem cannot be solved by methodological rules which should allow for the selection of the right theory. But it is also obvious that, even if there is no complete homogenization, there should be a high degree of it, and more importantly, this should be the result of a collectively shared ideal of a homogeneous community of scientists. This is demonstrated by the fact that in science disagreement can be overcome in

---

45 This understanding of the norm of universalism as a specific modern value can be contrasted with the classic understanding of that norm that was established by Parsons: "definitions of role-expectations in terms of a universally valid moral precept, e.g., the obligation to fulfil contractual agreements, an empirical cognitive generalization, or a selection for a role in terms of the belief that technical competence in the relevant respects will increase the effectiveness of achievement in the role, are universalistic definitions of roles. On the other hand definitions in such terms as 'I must try to help him because he is my friend,' or of obligations to a kinsman, a neighbor, or the fellow-member of any solidary group *because of this membership as such* are particularistic" (emphasis in the original). Talcott Parsons, *The Social System* (London: Routledge, 1954), 62.

favor of an unconstrained consensus within the relevant community of the scientist.[46]

The institutionalization of the democratic nation-state starts markedly later then the institutionalization of modern science. With the French revolution a model of politics emerges that is based on a new pattern of legitimation and which in a long historical process leads to new political institutions.[47] In this respect the democratic nation is not just based on the new legitimatory pattern of the sovereignty of the people in contrast to the dynastic legitimation of power. It is also framed by the general idea of progress and the rebuilding of society.[48] An important aspect here, too, is community building, which is experienced by those subjected to a particular authority and transforms them into political subjects. It is on this basis that revolutionary claims for political participation are made and new political institutions based on the public control of government arise.

In this connection a public—a political public—emerges that is very different from the public within science. These publics are different with respect to the subjects of discussion, but also with respect to its character. On the one hand questions about the structure of the world are discussed which are more or less detached from the conflicting interests of ordinary life and should be answered by accomplishing objective knowledge, while on the other hand the focus is on questions of political decision-making, which are connected with conflicting interests and are settled by the use of compromise in political procedures. Nevertheless, in both cases a similar problem has to be solved: particular solidary groups and local life-worlds have to be broken up and a process of homogenization has to take place. In science as well as in politics a shared perspective has to be established: on the one hand a shared perspective on rather unworldly scientific prob-

---

46 This leads to core problems in the sociology of the scientific community. In science, community-building begins when methodology ends. Since Kuhn, it has been common knowledge that the "choice of theories" is underdetermined by methodological rules. Kuhn stresses the importance of the scientific community and its values. See Kuhn, *Structure of Scientific Revolutions*, 167. But Kuhn's focus was on the value of unanimity that prompts the community to avoid conflict and to turn to a shared system of knowledge, i.e. "puzzle solving." Cf. Thomas S. Kuhn, *Die Entstehung des Neuen: Studien zur Struktur der Wissenschaftsgeschichte* (Frankfurt: Suhrkamp, 1978), 381–82.

47 Cf. Stefan Kutzner, *Die Autonomisierung des Politischen im Verlauf der Französischen Revolution: Fallanalysen zur Konstituierung des Volkssouveräns* (Münster: Waxmann, 1997).

48 Dieter Langewiesche, "'Nation', 'Nationalismus', 'Nationalstaat' in der europäischen Geschichte seit dem Mittelalter—Versuch einer Bilanz," in Langewiesche, *Nation, Nationalismus, Nationalstaat*, 14–34, esp. 22, 32.

lems and their solution, and on the other hand a shared perspective on pressing political problems and their solution. But at the same time, the processes of breaking up local life-worlds in favor of a more or less homogeneous community in both cases are quite different. In the case of science a common space of communication is established by scientific publication in special journals, the reproduction of evidence and regular meetings in which affiliation to a special working community of scientists is demonstrated, a community that shares a specific branch of knowledge, but also produces such a shared knowledge in an ongoing discursive process. The members of such a working community share a very specific interest in the advancement of a particular branch of knowledge, and such a community can integrate people with very different cultural affiliations. In the case of politics a common space of communication is established by the public press, opinion making and rituals of demonstrating affiliation to a quite huge group of people, the nation, but also to particular parties representing different interests, and such a community is limited in its capacity to integrate people with different cultural affiliations. In the case of science the established space of communication is a space of rational and consensus oriented discourse. In the case of politics a discourse is established in this space that is characterized by perpetual controversy and dispute. Nevertheless, decisions have to be made, which calls for political procedures (if controversy and dispute is not suppressed by creating inner and outer enemies). Since the institutionalization of science depends on politics, due to the different structure formations in both spheres that can likewise be ascribed to the norm of universalism, a constellation emerges in which nation-states compete for the advancement of science by creating national systems of science.

Against this backdrop, Virchow's speech, even if considered to be an expression of a process of universalization in the sphere of politics, seems to be a deviation from the norm of universalism at least with regard to a scientific community committed to this very norm. This deviation seems to be inherent in the concept of German science and is connected to Virchow's implicit theory of the emergence of modern science. As we have seen, according to Virchow modern science requires that scientists achieve intimate contact with the life of their own nations and develop their own perception of nature. Such a perception should be grounded in the peculiar nature of the respective nation itself. If so then this implies that there must be national styles of research grounded in the peculiar character of a nation.[49]

But does this really imply a violation of the idea of an overall community of scientists participating in a worldwide structure of communication? As we have seen we can also find hints in the speech itself of a strong concept of a universally valid scientific method. Virchow is claiming that Protestantism led to the emergence of a correct and somehow natural method of thinking. How can the idea of a universally valid scientific method be considered compatible with the assumption of the national sources of scientific productivity? Now, on the basis of the old distinction in the philosophy of science between the context of discovery and the context of justification, it is of course quite easy to connect these ideas.[50] In Virchow's speech we find no hints of the claim that there are different national standards of justification. It can be read solely as a theory of the nation as an indispensable source of the generation of new ideas. And if we take into consideration Virchow's claim that German science has a duty to contribute to the progress of science and the culture of mankind then a universalistic concept of justification even seems to be mandatory. Thus, Virchow's concept of German science can be said to be in no way a deviation from the norm of universalism in science. Nevertheless, it has implications for the distinction between national and international cooperation. For Virchow the working unit of science is the national community of scientists, and in this respect the German community of scientists should be of outstanding importance. That of course seems to be incompatible with the idea of intensive international cooperation, but it is compatible with an overall exchange and evaluation of ideas.[51] In this sense, the idea of science that is formed in Virchow's speech indeed has implications for the order of research, but these implications seem to be far from what such a formulation as the "nationalization of science" might suggest, especially when it is contrasted with a former cosmopolitan science committed to the norm of universalism.

49 Goschler, "Deutsche Naturwissenschaft und naturwissenschaftliche Deutsche," 109.

50 Hans Reichenbach, *Experience and Prediction: An Analysis of the Foundations and the Structure of Knowledge* (Chicago: Univ. of Chicago Press, 1938).

51 On physics in Germany, see Gabriele Metzler, "Nationalismus und Internationalismus in der Physik des 20. Jahrhunderts."

5. Shifting Alliances of Modernization: Concluding Remarks

What can be taken from the preceding considerations as a basis for the comparison between the nineteenth and twentieth centuries aspired to in this volume? To what extent might documents such as the speech analyzed here give hints to differences that matter with respect to such a comparison? Here, too, the concept of societal alliances as they are formed in the discourses on the order of the modernity can be brought to fruition. At first one has to state that the alliance looming here is bound to a transient historical constellation. In this constellation science is able to present itself not only as a source of scientific education but also as an intellectual power that shapes the emerging nation. However the societal role claimed in this context, i.e. to render the method of scientific thinking part of the common way of thinking, becomes unavoidably obsolete with the success of this program. And not only this: Today the objectives framed in the speech, which are crossed by anti-authoritarian motives and pressure for a democratization of society, exhaust precisely the kind of science that shall be enthroned here. The natural sciences as well as their distinctive claims to knowledge and ideas of autonomy have themselves become part of a criticism concentrated on the disenchantment of science itself and its societal control, which feed on anti-authoritarian motives and is intertwined with democratization efforts.

But in reference to the comparison between the nineteenth and twentieth centuries the question also arises, to what extent have the societal alliances shifted fundamentally that form under the banner of modernization, especially when the natural sciences as a major driving force of modernization are complemented in this role by the social sciences? On the one hand an older constellation is found, in which the representatives of a scientific community committed to the ethos of the modern experiment-based sciences and formed on a national level, succeeded in presenting themselves as an important embodiment of the nation alongside the state (or, as in the case of Virchow's speech instead of the state), by defining the discourse on science held in public and in establishing patterns of interpretation compliant with their own understanding of science. These patterns then set the agenda for the institutionalization of science. On the other hand a newer constellation is found, in which the interpretation of the role of science in society takes place in a discourse that is closely linked to the discussions in the social sciences in particular. What this implies for the standing of science in society is made clear, especially if one looks at the role of the scientific profession in each of the two constellations. In one the scientific profession was able to present itself and the particular understanding of science that is represented by it as a motor of modernization. In the other science becomes the object of a discourse of modernization, in which the scientific profession is first of all considered an interest group and the particular understanding of science that is protected by this profession is branded an ideology that hinders societal progress.[52]

The break in structure implied here becomes even more contoured when one takes into account the central topics of the discourse of modernization which can be found in the social sciences and in which the societal position of modern science is negotiated. In this discourse of modernization a criticism of the established order of science is formulated, the epistemic authority of science is deconstructed, and the knowledge for a "new governance of science" is produced. With respect to the first point, besides the aforementioned criticism of the professions, be it as relics of a feudal society or as powerful interest groups that create monopolized markets, two strands of discussion are found here which are relevant for the formation of the societal role of science. The first is the criticism of the idea of *Bildung* through science as it was framed in Germany especially by German idealism. In the context of this criticism science tends to be reduced to the production of expert knowledge. Thereby it is denied the potential of a *Bildung* that exceeds the acquisition of expert knowledge.[53] Second, the criticism of the idea of a scientific-technical progress directed by scientific experts has to be mentioned. Here the lack of democratic legitimation of scientific experts and the problem of non-intended side-

---

52 This interpretation of the professions became dominant in the 1970s. Famous examples of a power-approach in the sociology of professions include Terence J. Johnson, *Professions and Power* (London and Basingstoke: Macmillan, 1972); Magali Sarfatti Larson, *The Rise of Professionalism: A Sociological Analysis* (Berkeley: Univ. of California Press, 1977). Criticism of the professions is shared by sociologists of science who consider institutions of modern science to be relics of an estate-based society: "Die Institution der Wissenschaft trägt nach wie vor die Züge eines Standes." Peter Weingart, *Die Wissenschaft der Öffentlichkeit*, 7.

53 Helmut Schelsky, *Einsamkeit und Freiheit: Idee und Gestalt der deutschen Universität und ihrer Reform* (Düsseldorf: Bertelsmann, 1971). Schelsky's position stands out for its technocratic radicalism. In his view all political judgement in the end will be reduced to expert judgement. See Helmut Schelsky, *Der Mensch in der wissenschaftlichen Zivilisation* (Cologne: Westdeutscher Verlag, 1961).

effects are at the center of discussion.[54] With this line of criticism the fact is added that a science that is reduced to the production of expert knowledge shall consequently be established as an object of societal regulation that is guided by criteria of societal relevance and reflects the consequences of the technological developments based on scientific knowledge. The common direction of the impact of both lines of thought can be seen in the fact that they are equally directed against an understanding of science based on an idea of "pure knowledge." At the same time more utilitarian functions of science in society became the focus of attention. All this is paralleled by an attempt at demystification embracing all aspects of science. Modern science is described as a mere product of historical constellations of power, in which political interests are purported to be the decisive factor, and its knowledge is described as the result of a production process, in which the formative principle is not the scientific community and its regulatory ideals, but ideological commitments, random configurations in laboratory equipment and ordinary everyday practices.[55] Finally, this destruction of the understanding of science which was characteristic of the age of "classical modernity" corresponds to an ever broader knowledge of modernization provided by the social sciences themselves. This knowledge allows for the reorganization of academic life according to criteria of societal relevance viewed from a "rational" perspective.[56] As an implication the organization of scientific action is characterized less and less by institutionalization processes, in which the knowledge from the scientists' own personal experience working in the respective institutions takes effect. Instead it results in the application of the currently available knowledge of modernization.[57]

---

54 Protagonists of such criticism include Jurgen Habermas, *Technik und Wissenschaft als 'Ideologie'* (Frankfurt: Suhrkamp, 1968); Ulrich Beck, *Risikogesellschaft: Auf dem Weg in eine andere Moderne* (Frankfurt: Suhrkamp, 1986).

55 For sociologies of demystification, see the sociology of scientific knowledge developed in David Bloor, "The Strong Programme in the Sociology of Knowledge," in David Bloor, *Knowledge and Social Imagery* (Chicago and London: Univ. of Chicago Press, 1991), 3–23; Bruno Latour and Steve Woolgar, *Laboratory Life: The Social Construction of Scientific Facts* (Beverly Hills, CA: Sage, 1979), i.e. so-called "laboratory studies."

56 Examples include Henry Etzkowitz and Loet Leydesdorff, "The Dynamics of Innovation: From National Systems and 'Mode 2' to a Triple Helix of University—Industry—Government Relations," in *Research Policy* 29 (2000): 109–24.

57 Richard Münch, *Globale Eliten, lokale Autoritäten: Bildung und Wissenschaft unter dem Regime von PISA, McKinsey & Co* (Frankfurt: Suhrkamp, 2009).

The result of all these efforts at intellectual destruction and organizational modernization is that science takes place in a completely changed frame. This frame is created on the basis of a global knowledge of modernization with the intention of constructing an effective innovation system from a rational point of view. The concern of politics is primarily to condition the respective national science system to the global competition that takes place on the basis of the application of a global knowledge of modernization.[58] That implies that the science policy of a particular nation-state is no longer based on an alliance with the respective national profession of science, which then strives to distinguish itself through outstanding research achievements. This is replaced by an alliance of national politics with a globalized modernization industry. The scientific profession is forced to merge with the modernization processes that are activated on this basis. It can be assumed that this will also fundamentally change its character as a profession.

---

58 See Tobias Werron, "Universalized Third Parties: 'Scientized' Observers and the Construction of Global Competition between Nation-States," in this volume.

Science in an Emerging Nation-State: The Case of the Antebellum United States

Axel Jansen

1. Introduction

At the beginning of the nineteenth century, experiment-based research was a well-established mode of truth-seeking in many European countries and in the United States, but a belief in science and its explanatory potential prompted different responses in different national contexts. At a time when Napoleonic France considered science to be an important symbol of national cultural achievement, and while England relied on the Royal Society in London as well as emerging associations to provide a common intellectual focus for the pursuit of knowledge, American advocates of science faced a rather different context for legitimizing their work. In order to develop an appreciation for science within the growing number of American states, those interested in the natural sciences also had to help secure their country's political cohesion and viability. As scientists, they wanted their nation-state to appreciate and support their work; as US citizens in a century of nation-building, they were convinced that their country's future and their role as scientists hinged on the adoption of a rational perspective represented by their work. The scientific profession during that century, in other words, sought to help build an American nation-state.[1]

---

1 The term "profession" is used in line with a revised theory of the profession sketched in the introduction to this volume. This approach highlights the role of a particular habitus that evolves in response to the specific problems engaged by scientists, i.e. advancing theory through tackling problems of explanations whose resolutions promises to advance a field. The revised theory of professionalization proposes that such "vicarious crisis management" goes on in other professions as well even if they focus on different types of crisis. For an older perspective on professionalization in science that focused on "full-time professionals" vs. "amateurs", see Nathan Reingold, "Definitions and Speculations: The Professionalization of Science in America in the Nineteenth Century," *The Pursuit of Knowledge in the Early American Republic*, edited by Alexandra Oleson and Sanborn C. Brown (Baltimore: Johns Hopkins Univ. Press, 1976), 33–69.

This, at least, was the perspective of the leader of scientists in antebellum America. Alexander Dallas Bache (1806–1867), Benjamin Franklin's great-grandson, was the key figure in national scientific institutions in the United States during the 1840s and 50s, and Bache helped instigate and found several organizations including the National Academy of Sciences in 1863, becoming that institution's first president.

While Bache shaped key American scientific institutions on the regional and emerging national level, however, he may also be considered a token of the peculiar relationship between science and the evolving American national state during that period. An investigation of Bache's biography and institutional role elicits patterns relevant well beyond his own biography, not just because they reflect his times but because he was decisive in shaping them. For our purposes of looking for ways to grasp the intricate relationship between the universalistic endeavors of research science and their particular national context, Bache may well be our Representative Man as his biography helps grasp the evolving relationship between science and the national state in antebellum America.

2. Science in the American Federation

The setting for science in the United States was peculiar not least because the country was then only a few decades old and because it anticipated growing well beyond its contemporary boundaries. The country had been designed by its founders as an expanding federation, a federation that anticipated absorbing territory not yet under its control. The original thirteen states during the nineteenth century admitted new states to the union. In several successive steps, the federation came to include new states along the Pacific and in the continent's interior.

While the American states agreed on a Constitution in 1789, a regional and state-centered perspective carried over into the emerging federal arena. The Constitution established dual citizenship in both the states and the federal state, but national citizenship was a political project rather than the status quo.[2] The country lacked a coherent national public and a capital city

---

2 See both Dieter Langewiesche's and Peter Münte's essays in this volume for academic responses to a somewhat similar situation in German states during the nineteenth century.

that would serve as a cultural center.[3] In 1787, the Northwest Ordinance established a system of converting areas settled by European immigrants into territories and states, and the Louisiana Purchase in 1803 added a huge and largely unknown area to the country's settlement plans.

With the election of Thomas Jefferson to the presidency, a peaceful transfer of power from one political faction to another took place in 1800 for the first time, solidifying the federation's constitutional system. But the political framework was run by regional elites with a national political public that evolved gradually. Brian Balogh has recently argued that Americans throughout the nineteenth century had "consistently turned to the national government." But even Balogh had to concede in the title of his book that despite its role in foreign policy, military control of settlement areas, and a postal service before the Civil War, the nineteenth-century US government remained "A Government out of Sight."[4] James McPherson has suggested that the Civil War between 1861 and 1865 "fused the several states bound loosely in a federal *Union* under a weak government into a new *Nation* forged by the fires of a war in which more Americans lost their lives than in all of the country's other wars combined."[5] The Civil War has remained an important focus in American memory but it remains difficult to see how at the time, the war could be perceived as evidence for integration and nationhood. At the end of the war, however, few could doubt that an American state's membership in the United States was irreversible. Anticipating such developments and looking for ways to flesh out the American federation and its government (a government that was obviously claiming a much more active role than at any time prior to the war), in 1862 President Abraham Lincoln decided that the Union would abandon slavery and become a platform for negotiating issues of political justice. The Civil War

---

3 John C. Greene, "Science, Learning, and Utility: Patterns of Organization in the Early American Republic," *Pursuit of Knowledge*, 1.

4 I am subscribing to a "developmental vision" as presented by Peter S. Onuf and others. In his recent book, Brian Balogh suggests that the federal government had been much more authoritative in the nineteenth century than was presumed. But as the case of Bache shows, protagonists of American culture acted on the assumption that while the federal government did manage certain areas of life, the nation's integration lay in its future. Brian Balogh, *A Government out of Sight: The Mystery of National Authority in Nineteenth-Century America* (Cambridge: Cambridge Univ. Press, 2009).

5 James M. McPherson, *Battle Cry of Freedom: The Civil War Era* (New York: Oxford Univ. Press, 1988), viii.

had opened up the potential (rather than the political reality) of national integration through resolving problems on a common, national level.

These matters were reflected in the way science was organized in the United States in the early nineteenth century. Urban areas provided an amplifying context for scientific interests, with Philadelphia and Boston as centers of development. In the late eighteenth century, and in the first half of the nineteenth century, American universities, as institutions, had not yet dedicated themselves to research science.[6] Both Philadelphia and Boston laid claim to national leadership through the establishment of academies.

The American Philosophical Society (APS) had been founded in Philadelphia in 1743 and rejuvenated by Benjamin Franklin in 1769. The American federal government resided in Philadelphia before it moved to the newly built capitol city of Washington DC in 1810. During that period, the Philosophical Society benefitted from endorsement by cosmopolitans such as Thomas Jefferson. But by the 1820s, the society's *Transactions* were no longer published regularly. Boston had been an important center of learning throughout the colonial period and in 1780, the American Academy of Arts and Sciences had been founded there. Neither the APS nor the American Academy, however, could claim to represent science on a national level. While both organizations ambitiously aimed for American cultural independence from Europe, they remained regional affairs.[7]

Alexander Dallas Bache emerged as a leader among scientists in the US during the 1840s and 50s. He was the superintendent (director) of the United States Coast Survey, then the largest science-related organization within the federal administration, and with the guiding support of Bache and his closest colleagues and friends, scientific institutions began to emerge on the federal (national) level. Bache was instrumental in establishing and directing institutions such as the American Association for the Advancement of Science (AAAS) and the National Academy of Sciences. His background and thinking on the matter provides convenient shorthand

---

6 On the emergence of the American research university, see, for example, Roger L. Geiger, *To Advance Knowledge: The Growth of American Research Universities, 1900–1940* (New York and Oxford: Oxford Univ. Press, 1986).

7 Edward C. Carter, *One Grand Pursuit: A Brief History of the American Philosophical Society's First 250 Years, 1743–1993* (American Philosophical Society, 1993); Greene, "Science, Learning, and Utility," *Pursuit of Knowledge*, 1–32.

for an investigation of the trajectory of American science on an evolving federal and prospective national level.

3. Science and Technology in Philadelphia

While Bache's leadership was facilitated by his role in the federal administration, he was appointed to that post because his scientific peers had supported his candidacy. For his role in national science politics, his selection for the post of superintendent for the Coast Survey turned out to be decisive. In 1842, United States President John Tyler put Bache in charge of an enterprise in Washington that had a budget significantly bigger than that of the Smithsonian Institution and a staff larger than that of Harvard University. The Coast Survey employed a host of academically trained staff, such as surveyors and mathematicians, to map and calculate the federation's coastline, a task crucial for the country's shipping interests. Before moving into the late 1840s and 50s, however, it makes sense to take a step back and to investigate Bache's preparation for his federal role. This investigation serves a two-fold purpose: to sketch the qualifications that made Bache a successful candidate for the Coast Survey post in 1842, and to chart the scientific culture of one region—that of Philadelphia—at a time when science remained a regional affair.[8]

Alexander Dallas Bache was born in 1806 to a family very much aware of the limitations but also of the prospects of the American national enterprise. His early career was an extension of a highly successful family history that was intertwined with American independence and state-building. Through his father, Bache was a descendent of Benjamin Franklin, in many ways the quintessential eighteenth-century American. During his childhood and adolescence, however, his mother's side of the family exerted a much stronger and even more ambitious influence. His mother was the oldest daughter of Alexander James Dallas, a highly successful immigrant, who

8 For an analytical biography of Bache, see Axel Jansen, *Alexander Dallas Bache. Building the American Nation through Science and Education in the Nineteenth Century* (Frankfurt and New York: Campus, 2013). On Bache's candidacy for the Coast Survey post, see Hugh R. Slotten, *Patronage, Practice, and the Culture of American Science. Alexander Dallas Bache and the U.S. Coast Survey* (Cambridge: Cambridge Univ. Press, 1994), 61–75. On Philadelphia, see Robert V. Bruce, *The Launching of Modern American Science, 1846–1876* (New York: Knopf, 1987), 46–50.

had tied his fate to that of the rebellious American colonies in 1784. In Pennsylvania, Dallas was a leading figure in America's emerging opposition movement around Thomas Jefferson.

Considering his family's background in state-building, it comes as no surprise that when fourteen-year-old Alexander Dallas Bache graduated from school in Philadelphia, he chose to attend the United States Military Academy at West Point. At that age, of course, he probably followed his family's advice on this matter. The navy would perhaps have been the most consistent choice but it would have associated this family of Democratic-Republican persuasions with Federalism. Harvard also stood for a kind of elitism his family abjured. West Point, however, suggested the building of a federal elite on a different set of premises. The Academy had been founded by Jefferson in 1802 to signal that the Republican president would continue to support federal development championed by his political opponents, but that he would do so on his own terms. It matched Jefferson's expansionist vision of America by producing for the young republic a federal elite under the supervision of the US Army Corps of Engineers. Separated from the regular army, West Point graduates were to provide militias with leadership in times of crisis and the technological expertise in developing the federation's infrastructure. And in these ways, the elite institution was to help represent and diffuse the national idea among the country's citizenry.[9]

Bache excelled at West Point, graduating at the top of his class. His achievements reserved him a position in the US Army Corps of Engineers. But Bache soon moved on. Rather than pursuing a lucrative career as a civil engineer at a time when they were in high demand, the then 23-year-old chose to accept a post as Professor for Natural Philosophy and Chemistry at the University of Philadelphia. Returning to his home town in 1828 offered him an opportunity to join the city's evolving culture of science and technology. During the next few years, Bache would develop this culture in line with his family's national perspective reinforced by his West Point education.

---

9 Stephen E. Ambrose, *Duty, Honor, Country: A History of West Point* (Baltimore: Johns Hopkins University Press, 1966); Theodore J. Crackel, "The Military Academy in the Context of Jeffersonian Reform," and "West Point and the Struggle to Render the Officer Corps Safe for America, 1802–33," both in *Thomas Jefferson's Military Academy: Founding West Point*, edited by Robert M. S. McDonald (Charlottesville: Univ. of Virginia Press, 2004), 99–117 and 154–181, respectively.

Had there been an opportunity to develop science and technology on a national plane, we may speculate that Bache may perhaps have chosen to join such efforts in 1828 when he completed his West Point training. As things stood, however, Bache sought such opportunities in Philadelphia.

As a professor at the University of Philadelphia, Bache could have confined himself to handling his significant teaching load, but he didn't. Soon after his return to the city, he was elected a member of the American Philosophical Society, an organization his great-grandfather had shaped in the late eighteenth century. But Bache chose to invest his spare time in helping develop another emerging Philadelphia institution, the Franklin Institute of the State of Philadelphia, for the Promotion of the Mechanic Arts. Both the APS and the Franklin Institute provided platforms for research that American universities did not yet furnish. But while the Philosophical Society was Philadelphia's attempt to provide an emerging nation with a scientific organization on par with its European rivals, the Franklin Institute's founding was a response to the increased significance of technology in the context of industrialization.[10] At the Franklin Institute, Bache joined, and soon led, efforts to bolster the region's evolving large-scale manufacturing of steel-based production such as steam engines for locomotives and steam-powered ships. He was instrumental in drafting two reports during the early 1830s, and they characterize the role he envisioned for the Franklin Institute.

One is Bache's "General Report on the Explosion of Steam-Boilers" published by the Institute in 1836,[11] the other his expertise on the issue of weights and measures, also written as the chairman of a committee at the Franklin Institute two years earlier.[12]

The explosion of steam boilers on steam boats was a notorious problem that cost many lives. Bache headed a panel of 17 to investigate the

---

10 Bruce Sinclair, *Philadelphia's Philosopher Mechanics: A History of the Franklin Institute* (Baltimore: Johns Hopkins Univ. Press, 1974).

11 The Franklin Institute, "General Report on the Explosions of Steam-Boilers (1836)," in *Early Research at the Franklin Institute: The Investigation into the Causes of Steam-Boiler Explosions, 1830–1837*, edited by Bruce Sinclair (Philadelphia: Franklin Institute of the State of Pennsylvania, 1966).

12 Alexander Dallas Bache, "Report of the Committee on Weights and Measures. Appendix to the Report of the Committee of the Franklin Institute on Weights and Measures. Abstract of the Reports on Weights and Measures Which Have Been Submitted to the Congress of the United States, or to the Legislature of Pennsylvania," *Journal of the Franklin Institute* 8, no. 4 (April 1834): 232–47.

matter. When the US Secretary of the Treasury learned that the Franklin Institute was pursing such work, he commissioned the Philadelphia institution to conduct experiments on his behalf, and he also agreed to pay for it. Robert V. Bruce has inferred that Bache's research on steam boilers produced much data and little else, but this view underestimates another important dimension of the committee's work.[13] Under Bache's direction, the Franklin Institute pursued the first research project that was funded by the federal state. In his concluding report on the Institute's investigations, furthermore, Bache did not seek to use the federal government's prestige as leverage for the Franklin Institute's standing; quite the opposite. As the lead author on the *Report on Steam Boilers*, the now 30-year-old Bache by inference used the Franklin Institute's authority, an authority based on science, to bolster the federal government's political status.

Bache suggested that the federal government pass a law to improve the security of steam boilers. "In submitting this project," he wrote, "the Committee obviously do not entertain a doubt of the competency of the [United States] Congress to legislate on the matters embraced on it." The committee's reassertion was intended to help dispel such doubts and to propose federal action on behalf of the safety of American citizens. It would take another 16 years, however, before the United States government acted on the committee's recommendations. A federal system for licensing and inspecting steam boilers was not introduced until 1852.[14]

Bache's work on weights and measures suggested a similar perspective. In 1833, the Franklin Institute was asked by the Pennsylvania House of Representatives to present ideas for standardized weights and measures for the state. During the preceding years, this topic had received some attention across the United States. In his contribution to the Franklin Institute's report, Bache favored a pragmatic approach. He preferred an international solution and standard but considered such a solution unrealistic. Instead, he suggested to the Pennsylvania legislature that a federal solution be sought that would apply to the entire United States. Bache found the metric system unsuitable for his country because it was based on experiments along a meridian that ran through France. He preferred a solution based on an experiment conducted on US territory, and thus suggested keeping

---

13 Robert V. Bruce, *The Launching of Modern American Science, 1846–1876* (New York: Knopf, 1987), 16–18.
14 Louis C. Hunter, *Steamboats on the Western Rivers: An Economic and Technological History* (New York: Dover, 1994), chap. 13.

certain weights and measures (such as the yard) and to verify them by experiment in the United States.[15] The location of such experiments should be fixed by law so as to make possible "the recovery of the standard if lost, or its verification if required."[16] But the Pennsylvania government ignored suggestions for a federal solution, chose to act alone, and assigned to the Franklin Institute the official role of securing the standards for the state.[17]

Both in his work on steam boilers and in his expertise on weights and measures, Bache took for granted that he and his colleagues represented a self-reliant authority independent of political endorsement and support, an authority he sought to mobilize to bolster the political project of building America. Even though today we would perhaps expect him to boost the Franklin Institute and its committees by highlighting federal support for their project, he did no such thing. Instead, his work in Philadelphia during the 1830s suggests that he sought to mobilize standards of rationality represented by himself (as well as his science- and technology-interested peers) to help guide his country's development at the federal level, if possible, or at the state level, if necessary. During this period, federal support was scarce and a matter reserved for the future, but Bache's regional perspective anticipated a federal and national development. Such a perspective was reflected in his educational work as well.

That work, in fact, derived from the same set of standards and premises that shaped his role in providing technical expertise. As he came to assume a more prominent role at the Franklin Institute, Bache—as chair of the Committee on Instruction—suggested a new format for the Institute's meetings. While questions of science and technology had been discussed in monthly meetings, Bache suggested holding separate "conversation meetings" where individual members could raise questions and ideas on scientific and technological developments.[18] While Bache was that circle's most

15 "Report in Relation to Weights and Measures in the Commonwealth of Pennsylvania," Journal of the Franklin Institute 14, no. 1 (July 1834): 6–14. Concerning the experimental basis of weights and measures, see Paolo Agnoli and Giulio D'Agostini, "Why Does the Meter Beat the Second?," physics/0412078, December 14, 2004, http://arxiv.org/abs/physics/0412078.

16 Bache, "Abstract of the Reports on Weights and Measures Which Have Been Submitted to the Congress of the United States, or to the Legislature of Pennsylvania," 241.

17 Sinclair, Philadelphia's Philosopher Mechanics, 193.

18 Franklin Institute, "Annual Meeting of the Franklin Institute," Journal of the Franklin Institute 11, no. 2 (February 1833): 85–86; Franklin Institute, "Annual Report of the Board of Managers," Journal of the Franklin Institute 11, no. 2 (February 1833): 89–90.

active member, he considered it crucial for its success that meetings retained "[an] absence of form."[19] When Bache became chairman of the Franklin Institute's Board of Managers in 1834, he was instrumental in setting up a Committee on Science and the Arts. The committee was in charge of reviewing inventions, suggesting improvements, and reporting developments to members of the Institute and to the public. What Bache insisted on was that membership in this committee was opened up and that any member of the Institute who deemed himself capable could join in the reviewing and discussion of technological gadgets and scientific ideas. "It is believed," argued Bache,

that there are many of our younger fellow members, who, having been during years past in attendance upon the lectures and schools of the Institute, are now ready to repay with interest from their acquired stock of knowledge, the benefits which they may have received.[20]

Bache thus sought to implement institutional opportunities for a culture of self-directed criticism. At the same time, Bache aimed for a development of a rational discourse that would help improve life in America.

4. Girard College and Central High School, 1836–1842

In 1836, Bache left his professorship at the University of Pennsylvania for the post of president of the Girard College for Orphans in Philadelphia. Stephen Girard had been the Bill Gates of his times. Upon his death, he left seven million dollars (about 173 million dollars today) to build a college in his city. He had left detailed plans for a building that was to be constructed of fireproof marble. The building was both for the school and the internment of his diseased self in a sarcophagus. Bache's new position came with a generous pay-check. Even though construction of the college buildings was only just getting underway, Bache was hired and sent on a two-year trip to Europe, all expenses paid, to investigate the continents'

---

19 Franklin Institute, "Annual Report of the Board of Managers," Journal of the Franklin Institute 13, no. 4 (April 1834): 228.
20 Ibid., 229–30.

systems of education, a tour that, as Robert Bruce has noted, helped shape Bache's vision of science in America.[21]

Given these benefits, this was too good an opportunity to let pass. Between 1836 and 1838, Bache travelled through Europe. He attended lectures by Michael Faraday in London, met with François Arago in Paris (with whom he was particularly impressed) and with Alexander von Humboldt in Berlin. He investigated grammar schools as well as universities in order to familiarize himself with all dimensions of Europe's educational systems. After his return to Philadelphia, he wrote a 660-page report that became a standard for anyone interested in education.[22] In his letters to the Girard college trustees, Bache showed a keen interest in the interplay between research and education, an interest he had already shown in his work at the Franklin Institute.

In one of his letters to the Girard College board, for example, Bache proposed that all of the school's teachers be sent abroad on an educational tour. Bache not only sought to turn Girard College into a model institution in the United States for the most recent European insights into pedagogy (such as those by Philipp Emmanuel von Fellenberg in Hofwyl, Switzerland), but he also aimed to improve European models. In one of his letters, he referred to institutions there "which approach nearest to the model of what I think an elementary school ought to be."[23] One improvement Bache sought to make was to base the new college on the principle of collegiality. While Bache spoke of pedagogical training when referring to the "elementary school," he referred to teachers of older students as "professors." After his return to Philadelphia in 1838, he translated his ideas by erecting a laboratory on college grounds to facilitate regular measurements of the earth's magnetic field. With the help of students, measurements were recorded and published in several bound volumes.[24]

21 Bruce, *Launching of Modern American Science*, 18.

22 Alexander Dallas Bache, *Report on Education in Europe: To the Trustees of the Girard College for Orphans* (Printed by Lydia R. Bailey, 1839).

23 Alexander Dallas Bache to Nicholas Biddle, October 9, 1837, box 3, vol. 3, Bache Papers, Smithsonian Institution Archives, Washington DC.

24 Girard College Magnetic and Meteorological Observatory, Alexander Dallas Bache, and United States Army. Corps of Engineers, *Observations at the Magnetic and Meteorological Observatory, at the Girard College, Philadelphia, Made Under the Direction of A. D. Bache, LL. D. and With Funds Supplied by the Members of the American Philosophical Society, and by the Topographical Bureau of the United States* (Washington: Gales and Seaton, printers, 1847).

5. Federal Science

Bache's ejection from his educational projects in Philadelphia coincided with an opportunity to take his ideas to Washington DC. In a concerted campaign orchestrated by Joseph Henry, the Smithsonian's secretary (director), the scientific community supported Bache's appointment as superintendent of the US Coast Survey in 1842.[25] Bache moved to the federation's fledgling capital city the following year. As we have seen, he brought to Washington a distinct perspective that considered a scientific discourse to be the basis for technology-development and education. Bache's arrival in Washington coincided with a national development of science and its institutions in the United States, a development that Bache helped to initiate and direct.

But did the United States lack federal institutions of science prior to 1842? I have already mentioned the regional role of the American Philosophical Society in Philadelphia and the American Academy in Boston. Under Bache's direction, the Franklin Institute during the 1830s became an institution of science-based technology development that was relevant well beyond the Philadelphia region. On the federal level, there had been some initiatives prior to 1842. Thomas Jefferson had sent an expedition to explore the vast expanse of land the country had bought from France in 1803 through the Louisiana Purchase. The founding of the United States Military Academy at West Point had provided important educational opportunities for a science-oriented engineering curriculum. In 1846, after a long debate, Congress chose to accept a bequest from a quirky Englishman to found the Smithsonian Institution and to become the caretaker of this independent, transnational institution dedicated to science. And in 1807, the founding of the United States Coast Survey had fallen in line with similar interests to provide the federal government with the tools to understand the territory it had acquired and to aid the federation's merchants in navigating its coastline. But even though West Point was a permanent establishment with a strong interest in science-based engineering, it was not a scientific institution. The Smithsonian was dedicated to science but it was an international, not a national American institution. And while the Coast Survey relied on science-based technologies to chart the country's vast and growing coastline, its task remained limited. Congress anticipated

---

25 Slotten, *Patronage, Practice, and the Culture of American Science*, 61–75.

the survey would complete its charting of the US coastline and thus fulfil its mission at some point in the future. In line with its limited task, the Coast Survey was a temporary measure. A review of science-related institutions in the United States in the 1840s and 50s shows, therefore, that no permanent national scientific institution existed, and thus no institution that would suggest the US had committed itself to science.[26]

Bache had long been interested in developing, at a national level, experiment-based research science as a basis for education and technology-development. Prior to the 1840s and 50s, it seemed hardly possible to do so because of the federation's lack of integration. Bache's arrival in Washington DC coincided with emerging prospects for fleshing out the evolving national context in science and the arts. Given his family background, his Philadelphia experience, and his new role as superintendent of the country's largest science-related organization, Bache was in a good position to assume leadership of the emerging process of building national science institutions and one institution in particular: a national academy of sciences that would expand American cultural independence from Europe by representing the American people's dedication to the universalistic principles of science.

Emerging scientific associations provided the platform to discuss such issues among scientific peers. A national organization by geologists in 1848 gave rise to the American Association for the Advancement of Science. Following its British model, the AAAS organized annual meetings and published proceedings. It complemented the *American Journal of Science and the Arts* that had been published in the US since 1818, by providing opportunities for establishing a national reputation, while subcommittees within the AAAS responded to an increasing need for specialization within the natural sciences.[27] The AAAS was the first comprehensive national scientific organization in the United States, but it did not represent a public commitment to experiment-based science in the United States.

In the eyes of Bache and his closest colleagues, such a commitment remained insecure indeed, and Bache outlined his concerns in several speeches he gave in the late forties and through the fifties.

26 A. Hunter Dupree, *Science in the Federal Government: A History of Policies and Activities to 1940* (New York and Evanston: Harper Torchbooks, 1986 [1957]), particularly 66–90, 100–05.

27 Sally Kohlstedt, *The Formation of the American Scientific Community: The American Association for the Advancement of Science, 1848–1860* (Urbana: Univ. of Illinois Press, 1976).

The common denominator of these speeches was Bache's concern for a national foundation for science in America. Bache highlighted how the consolidation of the American federation into an American nation continued to hit political roadblocks, and that the country was losing valuable time before it could make a commitment to the kind of cultural progress that he associated with research science. Without such a commitment and endorsement, he felt that the responsibility to enforce standards within the scientific profession evolved on him and others in positions of power, for they had to shield claims for interpreting nature from those ("charlatans") who sidestepped peer review by appealing to the public.[28]

In 1851, in the most prominent of his public addresses, for the first time Bache publicly proposed the founding of a national academy.[29] In this "magisterial discourse on the state and needs of American science" that, according to Robert V. Bruce, "confirmed him as the leading spokesman and mentor of the American scientific community," Bache took a sweeping view of science in its evolving American setting as he and his closest friends conceived of it.[30] "While science is without organization," Bache pointed out, "it is without power: powerless against its enemies, open or secret; powerless in the hands of false or injudicious friends."[31] This much-quoted phrase seemed to confirm Bache's power politics.

But did it? What Bache was saying here was that science required protection, and in the context of his speech he was pointing to the integrity of science rather than its influence in other spheres of society. Bache considered the organization of science important to protect its standards, something the professional community could not do without an established structure. Bache made this point in the context of observations on science in America. "Separated by vast distances, scattered in larger or smaller communities," he said to his colleagues,

---

28 For an example of Bache's involvement in a public controversy, see Mary Ann James, *Elites in Conflict: The Antebellum Clash over the Dudley Observatory* (New Brunswick: Rutgers Univ. Press, 1987).

29 Dupree, *Science in the Federal Government*, 116–19; Merle Middleton Odgers, *Alexander Dallas Bache, Scientist and Educator, 1806–1867* (Philadelphia: Univ. of Pennsylvania Press, 1947), 168–70.

30 Bruce, *Launching of Modern American Science*, 264.

31 Alexander Dallas Bache, "Address of Professor A. D. Bache, President of the American Association for the Year 1851, on Retiring from the Duties of President," American Association for the Advancement of Science, *Proceedings* 6 (1852): lii.

the daily avocations of men of science in the United States keep us asunder. Our small numbers at any one point produces all the bad influences of isolation. We feel cut off from the world of science, and sink discouraged on account of the isolation; or having a position in the community about us, we become content to enjoy this, and forget that we owe a duty to the world outside; that we ought to increase, as well as to diffuse; to labor, as well as to enjoy the labors of others. Our country asks for other things from us than this.[32]

According to Bache, it took an extra effort and special dedication to pursue science in America, and his characterization implied that the dedication and sense of cohesion among scientists in the US must indeed have been strong as, unlike other countries, his colleagues could rely on little political or cultural support. Hence Bache considered organization "here" in the US "for good or for evil … the means to the end."[33]

Bache tied these reflections on the state of the professional community to the suggestion that the United States create a national academy of sciences. An *"institution of science,"* he said, *"supplementary to existing ones, is much needed in our country, to guide public action in reference to scientific matters."*[34] As an extrapolation of the role Bache had assumed as spokesman of science in Washington DC, he suggested a national academy as an institution to supplement and supplant the Coast Survey, an institution that during the fifties had evolved as a de-facto national academy with Bache as its "invisible president."

But Bache pointed to issues standing in the way when he moved on to reason that "[i]t is, I believe, a common mistake, to associate the idea of academical [sic] institutions with monarchical institutions." Bache, in other words, took up common American reservations towards anything that smacked of centralized political control at the federal level. He implies here that his contemporaries failed to see that political authority, in the United States, had a republican basis. He argued that in a republic it ought to be possible to create institutions dedicated to "aristocratic" cultural ideals represented by science.

But the absence of an aristocracy left a functional void. In his 1852 speech, Bache cautiously suggested to his colleagues that in lieu of an aristocracy, the ideals of science should eventually be endorsed by the democratic "sovereign" in order to secure science in American society and to

---

32 Ibid.
33 Ibid.
34 Ibid., xlvii–xlviii, Bache's emphasis.

secure the benefits of scientific rationality for American national development. Before his country was ready to dedicate itself to scientific principles, Bache considered himself and his colleagues, the scientific profession, to be responsible for "guarding the palladium" as he called it.

In the 1850s, in short, Bache and those around him hoped that a national academy would eventually be founded, an event that would signal that the country endorsed science and its principles. But they were reluctant to push the matter because they thought that America was not ready. Given a political climate in which the founding of such an academy would have implied an expansion of federal powers (something many politicians in the South and elsewhere wished to avoid in order to forestall federal activity in sensitive policy areas such as slavery) the proposal for a federal academy of science would not have gone anywhere. Bache had become the key leader of scientists in the United States and while he bemoaned the absence of an academy, he also controlled attempts to create one. He was the *primus inter pares* of a small group of scientists that called itself the "Lazzaroni", a term frequently inferring idle Neapolitan beggars who had sided with the king against Napoleon.[35] With Bache as their informal "president", they wielded considerable influence. Among them were Harvard mathematician Benjamin Peirce (father of philosopher Charles Sanders Peirce), Joseph Henry (the Smithsonian Institution's secretary), Louis Agassiz (a Swiss-born biologist at Harvard with considerable public visibility), Charles Henry Davis (a former naval officer who worked for the Naval Observatory's *Nautical Almanac*), and mathematician Benjamin Apthorp Gould (who had declined a Göttingen professorship to dedicate himself to building up science in America). During the late forties and early fifties, the group met informally for lavish dinners and conversation, and coordinated their professional politics. They were criticized for their high-handedness, but they felt justified to preserve control where they had it because of their concern for the fate of science in America.

---

35 Lazzaroni were stevedores with brief work shifts rather than beggars. John A. Davis, *Naples and Napoleon: Southern Italy and the European Revolutions, 1780–1860* (Oxford Univ. Press, 2006), 27.

6. Founding of the National Academy of Sciences

When the Civil War broke out in 1861, the Lazzaroni had been dysfunctional for some time, but they came back together one more time, in order to found a national academy. Group cohesion evaporated after the Lazzaroni had created a generalized copy of itself, with Bache carrying over into the office of president. But they took the initiative under circumstances quite different from the ones they had hoped for. The National Academy of Sciences was created by an act of Congress on March 3, 1863, with President Abraham Lincoln signing it into law late the same day.

On the occasion of another important piece of Civil War legislation, the 1862 Morrill Land Grant Act, Daniel Kevles has recently invoked the established view on the academy's founding.[36] While an active federal policy was blocked by southern resistance before the Civil War, he wrote, secession provided northern and western representatives in Congress with an opportunity to implement long-standing plans for expanding federal power. They "formed an imposing alliance determined to satisfy pent-up demand for federal promotion of Western settlement and national economic development under the leadership of the recently formed Republican Party." And with respect to the Academy, Kevles suggests that Bache, "the longtime federal scientist, felt the need for an institution of authoritative scientists that would safeguard public policymaking in an increasingly technical age from charlatans and pretenders."[37] Bache left no letters in which he explained his rationale for acting on the academy idea, but we may infer his motives from correspondence that reveals his assessment of the war, while a close reading of some of these letters shows that though Bache was an institutional realist, he was also a cultural romantic.

During the war, Bache struck a friendship with Francis Lieber, a Prussian émigré and a political scientist—a strong public advocate of a new, consciously national perspective on American politics, a perspective concerned about the loyalty of Americans in both the South and the North. Lieber had three sons, all of them in the war, one of them in the Confederate army fighting the other two.

---

36 Kevles, "Not 'A Hundred Millionaires'. Federal Science in the Civil War and the Gilded Age," Lecture at the Library of Congress, July 7, 2012, http://issues.org/wp-content/uploads/2014/05/kevles_lecture.pdf.
37 Ibid., 5–6.

Fig. 1. Alexander Dallas Bache, between 1855 and 1865.

In one of his letters to Lieber, Bache poked fun at a union soldier who had found a marker set near a field by a Coast Survey team but who mistakenly thought that the marker had been set, not by the Coast Survey, but by the Confederacy. Instead of recognizing the acronym "U.S.C.S." for "United States Coast Survey," the union soldier had made it out as suggesting a boundary between the "United States" and the "Confederate States" ("U.S." plus "C.S."). For Bache, the soldier's error and ignorance represented his country's dismal situation. Bache conceived of the Coast Survey as a token of the country's ambitious future. But the soldier had remained oblivious to such political and cultural aims. What hope could there be if the evolving nation's ambitions remained invisible to the American people?

In the same letter, Bache wrote that he refused to help a confederate soldier's wife whose property was about to be confiscated by the US government. "Her husband," Bache wrote bitterly, "is a double distilled traitor + her appeal is founded on the fact which doubly condemns him, that he was once in the employ of the Coast Survey!"[38] The woman's husband was a confederate soldier and this made him a traitor. What made him "double distilled", however, was that having once worked for the Coast Survey his disloyalty was even more unforgiveable as Bache considered the Coast Survey to be not just one federal agency among others, but nothing less than an advance representation of the emerging American nation.

Exasperated with his country's fratricidal war, in his letter Bache condemned both the ignorant soldier in the North and the wife of a former employee in the South. Even though he despised the attack by both sides on the country's integrity, Bache supported the union effort as the only way to preserve the country's future.

In the fall of 1862, union armies faced severe setbacks in the field. Their success and the country's future seemed all but certain. This is when Bache and his comrades decided to act on the old academy idea. The absence of Southerners from Congress provided an opportunity but not the motive. At least with respect to Alexander Dallas Bache (without Bache, the academy would have lacked authority) a key aim was to erect a standard for the union, to provide the country with a signal that there was something worth fighting for and that the American union would eventually emerge as an American nation, ready to pledge itself to the universalistic discourse of science.

7. Legitimizing Science in America

In his well-known 1851 speech, Alexander Dallas Bache had expected and hoped for the US sovereign to implement a national academy so as to endorse and mobilize the rationality of science and to secure the profession's standing in America. The founding of the Academy during the Civil War represented his effort at mobilizing the scientific profession's standing on behalf of his country, which he feared was breaking apart. Bache's republican hopes for integrating science with the national project would eventually evolve along unexpected lines.

In the history of legitimizing American science, the key development in the late nineteenth century was the coordinated growth of research univer-

---

38 Alexander Dallas Bache to Francis Lieber, May 15, 1862, box 1, Lieber Papers, Huntington Library, San Marino, CA.

sities and professional associations. This expanded backbone to the rise of American science corresponded to the tacit acknowledgment that science provided an essential content for the education of emerging elites.[39] The rise of science-based education after the Civil War opened up teaching careers to researchers who were themselves trained at universities. Science was also endorsed by an unmatched private provision and support for research by wealthy donors such as Andrew Carnegie and John D. Rockefeller. Their foundations further strengthened the universities as centers of research during the 1920s when these foundations decided to support them instead of funneling funds into research institutes without academic ties. Such support represented "private, elite authority over American science during the 1920s."[40] Before 1940, therefore, American basic science was very much a project supported by wealthy businessmen endorsing scientific research, and by urban elites endorsing educational programs offered by the country's most prestigious and research-oriented institutions of higher education.

After the Civil War, Bache's perspective carried over into the new period. On the widening scale of an American elite's commitment to science and its institutions, and in the context of an emerging recognition of the country's intellectual contribution, the impetus compelling recognized scientists to bolster American nationhood and the federal state took on new forms. Its most visible attempts took place during times of crises or when such crises provided opportunities to bolster the state with authority that would in turn bolster the reputation of its scientists. Examples include those arising from Bache's own circle, such as the idea by mathematician and longtime Lazzarone Benjamin Peirce in the 1870s that the American Social Science Association (ASSA) merge with Johns Hopkins University.[41] At a time when the joint role of research universities and research associa-

---

39 Burton J. Bledstein, *The Culture of Professionalism: The Middle Class and the Development of Higher Education in America* (New York and London: W. W. Norton, 1976). Andrew Jewett, *Science, Democracy, and the American University: From the Civil War to the Cold War* (Cambridge: Cambridge Univ. Press, 2012), 39–44. Jewett focuses on educators ("scientific democrats") who after the Civil War wished to replace the institutional Christianity of denominational colleges with science-based education (p. 39).

40 Geiger, *To Advance Knowledge*, 99.

41 Thomas L. Haskell, *The Emergence of Professional Social Science: The American Social Science Association and the Nineteenth-Century Crisis of Authority* (Urbana and Chicago: Univ. of Illinois Press, 1977). Dorothy Ross, *The Origins of American Social Science* (Cambridge: Cambridge Univ. Press, 1992).

tions had not yet snapped into place, Peirce hoped to turn Johns Hopkins into the cultural hub for educators across the United States. In 1916, and well before the US would enter World War I, astrophysicist George Ellery Hale, on behalf of the National Academy of Sciences, sought to create a national role for scientists by offering the services of science to the federal government, helping pave the way for a political endorsement of a new national role he envisaged for his government. Within the National Academy, the new National Research Council was not intended to advance basic research but to enshrine the work of scientists with public relevance, and to take on an opportunity for public service. Despite its public character, the NRC was not financed by the federal government, but from private sources such as the Carnegie Foundation.[42] It sought to renew legitimacy for a system of private science that had evolved outside the federal establishment.

American science came into its own in the late 1920s when research in such fields as physics caught up with developments in Europe. The emergence of the United States as a leading force in science then took place within its private system of research universities in conjunction with professional associations to enable, coordinate, and evaluate work in the various fields.[43] During the 1930s, MIT president Karl T. Compton, head of President Franklin Roosevelt's Science Advisory Board, led efforts to mobilize federal funds for research. But Compton and his board wished to use such funds for basic research without political stipulations. The board refused to earmark money for new hires to help fight the war on unemployment.[44] Basic research organized at universities and in associations would eventually be supported by the federal government but in keeping with its nineteenth-century roots in such fields as geology and geodesy, the legitimacy for such funding would derive from science-based technology rather than from science itself.

During the buildup to the American entry into World War II, science once again reached out to the federal state. During World War I, the National Academy had supported the federal government through its National Research Council, but funding for these efforts had been private. In the late thirties, Karl Compton brought Vannevar Bush to Washington,

---

42 The Carnegie Foundation in the 1920s paid for a building on the Mall in Washington DC that became the home of the National Academy of Sciences

43 Geiger, *To Advance Knowledge*.

44 Kevles, *The Physicists*, 252–66.

an MIT engineering professor. Instead of looking for federal funds to support basic research, Bush looked for ways to mobilize research for the American military. In 1941, the Office of Scientific Research and Development (OSRD) became the umbrella organization for developing and producing technology including radar and the nuclear bomb. Physics thus took the lead in associating science-based technology with the federal state. By the time the US entered World War II, more than 1,700 physicists worked in defense technology development, a quarter of the entire profession.[45] By the time the war ended, that number would be dwarfed by the more than 125,000 administrators, scientists, engineers, and construction workers employed under the rubric of "science" in the 37 installations of the Manhattan Project.[46]

The significance of science-related technology for American political hegemony after World War II provided the scientific community with significant political leverage vis-à-vis emerging federal funding structures and in American culture at large.[47] The prominent role of American science during the Cold War has been intertwined with the recognition by the federal state of science-based technology as an essential component of power. This perhaps obscured the role that some scientists from within the profession's ranks sought in bolstering the consolidation of the American nation-state as a basis for the recognition of their work.

---

45 Ibid., 320.

46 Ira Katznelson, *Fear Itself: The New Deal and the Origins of Our Time* (New York: Liveright, 2013), 349.

47 Roger L. Geiger, *Research and Relevant Knowledge: American Research Universities since World War II* (New York: Oxford Univ. Press, 1993).

The State-Technoscience Duo in India: A Brief History of a Politico-Epistemological Contract

Shiju Sam Varughese

1. Introduction

The historical transformation of the social contract of science is a central concern for contemporary Science, Technology and Society (STS) Studies scholarship, given the anxieties that exist about the emergence of mode II knowledge production and its capacity to radically alter the way science relates to the nation-state in today's liberal democracies.[1] "Science and neoliberalism" is similarly an emerging concern for scholars in the field.[2] In the neoliberal context, academics have argued that there are new, multiple spaces where this social contract is operationalized and the public can engage with science.[3] In order to democratize science to be "socially robust," more public deliberation as well as new institutional mechanisms to increase citizen participation in the governance of science is proposed by scholars from different theoretical traditions. Social historians of science have pointed out that the contract between science and society operates differently in diverse social contexts, thanks to the unique historical trajectories of these societies. From this vantage point, exploring the non-European contexts of the professionalization of science is highly rewarding, as the process is deeply intertwined with the region-specific characteristics of the contract. The case of South Asia in general and India in

I thank Itty Abraham for his comments on an early draft of this chapter.

1 See Michael Gibbons et al., *The New Production of Knowledge: The Dynamics of Science and Research in Contemporary Societies* (Los Angeles: Sage Publishers, 1994); Helga Nowotny, Peter Scott and Michael Gibbons, *Rethinking Science: Knowledge and the Public in the Age of Uncertainty* (Cambridge: Polity Press, 2001).

2 See, for example, the special issue on 'STS and Neoliberal Science' in Social Studies of Science 40(5), October (2010). See also Kelly Moore et al. "Science and Neoliberal Globalisation: A Political Sociological Approach," *Theory and Society* 40 (2011): 505–32.

3 See Nowotny, Scott and Gibbons, *Rethinking Science*.

particular, as this chapter will demonstrate, offers new insights into the professionalization of science that were coupled with the evolution of the postcolonial state in a very different fashion. In this chapter, my analysis focuses on the historical evolution of the relationship between the Indian state and science from the 1950s to contemporary times to suggest how academic science was marginalized and became subservient to a hegemonic, mission-oriented "technoscience"[4] under the aegis of the state. My arguments take their cue from the theoretical proposition regarding the formation of a politico-epistemological contract between the state and technoscience that shapes the wider relationship between science and multiple publics.[5]

The chapter follows the evolution of the contract through three broad phases of Independent India.[6] The first phase is from the 1950s to the early 1970s, a period of great optimism in science and its potential to serve the infant nation-state. This period is generally known as "the Nehruvian era" after the first Indian Prime Minister Jawaharlal Nehru (1947–64) who was deeply influenced by the USSR model of centralized economic planning. The second phase begins in the late 1970s with the emergence of a diversity of new social movements all over the country which offered a strong epistemological critique of modern science and its close coupling with the nation-state. The third phase corresponds with the adoption of the New Economic Policy by the Indian government in 1991 that opened

---

4 The term technoscience is used here not in the Latourian sense "as a fusion of science, organization and industry" but denotes the particular function of science and technology as part of the development machine of the state. Bruno Latour, *Pandora's Hope: Essays on the Reality of Science Studies* (Cambridge: Harvard Univ. Press, 1999), 203. In this sense, technoscience is different from academic science, although academic science can be reductively appropriated and/or overshadowed by technoscience as the chapter demonstrates.

5 See Shiju Sam Varughese, "Where are the Missing Masses? The Quasi-Publics and Non-publics of Technoscience," *Minerva* 50, no. 2 (2012): 239–54. The term "public" is generally used in plural as "publics" in the chapter, due to the fact that the state-technoscience relationship has produced diverse formations of the public in India, thanks to the emergence of multiple public spheres as well as complex political negotiations to claim citizenship rights, despite universal adult franchise. The public spheres in India show great diversity in their organization and deliberative function, resulting in the formation of heterogeneous publics. The Habermasian notions of the public and public sphere hence cannot be employed in its original form to examine the Indian context.

6 India became decolonized in 1947, followed by the national struggle for independence from the British Empire.

up India to the challenges and possibilities of neoliberal globalization. Through this brief analysis of the social history of the politico-epistemological contract between science and the state in India, I will demonstrate how a symbiotic relationship between the nation-state and the "technoscientific complex" was established in accordance with the contract between state and technoscience, and how academic science in the country became marginalized in this process. The development of this contract will be analyzed in relation to two specific sectors: nuclear energy and agriculture. These sectors are of special interest because of their emergence as sites of enunciation of the sovereign power of the infant nation-state that sought legitimacy from modern science to establish itself. The nuclear energy sector in the country is kept detached from any democratic intervention, shrouded in the rhetoric of national security because of the state's nuclear weapon development program. The developments in the agriculture sector were connected to the central problem of increasing food production with the help of advanced technological input. In recent decades, these two sectors reflect the impact of neoliberalism on science and its relationship with the state.

2. Academic Science before and after Decolonization

Although there were several views available before 1947 about what role science and technology ought to play in the newly formed Indian nation-state, it was the Nehruvian-Bernalist vision of development based on science and technology as the "epistemological engine of progress" that became the official ideology.[7] The Gandhian critique of western science and technology, and his emphasis on local villages, remained a marginal concern for the developmental aspirations in the Nehruvian period.[8] A strong endorsement of this developmentalist view was made during the second

---

7 Dhruv Raina, "Evolving Perspectives on Science and History: A Chronicle of Modern India's Scientific Enchantment and Disenchantment (1850–1980)," *Social Epistemology* 11, no. 1 (1997): 9. Professor J. D. Bernal's views on the social function of science had a great influence on Nehru. Bernal had visited India several times in the 1940s and 1950s and his suggestions were central to the shaping of the national S&T system.

8 The Gandhian perspectives found resonance only in the "Community Development Schemes" that focused on rural development since 1952. See, Benjamin Zachariah, *Nehru* (London and New York: Routledge, 2004).

Five-Year plan of the country (1956–61), when heavy industrialization was promoted.[9] The quintessential need for centralized planning of the national economy was a strong discourse in the 1950s and 60s and the advent of development economics as a strong academic field enhanced this.[10] The Nehruvian paradigm of development in the 1950s "incorporated three elements; centrality was accorded to a particular kind of knowledge, a well-defined strategy for its deployment in social transformation, and a path of action: these being modern science, planning and industrialization, respectively."[11] A strong belief that modern science and technology, wedded with the centralized planning process, could help the country modernize more rapidly became the hallmark of the period.

The Nehruvian era was marked by the formation of a unique coupling between science and politics. The emphasis on centralized planning and expectations about the ability of science and technology to find solutions for societal problems led to the creation of an elite group of scientists who were given the dual task of guiding the policy-making process and developing the national science and technology (S&T) system. Expert advice was actively sought by the state to make policy decisions and it was believed that a self-reliant national S&T system was quintessential to the nation's progress. Scientists such as Homi J. Bhabha, who established the nuclear energy program of the country, and Shanti Swaroop Bhatnagar, founder director of the Council for Scientific and Industrial Research (CSIR)[12] and the first chairperson of the University Grants Commission (UGC),[13] worked closely with political leaders such as Jawaharlal Nehru. At this juncture of the emergence of the institutional frame of big science, the

---

9 The first three five year plans in India correspond to the first phase of the contract. The First Five-Year Plan (1951–56) focused on agriculture, and invested in big river-valley dam projects. The Third Plan (1961–66) reemphasized the modernization of the agriculture sector.

10 It is noted that "[a]lmost all major contemporary economists who took an interest in problems of development had occasion to interact with India's planners and policy-makers in the fifties and the early sixties." Sukhamoy Chakravarty, *Development Planning: The Indian Experience* (New Delhi: Oxford Univ. Press, 1987), 4.

11 Raina, "Evolving Perspectives," 9.

12 When World War II began, the British government established a Board of Scientific and Industrial Research in 1939 in India, modeled on the Department of Scientific and Industrial Research (DSIR) in Britain. This was reconstituted as the CSIR in September 1942 to coordinate the industrial R&D of the country. Today the CSIR has a network of 38 national R&D laboratories.

13 The UGC has been coordinating the university system in the country since 1956.

state became the major supporter of scientific and industrial research, creating the national system of S&T.

The net effect of the emergence of the state as the patron of science, however, initiated a process of "suppression of the university as the primary centre of scientific research."[14] The two research imperatives of the period under the tutelage of the state were the industrial research imperative and the nuclear research imperative which shaped the framework of the national S&T system.[15] These imperatives catalyzed the organization of the national S&T system around a technoscientific complex that served the developmental aspirations of the nascent nation-state. State support for separate research institutes outside the university system (under the CSIR) where short-term mission-oriented research for national development was carried out, led to a decline of university-based academic science,[16] and assessment endorsed by recent research on the early decades of nuclear physics in India. During the 1930s and 40s, Indian nuclear physicists had been at the frontiers of the field, collaborating with renowned laboratories in the West and engaging intensely with the international research community.[17] With the powerful emergence of the postcolonial state, however, the internationalism of Indian nuclear physics began to decline from the 1950s. The state limited nuclear research to nuclear energy and weapon development programs, and it neglected nuclear physics departments in the university system.[18]

Agricultural science in India reveals another side of this sad decline of academic science. Agricultural research went through a radical transforma

---

14 Dhruv Raina and Ashok Jain, "Big Science and the University in India," in *Science in the Twentieth Century*, edited by John Crege and Dominique Pestre (Amsterdam: Harwood Academic Publishers, 1997), 859–78.

15 Ibid.

16 Ibid.

17 Jahnavi Phalkey. *Atomic State: Big Science in Twentieth-Century India* (Ranikhet: Permanent Black, 2013). Until the beginning of World War II, a true internationalism was enjoyed by the scientific community all over the world, and the Indian academic community was also a part of it. According to Itty Abraham, "[i]n contrast to the halcyon days of the 1930s and before, when physicists from all over the Europe moved easily back and forth between each other's laboratories and countries, the 1940s marked the end of what we might call 'continental' physics in favor of a 'nationalized' physics [in Europe]." Itty Abraham, *The Making of the Indian Atomic Bomb: Science, Secrecy and the Postcolonial State* (Hyderabad: Orient Longman, 1998), 42. However, as I suggest here, the "nationalized science" in Europe was characteristically poles apart from its Indian counterpart.

18 Ibid.

tion in the immediate context of the Green Revolution (1965–75), directly linking it with the technoscientific task of enhancing food production. The colonial government did not show much interest in establishing agricultural research even after the central department of agriculture had been established in 1881. This picture changed in the early twentieth century with the founding of several agricultural and veterinary colleges.[19] The Imperial Agricultural Research Institute (in Pusa, Bihar) was founded in 1904, and many other central research institutes followed.[20] Although the Indian government's first Five-Year plan (1951–56) emphasized agriculture and rural development, it did not affect the way agricultural research functioned in the country. Rajeswari S. Raina has pointed out that in the 1950s and early 1960s, 60 to 75 percent of national research and development (R&D) resources were generated by research institutes sponsored by provincial governments. This decentralized research system enjoyed greater autonomy in terms of research priorities, and in seeking research funds.[21]

With the advent of the Green Revolution paradigm in the second half of the 1960s, however, the agricultural research system became reorganized under the Indian Council for Agricultural Research (ICAR).[22] The second plan's shift of focus to heavy industrialization had played havoc with the agricultural sector, and made famines an everyday reality for the rural poor.[23] As a consequence, the third plan aimed at increasing food production. High Yielding Varieties (HYVs) of wheat and rice were used for massive cultivation after their successful transfer from Mexico and the Philippines and their diffusion in India by the Ministry of Food and Agriculture and the newly reconstituted ICAR.[24] With the assistance of international donor agencies such as the Ford Foundation, the Rockefeller Foundation, and the United States Agency for International Development (USAID), agricultural universities modeled after US institutions were established in India from the early 1960s along with several research institutes

---

19 Rajeswari S. Raina, "Institutional Strangleholds: Agricultural Science and the State in India," in *Shaping India: Economic Change in Historical Perspective*, edited by D. Narayana and Raman Mahadevan (London, New York and New Delhi: Routledge, 2011), 99–123.

20 Ibid.

21 Ibid.

22 Ibid; Govindan Parayil, "The Green Revolution in India: A Case Study of Technological Change," *Technology and Culture* 33, no. 4 (Oct. 1992): 737–56.

23 Parayil, "The Green Revolution."

24 Ibid.

under the ICAR.[25] With the research system's centralization in later years, however, the immediate task of the agricultural research system, including the universities, turned out to be facilitating the transfer of technology under the Green Revolution program.[26] This made the national agricultural research system focus on mission-oriented short-term research on behalf of the state, while ignoring important long-term research projects. This reorientation prevented the State Agricultural Universities and regional institutes from adapting their research agendas to local concerns and demands. By 1965, the country's agricultural research system had been completely absorbed into the technoscientific complex. In the case of nuclear physics, however, academic research continued to be a separate, marginal activity in sharp contrast to the state's flourishing nuclear energy and weapons programs.

The industrial and nuclear imperatives of establishing the S&T system within the frame of big science in the 1950s and 1960s thus discriminated against university-based academic science in favor of the state-centric technoscientific complex. Economic development and national security were the dual objectives of that complex. "Self-reliance" was used as a rhetorical claim to garner public support for large technoscientific projects (both civilian and military) developed by the state with the help of the technoscientific complex. The coming of age of the technoscientific complex in the 1960s further marginalized the ill-developed academic research system and its potential to develop human resources through research-based education. As mentioned earlier, an elite group of scientists, engineers, technocrats and policy makers represented the technoscientific complex, frequently disguising technoscience *as* academic science, and they worked in close association with the country's political elite to actualize the state's developmentalist dreams. Hi-tech projects were promoted as peak achievements on India's path of rapid modernization.

These developments plainly suggest an important aspect of the Indian state. From its very inception, the technoscientific complex was an integral part of the country's sociopolitical constitution, performing ideological

---

25 Ibid. For example, Gobind Ballabh Pant University of Agriculture and Technology, the first agricultural university established during the period, was modeled on the University of Illinois. The assistance of the US in establishing agricultural universities came to an end by 1972. See Parayil, "The Green Revolution" for details.

26 Scholars have pointed out that the Green Revolution "was more an administrative feat using research results from the West than an achievement of Indian agricultural research." Raina, "Institutional Strangleholds," 114.

functions central to its stability. As the epistemic engine of progress, it provided the state with a raison d'être. It was an apparatus through which the state established and exercised sovereignty. In return, the technoscientific complex enjoyed special epistemological status without political interference and with protection from social audits.[27] This contract between the state and the technoscientific complex was both political and epistemological, and it was constitutive of sovereignty in the form of the state-technoscience duo.[28]

The professionalization of science in postcolonial India, therefore, was placed on a track altogether different from countries such as Germany or the United States because the postcolonial Indian state played a crucial role in harnessing S&T to serve developmental goals while the country's scientific community lost its international character and autonomy. A massive technoscientific complex replaced (rather than complemented) the country's academic research science. The sovereign state was founded through a politico-epistemological contract with a technoscientific complex masquerading as science proper.[29] The nation-state used the language of scientism to acquire legitimacy while abandoning academic science to dusty laboratories in universities that were little more than teaching and degree-awarding institutions.[30]

3. The Publics and the Contract

How did this peculiar institutionalization of science in India influence the rise of a public, or several publics? Although the contract is singular, mul-

---

27 Varughese, "Where are the Missing Masses?," 344.

28 Ibid.

29 Ibid.

30 Under the colonial administration, the university system was established in the country with the first three universities in 1857. These universities in Calcutta, Bombay and Madras Presidencies were teaching universities without postgraduate research departments. It was only in the early twentieth century that the research component was inserted into the university system. In India, the "Humboldtian" ideal of a university as a space for teaching and research took more than a century to take root. On contested references to Humboldt in nineteenth-century German academic discourse, see Dieter Langewiesche, "State–Nation–University: The Nineteenth-Century 'German University Model' as a Strategy for National Legitimacy in Germany, Austria, and Switzerland," in this volume.

tiple publics relate to the state-technoscience duo.[31] These publics have been generated in specific historical contexts in connection with the workings of the duo. Taking a cue from Partha Chatterjee's theorization of civil and political societies as two separate spheres of politics in South Asia,[32] I argue that in India, two different formations of public, the scientific citizen-publics and the quasi-publics, historically have been shaped in connection with the state-technoscience duo.

In the first phase of the contract between the state and technoscientific complex (the Nehruvian era), there was a clear distinction between elite citizen-publics and the wider population. Elite publics included an emerging urban middle-class that endorsed Nehruvian ideals of socialism and the powerful state to catch the train of modernity with the help of the technoscientific complex. These were the scientific citizen-publics[33] that the state originally considered as its authentic citizens. These national elite were appreciative of the technoscientific complex and believed in its

---

31 Ibid.

32 Partha Chatterjee points out that "[i]n terms of the formal structure of the state as given by the constitution and the laws, all of society is civil society; everyone is a citizen with equal rights and therefore to be regarded as a member of civil society. The political process is one where the organs of the state interact with members of civil society in their individual capacities or as members of associations." However, this dominant understanding of the state-citizen relationship is just half the story in postcolonial contexts like India. For Chatterjee, in practice, there is a political society coexisting with the civil society where the subaltern population groups have a quasi-legal existence: "Most of the inhabitants of India are only tenuously, and even then ambiguously and contextually, rights-bearing citizens in the sense imagined by the [Indian] constitution. They are not, therefore, proper members of civil society and are not regarded as such by institutions of the state. But it is not as though they are outside the reach of the state or even excluded from the domain of politics. As populations within the territorial jurisdiction of the state, they have to be both looked after and controlled by various governmental agencies. These activities bring these populations into a certain *political* relationship with the state. But this relationship does not always conform to what is envisaged in the constitutional depiction of the relation between the state and members of civil society." Partha Chatterjee, *The Politics of the Governed: Reflections on Popular Politics in Most of the World* (New York: Columbia Univ. Press, 2004), 38. Chatterjee's emphasis. See also Partha Chatterjee, *Lineages of Political Society: Studies in Postcolonial Democracy* (New York: Columbia Univ. Press, 2011).

33 The scientific citizen-publics are not a homogenous category, given the diversity of civil society politics in a country like India. Within civil society, there are multiple formations of the public based on a variety of political practices and sites of political deliberation. Hence the term is used in its plural as scientific citizen-publics. The same logic is applicable to the use of quasi-publics too.

powers for social transformation. They supported the state's Nehruvian-Bernalist vision of science and internalized the need to cultivate a scientific temper for social transformation and progress.[34] The elite minority of citizens provided a collective voice that determined public opinion at large while they were the beneficiaries and supporters of the contract between the state and the technoscientific complex.

While the scientific citizen-publics endorsed the state-technoscience duo, the wider population was deemed illiterate, ignorant, and hence under the state's paternalistic wings. This wider subaltern population had to be educated and enlightened through state-mediated science popularization and informal science education programs. In contrast to its trust in knowledgeable and scientifically tempered citizen-publics who are expected to don the mantle of national progress, the state-technoscience duo demanded from the wider population both a willingness to be governed as population groups, and their sacrifice for the nation upon request. This perspective was epitomized by Jawaharlal Nehru's speech to villagers who were to be displaced by a massive dam project in Hirakud in the eastern Indian province of Orissa in 1957. Nehru demanded from the villagers (mostly indigenous tribes dependent for their livelihood on the fertile land to be submerged by the reservoir) that "[i]f you suffer, you should suffer in the interest of the country."[35] In practice, therefore, the state-technoscience duo did not consider equivalent to the scientific citizen-

---

34 Nehru pointed to the indispensability of a scientific temper in 1946 while arguing that "we have still to hold on to [reason and the scientific method] with all our strength, for without that firm basis and background we can have no grip on any kind of truth or reality. [...] The applications of science are inevitable and unavoidable for all countries and peoples to-day. But something more than its application is necessary. It is the scientific approach, the adventurous and yet critical temper of science, the search for truth and new knowledge, the refusal to accept anything without testing and trial, the capacity to change previous conclusions in the face of new evidence, the reliance on observed fact and not on pre-conceived theory, the hard discipline of the mind—all this is necessary, not merely for the application of science but for life itself and the solution of its many problems." Jawaharlal Nehru, *The Discovery of India* (Delhi: Oxford Univ. Press, 1989), 512. The idea of a scientific temper in 1976 was incorporated in the Indian Constitution. Article 51A/h states that it is a fundamental duty of citizens "to develop the scientific temper, humanism and the spirit of inquiry and reform."

35 Quoted in Arundhati Roy, *The Algebra of Infinite Justice* (New Delhi: Penguin, 2002), 35. The dam displaced 22,000 families of which 12,700 families belonged to scheduled castes and scheduled tribes. Most of the villagers were never resettled and a strong social movement of the displaced is active even today. See http://www.merinews.com/article/hirakud-dam-displaced-families-seek-rehabilitation/136009.shtml.

publics the massive proportion of the Indian subaltern population that was expected to make enormous sacrifices for the nation's progress. Dalits[36] and indigenous communities in rural India never enjoyed the full citizenship granted to them by the constitution and had to sacrifice their livelihood for massive technoscientific development projects such as dams, coal and mineral mines, and nuclear reactors. While the state-technoscience duo had great confidence in the social and political agency of the scientific citizen-publics as beneficiaries of national progress, the wider population was treated as having no agency and voice, and hence as requiring the state's parental care through welfare mechanisms and technologies of population management.[37]

3. Emergence of a Critical Discourse on Science

By the 1970s, however, public support for science began to recede among the scientific citizen-publics, and Nehruvian developmentalism started to face challenges from new social movements. India's first nuclear experiment in 1974 and the declaration of a state of national emergency from 1975 to 1977 under the regime of Prime Minister Indira Gandhi (1966–77), prompted skepticism about the stated objectives of the state-technoscience duo.[38] During the state of national emergency, civil liberties were suspended, revealing "the state of exception,"[39] a face of the sovereign state that had not been revealed on a national scale.[40] The tyranny displayed during

---

36 Dalits are the "lower castes" in the caste hierarchy of Indian society.

37 See Chatterjee, *The Politics of the Governed*; Chatterjee, *Lineages of Political Society*.

38 Following instructions from Indira Gandhi, a state of national emergency was declared for a period of 21 months by Indian President Fakhruddin Ali Ahmed under Article 352(1) of the Constitution in response to "internal disturbance" in the wake of escalating political opposition to her rule. Mrs. Gandhi's political opponents were tortured and jailed. Police suppressed dissenting voices and censored the press.

39 Giorgio Agamben. *Homo Sacer: Sovereign Power and Bare Life* (Stanford: Stanford Univ. Press, 1998).

40 States of exception had been declared on a smaller scale since the inception of the Indian nation-state, even under Prime Minister Jawaharlal Nehru. The North-East provinces of India, for example, became a fuzzy frontier that the state sought to control through military interventions. "Indian attempts at 'nation-building' by force of arms, with the Indian 'defense forces' in culturally alien territory indulging in large-scale killing and rape, were hardly the best ways of demonstrating to the Nagas [an indigenous

this period accelerated the functional capabilities of the technoscientific complex to serve a Fascist state. Just before the announcement of emergency rule, India had attained the status of a nuclear power. During the state of emergency, the technoscientific complex actively helped the state implement programs of population control through forced contraception among the poor and of sanitizing urban areas through demolishing slums. The twin projects of family and urban planning were "exercises in the authoritarian application of science,"[41] and emergency rule revealed the duo's potential for exercising tyrannical force across the entire citizenry.

As Shiv Viswanathan has pointed out, "[t]he Emergency and its aftermath became text, context and pretext of an efflorescence of civil society debates on knowledge."[42] This was also a period that saw the rise of Science Studies as a separate intellectual activity from science policy-making in the country, contributing to the ongoing debates on knowledge.[43] The global rise of an environmental consciousness in the 1970s as well as women's movements and indigenous movements catalyzed the urge to develop a critique of modern science and its support for the state. The rural population also became organized against the state-technoscience duo and demanded that its activities be stopped, a trend that gained momentum in the 1980s and 1990s.[44] These social movements, however, were guided by members of the critical scientific citizen-publics who translated the population's concerns into a language understandable in elite public spheres. These social protest movements advanced alternative visions against the developmentalist ideology of the state.

---

community in the North-Eastern province] the warm and enveloping joys of belonging represented by Indian nationhood. But Nehru's centralized state could not afford to have fuzzy edges. It was in the north-east of India that the Nehruvian vision took on its most brutal and violent forms." Zachariah, *Nehru*, 210–11. A more recent case is the 2002 mass killing of the minority population of Muslims in Gujarat in western India.

41 Shiv Viswanathan, "Beyond the Social Contract: Science, Knowledge and the Democratic Imagination in India," *Harvard Asian Quarterly* (n.d.): 4–11, 6.

42 Ibid., 6.

43 Ibid. The spirit of the period is captured in an edited volume that presented a strong critique of the epistemology of science: Ashis Nandy, ed., *Science, Hegemony and Violence: A Requiem for Modernity* (New Delhi: Oxford Univ. Press, 1988).

44 A major social movement by people affected by a project in the 1980s was the Narmada Bachao Andolan (Save-Narmada-Movement) under the leadership of Medha Patkar, in the context of the Sardar Sarovar dam project in western India.

*Fig. 2. Prime Minister Jawaharlal Nehru inaugurates the Bhakra dam at Bilaspur,
October 1963.*

(Courtesy of Press Information Bureau, Government of India)

This was a period when several new science movements radically reshaped the attitude of the scientific citizen-publics. These movements became stronger in the 1980s and their perspectives varied from high lighting the social function of science to a strong epistemological critique of western science and technology.[45] The Bhopal disaster on December 3, 1984 vindicated arguments by civil society movements against heavy industrialization and massive technoscientific projects for their environmental and social impacts. By the late 1980s, effects of the Green Revolution of the 1960s became visible when agricultural land turned out to be less fertile in places where the program seemed to have been successful. This was due to the program's use of fertilizers and its overuse of pesticides.[46] By that

---

45 People's science movements such as the Kerala Sastra Sahitya Parishad emphasized a Marxist-Bernalist vision of science for social revolution. The appropriate technology movement and the alternative science movements were more Gandhian in their perspectives. For a detailed discussion, see Raina, "Evolving Perspectives."

46 For a critique of the Green Revolution, see D.N. Dhanagare, "Green Revolution and Social Inequalities in Rural India," *Economic and Political Weekly* 22 (19–21), May (1987):

time, agriculture was in great need of capital for better returns from high yielding varieties of seeds as large areas had become unviable for small and marginal farmers.

Developments in the 1970s and 1980s had the cumulative effect of changing the intellectual and political atmosphere, and the state-technoscience duo lost its earlier attraction and endorsement from the scientific citizen-publics, followed by the emergence of a strong epistemological and political critique of its functioning. The second phase of the contract, the decades of the 1970s and 1980s, thus witnessed the gradual transformation of the scientific citizen-publics into a critical voice, which had lost their trust in the state-technoscience duo as they became disillusioned with the statist agenda of development and national progress. This was also the period when population groups began to become organized as quasi-publics, with the intellectual assistance and political support of the scientific citizen-publics. The politicization of both the scientific citizen-publics and quasi-publics in this period of "disenchantment with science"[47] created an interim crisis for the state-technoscience duo, as some developmental projects, in the midst of public criticism, had to be either scrapped or reconfigured to align them with better governance practices.[48]

4. Neoliberalism and the Contract

At this juncture of a growing public distrust of technoscientific developmentalism, the Indian economy's shift to neoliberal policies was a major turn of events in the 1990s. The adoption of the New Economic Policy and the concomitant shift from Nehruvian welfarism to neoliberal market-orientation helped the state-technoscience duo circumvent the crisis. The state has also expressed its desire to integrate the national S&T system with its global counterpart. In the post-1990s, large scale technoscientific projects returned to the state's developmental agenda, this time with the sup-

AN 137–AN 144; Vandana Shiva. *The Violence of the Green Revolution: Third World Agriculture, Ecology and Politics* (Goa: The Other India Press, 1991).

47 Raina, "Evolving Perspectives."

48 A major victory for the civil society groups was the verdict against the construction of a dam in the ecologically sensitive Silent Valley region of the Western Ghats in South India. See Darryl D'Monte, *Temples or Tombs? Industry versus Environment: Three Controversies* (New Delhi: Centre for Science and Environment, 1985).

port of multinational companies. The Indian government opened up agriculture to greater corporatization, and for its nuclear energy sector it sought new technological cooperation with developed countries. The Nehruvian welfare state's rhetoric of self-reliance was abandoned, even in strategic areas such as nuclear power. In recent decades, urban development projects and the opening up of farmland for massive industrialization have also received an impetus from the state's conversion to neoliberalism. In the Nehruvian period, the technoscientific complex functioned within a public system of S&T but it has since been reconfigured to suit the state's new aspirations. Recent developments in agriculture and nuclear energy indicate that government has become open to private R&D and technology transfers.

Public critique of the Green Revolution ironically helped the state-technoscience duo create favorable conditions for its transformation to neoliberalism. Using the rhetoric of an "ever-green" revolution, it aims to boost the role of corporate R&D in agriculture. This "second green revolution," according to its proponents, will solve problems created by the Green Revolution, such as increased use of chemical fertilizers and pesticides, especially with a shift to genetically modified crops. Similarly, in the nuclear energy sector, there is a clarion call for increasing the generation of nuclear energy claimed to be "clean and green" unlike the previous era's non-renewable energy and hydroelectric power. India aims to shift 25 percent of its energy capacity to nuclear power by 2050, and six new reactors currently are under construction. During the Nehruvian era, India emphasized technological self-reliance and developed its own reactors. With the inception of its neoliberal phase, India has begun to buy reactors from Russia, France, and the US through civil nuclear cooperation agreements. Although the slogan of self-reliance has been relinquished, the nuclear energy sector continues to enjoy protection from social audits in the name of national security.

A quick glance through the scholarly literature on the neoliberal phase of public engagement with science and technology reveals that this is a period of increased skepticism and strengthened public participation in the governance of science.[49] Although a similar phenomenon of increased

49 See Nowotny et al., *Rethinking Science*; Alan Irwin, *Citizen Science: A Study of People, Expertise and Sustainable Development* (New York: Routledge, 1995).

public involvement in science has been noted in India by many scholars,[50] such an enquiry could be misleading unless the processes involved are analyzed with reference to transformations in the contract between the state and the technoscientific complex.

The decision to introduce genetically modified (GM) crops in India began in the mid-1990s, although the first successful commercial release of genetically modified cotton plants took place only in March 2002.[51] (Bt Cotton has a greater resistance to bollworms and *Lepidoptera* larvae due to a gene transferred from *Bacillus thuringiensis*, a soil bacterium.) Following the release, public protests against the corporatization of agriculture erupted across India. Protesters targeted the commercial interests of Monsanto, a US-based transnational seed company that had developed the GM-crops. Scholars have noted that protests reflected a wide array of issues including environmental risks, health hazards, farmers' rights, food security, biodiversity, globalization, trade agreements, and linking the local to the global.[52] The controversy was reinvigorated when the Genetic Engineering Approval Committee (GEAC)[53] approved field trials for Bt brinjal, a GM-version of the eggplant *Solanum melongena*,[54] but strong public protests and public consultations in seven cities across the country in 2010 forced the Ministry of Environment and Forests (MoEF) to declare a moratorium on its release.[55] As I have pointed out elsewhere, the public consultation orga-

---

50 For example, Ian Scoones in a comparative study of civil society mobilization against genetically modified (GM) crops in India, South Africa, and Brazil has argued that "these anti-GM groupings are all examples of hybrid network forms of social activism, linking people, issues and politics in new ways around globally-defined issues, but always co-constructed in local contexts and through local political processes." Ian Scoones, "Mobilizing against GM Crops in India, South Africa and Brazil," *Journal of Agrarian Change* 8, nos. 2 & 3, April–July (2008): 317.

51 Ibid.

52 Ibid, 326.

53 Recently it has been renamed the Genetic Engineering Appraisal Committee (GEAC).

54 "Bt" in Bt Brinjal stands for *Bacillus thuringiensis*, as in Bt Cotton.

55 See Varughese, "Where are the Missing Masses?" for a detailed analysis. See also, Ronald J. Herring, "Reconstructing Facts in Bt Cotton: Why Skepticism Fails," *Economic and Political Weekly* XLVIII, no. 33 (Aug. 17, 2013): 63–66; Chandrika Parmar and Shiv Visvanathan, "Hybrid, Hyphen, History, Hysteria: The Making of the Bt Cotton Controversy," unpublished paper presented in the IDS Seminar on Agriculture Biotechnology and the Developing World, Oct. 1–2, 2003; Scoones, "Mobilizing against GM Crops"; Esha Shah, "What Makes Crop Biotechnology Find its Roots? The Technological Culture of Bt Cotton in Gujarat, India," *The European Journal of Development Research* 20,

nized by the ministry was the first of its kind, and it indicated the state-technoscience duo's willingness to include the scientific citizen-publics as a true partner in technoscientific regulation.[56]

While the scientific citizen-publics have been successfully recognized as critical partners in technoscientific decision-making, the duo's relationship with other population groups remains more complicated. As mentioned earlier, these groups by the end of the 1980s had successfully transformed themselves from scapegoats into political agents. They have been less susceptible to the welfare state's governing techniques and became "unruly publics" to enter a new political relationship with the state-technoscience duo. Their political negotiations with the duo in the shady zone of political society outside of civil society have been deeply linked to the negative impact of technoscientific projects on their livelihood. Such new politics has turned them into "quasi-publics" without full citizenship rights but with the means to negotiate with the state-technoscience duo.[57]

The quasi-publics and their negotiations, however, may remain invisible if they are not connected to debates and movements in civil society. This may explain why small and marginal farmers whose livelihood was affected by the Green Revolution received no public attention until in the last two decades thousands of them committed suicide due to the debt trap created by the capital-driven agricultural practices established under the Green Revolution paradigm[58] Once farmers' suicides started attracting media attention, agrarian distress became an important issue of public concern.

In contrast to the farming sector's invisible quasi-publics, nuclear reactor sites witness the emergence of strong political negotiations between the quasi-publics and the nuclear establishment. The local villagers' protests against the construction of reactors vindicate the emergence of the quasi-publics with unique political objectives and modes of negotiation in contrast to the scientific citizen-publics.[59] As in the case of GM crops, the issue of nuclear energy is generally deliberated in civil society by linking it

---

no. 3, September (2008): 432–47; Esha Shah, "'Science' in the Risk Politics of Bt Brinjal," *Economic and Political Weekly* XLVI, no. 31 (July 30, 2011): 31–38.

56 Shiju Sam Varughese, "The Public Life of Expertise," *Seminar* 654 (Feb. 2014): 21–26.

57 Varughese, "Where are the Missing Masses?" 247.

58 The highest number of farmer suicides was reported in 2004 with 18,241 deaths. A total number of 296,438 farmer suicides were recorded by the National Crime Records Bureau between 1995 and 2013.

59 Ibid.

with concerns of the scientific citizen-publics.[60] In sharp contrast, the quasi-publics are primarily concerned with the nuclear reactor's consequences for their livelihoods. Villagers of Kudankulam in southern India, for example, strongly protested against two Russian reactors that the Kudankulam Nuclear Power Project (KKNPP) planned to build there. Their resistance was prompted by risks created by the project for the livelihood of the local fishing community.[61] They successfully tied their demands to concerns of the scientific citizen-publics, thus increasing their bargaining power with the state-technoscience duo.[62] "The quasi-publics are concerned mainly with their livelihood and survival issues and trespass the boundaries of legality," thereby posing a serious threat to the duo's sovereignty.[63] Their strategies of negotiation can often only be countered by the state with pastoral care and punitive actions. In the case of KKNPP, protesting villagers were first offered welfare projects and employment opportunities. When they remained unpacified, legal action was brought against them (including charges of sedition) to defuse the movement. Despite escalating protests, the nuclear establishment and state authorities were unwilling to listen to the villagers' demands. This contrasts with the politics of civil society where the duo has been open to include the scientific citizen-publics as partners in the governance of technoscience. This is because the demands of scientific citizen-publics are mostly concerned with a general perception of risk and regulation. By contrast, the quasi-publics' perception of risk was deeply linked to their immediate livelihoods.

What does the constitution of these two different publics reveal about the contract between the state and the technoscientific complex in India?

---

60 These concerns included questions related to energy security, technological risks, differences between civilian and military uses of nuclear technology, the ecological impact of nuclear reactors, fear of nuclear disasters, skepticism towards safety measures, a lack of transparency and accountability in the nuclear establishment, the economic viability of nuclear energy, threats posed by the involvement of multinational corporations and agencies in the nuclear energy sector, economic corruption, and India's status as a nuclear state in geopolitics.

61 For a detailed case study, see Itty Abraham, "The Violence of Postcolonial Spaces: Kudankulam," in *Habitations of Violence*, edited by Kalpana Kannabiran (New Delhi: Oxford Univ. Press, 2015), forthcoming.

62 Varughese, "Where are the Missing Masses?" There is no assumption that the quasi-publics are always critical of the duo. Local villagers, for example, recently organized to demand employment opportunities at the nuclear reactor complex under construction in Kakrapar in western India (Gujarat).

63 Ibid., 249.

The state-technoscience duo's modes of operation have changed in neo-liberal times. These changes are reflected in governance mechanisms that have opened up the decision-making process only to a limited degree. Such changes, however, have had a rather weak impact on the original contract. The contract has been under serious attack from the quasi-publics, and this is why the state-technoscience duo uses its pastoral powers with the utmost care and force. By diversifying governance techniques, the duo thus safeguards the original contract, which is constitutive of its sovereign power.

5. Conclusion

I have argued that the contract between the state and technoscientific complex has remained intact throughout the history of the postcolonial nation-state in India. The technoscientific complex and the nation-state were mutually constitutive, leading to the formation of the state-technoscience duo. The emergence of the technoscientific complex during the early decades of national independence marginalized academic science and integrated itself into the state machinery of centralized national planning. The formation of the duo was endorsed by the national elite publics who shared the nation's developmentalist vision and supported the apparatus of the technoscientific complex.

The scientific citizen-publics existed in contrast to population groups from whom the duo demanded sacrifices and suffering in the name of national progress. This included the rural populace of the country, mostly subaltern social groups, which were thus displaced and uprooted by technoscientific projects. The scientific citizen publics, however, gradually moved away from a blind appreciation of the technoscientific complex and became skeptical of it during the 1970s and 1980s, a period that simultaneously witnessed the rise of an opposition within the subaltern population groups that was no longer prepared to sacrifice itself for the duo. This has led to the constitution of two divergent publics that engaged with the duo in radically different ways.

The adoption of neoliberal economic policies by the Indian state in the 1990s triggered off a new set of public protests and generated a dynamic civil society that demanded more democratic regulation of technoscience.

While the state-technoscience duo has gradually come to terms with the demands of the scientific citizen-publics, it had to negotiate with the quasi-publics in a different way so as to defuse threats to the contract between the state and the technoscientific complex. Despite being continuously threatened by the quasi-publics, the contract remains intact even under the neoliberal paradigm, as demonstrated by recent cases of public deliberations and collective action in connection with neoliberal transformations in the sectors of nuclear energy and agriculture.

Section III:
Legitimizing Fields of Investigation

Shifting Alliances, Epistemic Transformations: Horkheimer, Pollock, and Adorno and the Democratization of West Germany

Fabian Link

1. Social Change, Epistemic Transformation, and the Role of Knowledge in Stabilizing the Nation-State

Ludwik Fleck once pointed out that phases of general confusion cause profound transformations of scientific knowledge.[1] The seizure of power by the National Socialists in 1933 initiated such a phase. The emigration of intellectuals and researchers from Germany and, after 1938, from Austria to England, the US, and other countries brought profound transformations to the sciences and the humanities, changes that were especially rigorous in the social sciences[2] because many German and Austrian social scientists had a Jewish background or subscribed to leftist social reform.[3] Émigrés to America included researchers at the Institute for Social Research (*Institut für Sozialforschung,* IfS) in Frankfurt am Main, as the institute's director, Max Horkheimer, and his colleagues fled to Geneva, moved on to Paris, and then to New York City. Horkheimer would return to Frankfurt in 1949, followed by Frederick Pollock and Theodor W. Adorno, and shortly after, sponsored by the McCloy Fund and private donors, the IfS would reopen

---

1 Ludwik Fleck, *Entstehung und Entwicklung einer wissenschaftlichen Tatsache. Einführung in die Lehre vom Denkstil und Denkkollektiv* (Frankfurt: Suhrkamp, 1980), 124.

2 I use the term "social sciences" as a heuristic term, which describes social science in a strictly empirical, science-oriented manner as well as, social theory, social philosophy and social psychology.

3 See Konstantin von Freytag-Loringhoven, *Erziehung im Kollegienhaus. Reformbestrebungen an den deutschen Universitäten der amerikanischen Besatzungszone 1945–1960* (Pallas Athene 45) (Stuttgart: Franz Steiner, 2012), 132. See also Steven E. Aschheim, *At the Edges of Liberalism: Junctions of European, German, and Jewish History* (New York: Palgrave MacMillan, 2012).

in Frankfurt in 1951.[4] During the fifties and sixties, Horkheimer, Pollock, and Adorno, through empirical social research, radio and television broadcasts, public speeches, and newspaper articles, contributed to reeducation efforts and to the stabilization of democracy in the fledgling West German state.

This chapter explores the intellectual traces and shifting alliances prompted by the IfS leadership's transatlantic crossing and return, and asks how did Horkheimer and his colleagues contribute to West German reconstruction and democracy in the early Cold War period? I will propose that their transatlantic crossings were accompanied by profound epistemic transformations caused by changes in the social, political, and ideological environment of the IfS in exile. Alterations in the institute's network of scholars, politicians, and businessmen during the thirties, forties, and fifties influenced epistemic concepts developed by Horkheimer before 1931, and the key transformation was the separation of empirical social research from social philosophy. In Germany before the war, social philosophy had provided an essential intellectual framework for Horkheimer, Pollock, and Adorno, but in exile in the US, they had to focus on empirical research in order to sustain themselves. Methodologically, such research was based on mathematical principles and required significant private financial support. Empirical research projects designed to create useful knowledge were funded by American interest groups and as such they challenged the institute's scientific autonomy. As a result, social philosophy became for Horkheimer and his colleagues a last intellectual resort. The tension between methods of empirical social research and social philosophy continued after the institute's remigration in the late 1940s, but after their return to Frankfurt, Horkheimer, Pollock, and Adorno embarked on empirical research projects whose results were relevant for the contemporary democratization of West Germany. At the same time, Horkheimer and Adorno helped legitimize the Federal Republic of Germany through lectures, speeches, and radio broadcasts focusing on democracy and on Germany's National Socialist past.

From the eighteenth century, statistics, demography, and *Kameralwissenschaften* (cameralism) have sought to produce applied knowledge for the

---

4 See Martin Jay, *The Dialectical Imagination: A History of the Frankfurt School and the Institute of Social Research, 1923–1950* (Berkeley: Univ. of California Press, 1973 [1996]); Rolf Wiggershaus, *Die Frankfurter Schule. Geschichte. Theoretische Entwicklung. Politische Bedeutung* (Munich: DTV, 2008).

progressive improvement of society. With the rise of industrialization and pressing social questions, and in the context of their nineteenth-century professionalization and institutionalization, the social sciences developed a role in considering poverty, migration, political ideology, and economic problems that was aware of its practical significance.[5] Their reflective perspective ultimately aimed at an improvement of society. The nineteenth century witnessed the rise of new national projects, and the founding of new states went along with the expectation that the social sciences would help guide political efforts. Against this wider canvas, the twentieth-century case of the IfS is of particular interest because Horkheimer and Adorno contributed to the stabilization of the fledgling West German state after 1945, while Adorno also tried to destabilize German consciousness through his critique of the status quo. I will thus explore this seeming paradox.

To investigate the institute's transatlantic crossings and its reestablishment in Frankfurt after 1950, I will use a methodological combination of Imre Lakatos' methodology of scientific research programs and an approach from science studies, inspired by Bruno Latour's and Michel Callon's actor-network theory. This somewhat idiosyncratic tactic helps demonstrate how epistemic transformations are tied to shifts in financial support and in social and political alliances. In the case of Horkheimer, Pollock, and Adorno, their American experience prompted them to engage in democratic reeducation in postwar Germany, with a special emphasis on academic elites.

Intending to go beyond Karl Popper's theory of falsification and Thomas Kuhn's idea of scientific paradigm shifts, Lakatos developed the idea of scientific research programs in order to grasp the history and development of major epistemological directions. Such a research program consists of a "hard core" of assumptions and theoretical foundations, an "elastic protected area," and an "advanced apparatus of solving problems."[6] The "hard core" of scientific research programs is surrounded by

---

5 Theodore M. Porter, "Genres and Objects of Social Inquiry, from the Enlightenment to 1890," *The Modern Social Sciences*, edited by Theodore M. Porter and Dorothy Ross, vol. 7 of *The Cambridge History of Science* (Cambridge: Cambridge Univ. Press, 2008), 13–39, 26–32, 38–39; Dorothy Ross, "Changing Contours of the Social Science Disciplines," in *Modern Social Sciences*, edited by Porter and Ross, 205–37, see esp. 205–29.

6 Imre Lakatos, *Die Methodologie der wissenschaftlichen Forschungsprogramme* (Philosophische Schriften 1), edited by John Worrall and Gregory Currie (Braunschweig and Wiesbaden: Vieweg, 1982 [1978]), 4, 116–17.

auxiliary hypotheses, an "elastic protected area" that shields theoretical assumptions from attacks and attempts to falsify them. Attacks from outside are important, however, because auxiliary hypotheses will have to be adjusted in response to successful falsifications. Such processes improve the program, harden its "core", and make it successful.[7] Although Lakatos developed his epistemological concept with reference to physics and mathematics, its basic assumptions may also be applied to ideas developed by Horkheimer and the IfS in the early 1930s.

This epistemological approach must be connected with a perspective that focuses on the agency of scholars. Following Bruno Latour's and Michel Callon's actor-network theory, scholars have to establish networks and alliances with other scholars, politicians, businessmen, and journalists in order to mobilize resources for research and the acceptance of scientific knowledge in society. In order to establish new common interests, scholars need to explain their scientific ideas, and Latour and Callon call this mechanism "translation".[8] It is important to note as a key assumption in this chapter, therefore, that for economic support research depends on political and economic power.[9]

The historicity of knowledge implies ongoing transformations of knowledge.[10] In the case of the IfS, methods of empirical social research and particular theoretical approaches to social philosophy or scientific rhetoric were recontextualized and rearranged in the course of their transatlantic transfer. This resulted in profound transformations of these objects of translation.[11] Translations and transformations usually entail misconcep-

---

7 Ibid., 47–49.

8 Michel Callon, "Einige Elemente einer Soziologie der Übersetzung: Die Domestikation der Kammmuscheln und der Fischer der St. Brieuc-Bucht," in *ANThology. Ein einführendes Handbuch zur Akteur-Netzwerk-Theorie*, edited by Andréa Belliger and David J. Krieger (Bielefeld: transcript, 2006), 135–74, see esp. 135–36; Bruno Latour, *Science in Action: How to Follow Scientists and Engineers through Society* (Cambridge, MA: Harvard Univ. Press, 1987), 111–21.

9 Pierre Bourdieu, *Meditationen. Zur Kritik der scholastischen Vernunft* (Frankfurt: Suhrkamp, 2001 [1997]), 30.

10 Bruno Latour, *Die Hoffnung der Pandora. Untersuchungen zur Wirklichkeit der Wissenschaft* (Frankfurt: Suhrkamp, 2002), 36–95.

11 Helmut Lethen, "Der Gracián-Kick im 20. Jahrhundert," *Zeitschrift für Ideengeschichte* 7, no. 3 (2013): 59–76, 61. See Walter Benjamin, "Die Aufgabe des Übersetzers," in *Gesammelte Werke I. Berliner Kindheit um neunzehnhundert, Berliner Chronik, Einbahnstraße und andere Schriften* (Frankfurt: Zweitausendeins, 2011 [1923]), 383–93.

tions and failures[12] that may have the effect of a radical but sometimes unintended realignment of social scientific trajectories. The scholars of the IfS emigrated to another culture where social scientists spoke another language and maintained very different kinds of scientific thinking. In order to maintain their scientific careers, Horkheimer and his colleagues not only had to translate their epistemic approaches and practices, they also had to adapt their personalities and intellectual processes.[13]

2. Critical Theory and its Research Program

Founded in 1923, the *Institut für Sozialforschung* developed a somewhat closed scientific environment with a particular research program. After his election to the directorship, Max Horkheimer became the key figure in shaping the institute's epistemological foundation.[14] Although the institute's employees and collaborators maintained contacts to other scholars, such as Kurt Riezler, Paul Tillich, Adolf Löwe, or Karl Mannheim (whose Seminar for Sociology was located in the IfS building[15]), their scientific program differed from other approaches in the social sciences. Horkheimer broke with the sociology of science developed by Max Scheler and Mannheim, claiming that their approach was nothing more than an affirmation of the status quo. He rejected logical positivism as "metaphysics" due to its mathematical-logical reductionism. Horkheimer also criticized Lenin's blunt concentration on materialistic truth and considered unacceptable propositions for a new metaphysics by proponents of the philos-

---

12 Juri Lotman, *Culture and Explosion* (Berlin: Mouton de Gruyter, 2009), 22.

13 Hans-Jürgen Lüsebring, "'Lost in Translation'—Übersetzung und Exilerfahrung bei Eva Hoffman (Polen/Kanada/USA) und Jacques Poulin (Québec, Kanada)," in *Kultur übersetzen. Zur Wissenschaft des Übersetzens im deutsch-französischen Dialog*, edited by Alberto Gil and Manfred Schmeling (Berlin: Akademie Verlag, 2009), 97–106, see esp. 97.

14 Helmut Dubiel, *Wissenschaftsorganisation und politische Erfahrung. Studien zur frühen Kritischen Theorie* (Frankfurt: Suhrkamp, 1978), 193–97.

15 Amalia Barboza, "Das utopische Bewusstsein in zwei Frankfurter Soziologien: Wissenssoziologie versus Kritische Theorie," in *Soziologie in Frankfurt. Eine Zwischenbilanz*, edited by Felicia Herrschaft and Klaus Lichtblau (Wiesbaden: VS Verlag für Sozialwissenschaften, 2010), 161–78, esp. 163; Thomas Wheatland, *The Frankfurt School in Exile* (Minneapolis: Univ. of Minnesota Press, 2009), 97–98.

ophy of life and fundamental ontologists.[16] Horkheimer's intellectual and political commitments shaped the inner circle of the IfS. Besides Horkheimer, the institute comprised of his close friend Frederick Pollock as well as Erich Fromm, Leo Löwenthal, Herbert Marcuse, Otto Kirchheimer, Franz Neumann, Karl-August Wittfogel, and Henryk Grossmann. Scholars such as Theodor Wiesengrund (who would change his name to Theodor W. Adorno in the US), Siegfried Kracauer, and Walter Benjamin were loosely affiliated with the institute's intellectual endeavors.[17]

Horkheimer presented the institute's research program in his inaugural address in 1931. His interdisciplinary approach for the exploration of modern society evolved from a philosophical and critical perspective of dialectical materialism that was inspired by Hegel, Marx, and Freudian psychology. Philosophy should help analyze social inequality, which Horkheimer took to be the result of the suppression of human instincts by bourgeois society. Three objects of research were central for Horkheimer's concept: the economic life of society, the psychological development of individuals, and cultural change.[18] Horkheimer's approach contrasted with a "neutral" and "bourgeois" scientific concept. He wished to investigate mechanisms of social inequality and of ideology as outcomes of a social "totality", the "untrue" whole. The aim of his approach was the transformation of these mechanisms in order to create a better society.[19] This was, in Lakatos' terminology, the "hard core" of the institute's research program. Horkheimer planned to gather economists, empirical social scientists, social psychologists, historians, literature scholars, and art critics so that together they could use "the finest scientific methods" to explore "big philosophical

---

16 John Abromeit, *Max Horkheimer and the Foundations of the Frankfurt School* (Cambridge: Cambridge Univ. Press, 2011), 143–56, 249–51; Jay, *Dialectical Imagination*, 62.

17 See *Zeitschrift für Sozialforschung*, Doppelheft 1/2 (1932).

18 Max Horkheimer, "Die gegenwärtige Lage der Sozialphilosophie und die Aufgaben eines Instituts für Sozialforschung," in *Max Horkheimer. Sozialphilosophische Studien. Aufsätze, Reden und Vorträge 1930–1972*, edited by Werner Brede (Frankfurt: Fischer Taschenbuch Verlag, 1981 [1931]), 43; Hans-Joachim Dahms, *Positivismusstreit. Die Auseinandersetzungen der Frankfurter Schule mit dem logischen Positivismus, dem amerikanischen Pragmatismus und dem kritischen Rationalismus* (Frankfurt: Suhrkamp, 1994), 45.

19 Helmut Dahmer, "Faschismustheorie(n) der 'Frankfurter Schule'," in *Soziologie und Nationalsozialismus. Positionen, Debatten, Perspektiven*, edited by Michaela Christ and Maja Suderland (Berlin: Suhrkamp, 2014), 76–118, see esp. 79–80. See Emil Walter-Busch, *Geschichte der Frankfurter Schule. Kritische Theorie und Politik* (Munich: Fink, 2010), 47–52; Wiggershaus, *Frankfurter Schule*, 71.

questions."[20] While the interdisciplinary approach corresponds with Lakatos' idea of an "elastic protected area," the use of methods taken from different disciplines may be regarded as an "advanced apparatus of solving problems."

In Frankfurt, the institute staff began to work on this program. This encompassed qualitative critical studies of music, art, and film by Adorno, Benjamin, and Kracauer, and empirical projects such as Fromm's and Grossmann's large study of workers and employees in 1929/30. The latter drew on questionnaires that were also used by social researchers in the US at the time.[21] In order to explore the psychological mindset of workers and employees, Horkheimer proposed analyzing "published statistics, reports of political organizations, the material of offices of public service" as well as texts from the "media and belletristic." Empirical facts should then be integrated in these analyses. Horkheimer valued empirical facts as useful only "if one knows that inductive conclusions drawn from these empirical facts [...] are hasty." For the investigation of social problems, Horkheimer demanded that multiple methods be used.[22]

Even though the *Institut für Sozialforschung* had been founded by the industrialist Felix Weil and his father Hermann, the institute was affiliated with the Johann Wolfgang Goethe-University, Frankfurt am Main. Horkheimer could direct the IfS more or less as he wanted because he had the support of the Weil family.[23] Convinced of the economic benefits of socialism, Felix Weil had initially modeled the IfS after the Marx-Engels-Institute in Moscow, as an institution for the scientific exploration of Marxism and socialism. The first director of the IfS was Carl Grünberg who was known as a *Kathedersozialist* (academic socialist). Grünberg's research program was dedicated to the investigation of socialism and the workers' movement and this prompted him to establish a library and an archive.[24] Although the foundation of a research institute left of the political center was unique in Germany, its research program was neither new nor innovative. After Grünberg suffered a stroke in 1928, his successor,

---

20 Horkheimer, "Die gegenwärtige Lage," 41. See Dubiel, *Wissenschaftsorganisation*, 149; Jay, *Dialectical Imagination*, 21, 25.

21 Abromeit, *Max Horkheimer*, 282, 215.

22 Horkheimer, "Die gegenwärtige Lage," 44. My translation.

23 Dahmer, "Faschismustheorie(n)," 77–78; Rolf Wiggershaus, *Max Horkheimer. Unternehmer in Sachen 'Kritische Theorie'* (Frankfurt: Fischer Taschenbuch Verlag, 2013), 57–58. See Walter-Busch, *Geschichte*, 14–15.

24 Wiggershaus, *Frankfurter Schule*, 23, 33–46.

Horkheimer, broke with orthodox Marxism in his new and dynamic research program.[25]

3. Shifting Alliances and the Disruption of the Research Program of the IfS in Exile

From their empirical investigations of blue and white collar workers, Fromm and Grossmann concluded that fascism was likely to assume the reins of government in Germany.[26] By the time the Nazis seized power in 1933, Horkheimer and his closest colleagues had already left Frankfurt. After its last employee, Leo Löwenthal, departed Frankfurt in March 1933, the German police shut down the IfS. Categorized as "half Jew," Adorno remained in Germany because he was not in immediate danger. The institute's emigration from Germany did not change its overall financial situation as Weil and his father continued to provide financial support, and thereby secured independence for Horkheimer and his colleagues. With the help of his wide international network, Horkheimer founded IfS branches in Geneva and Paris.[27]

In Geneva, Fromm and his collaborators developed another empirical study of the relationship between authoritarian thinking and the social structure of families. This project showed that fascism, as a radical version of authoritarianism, was likely to assume political power in many European countries. As such, fascism was not a particular national phenomenon but had international dimensions.[28] It made sense to leave Europe for the US as Erich Fromm and Julian Gumperz, a German-American friend of the IfS abroad, both reported that democracy was more stable in America and that a fascist takeover was improbable.[29] Horkheimer was aware that even though the social sciences remained heterogeneous in the US, they were

---

25 Ibid., 46.

26 Leo Löwenthal to Heinrich Meng, 19 Jan. 1937, fol. 216, Na 1, 36-Korrespondenzen N, Archivzentrum der Universitätsbibliothek Frankfurt am Main (Archive Center, University Library, Frankfurt, hereafter cited as UBA Ffm).

27 Wiggershaus, *Frankfurter Schule*, 147–53.

28 Jay, *Dialectical Imagination*, 37; Wiggershaus, *Max Horkheimer*, 79.

29 Jay, *Dialectical Imagination*, 38–40; Wiggershaus, *Frankfurter Schule*, 161–65.

well-institutionalized at the country's universities.[30] Fromm and Gumperz contacted colleagues at Columbia University in New York, one of the prestigious Ivy League universities on the American east coast, and they managed to arouse the interest of Nicholas Murray Butler and Robert Lynd. Butler was Columbia's president and Lynd was a left-liberal sociologist who considered an affiliation of the IfS with the Faculty of Political Sciences at Columbia useful for strengthening the independence of sociology there.[31]

In New York City, Paul F. Lazarsfeld became an important transatlantic mediator between the IfS and social scientists at Columbia. Lazarsfeld was a social scientist from Austria who had successfully applied for American citizenship after he had arrived in New York as fellow of the Rockefeller Foundation in 1933.[32] In the early 1930s, Lazarsfeld had served as a research collaborator in empirical research projects at the IfS. At the time, Horkheimer had supported Lazarsfeld's emigration to the US.[33] When the IfS stood ready to cross the Atlantic, Lazarsfeld mobilized his reputation to campaign for the institute's affiliation with Columbia. He advertised to his colleagues that they could expect from the new affiliation a strengthening of empirical social research.[34]

After its emigration to New York, the institute's scientific program basically remained the same. The institute's financial supporters preserved its

30 Christian Fleck, *Transatlantische Bereicherungen. Zur Erfindung der empirischen Sozialforschung* (Frankfurt: Suhrkamp, 2007), 49, 186.

31 Eva-Maria Ziege, *Antisemitismus und Gesellschaftstheorie. Frankfurter Schule im amerikanischen Exil* (Frankfurt: Suhrkamp, 2009), 86; George Steinmetz, "American Sociology before and after World War II: The (Temporary) Settling of a Disciplinary Field," in *Sociology in America: A History*, edited by Craig Calhoun (Chicago: Univ. of Chicago Press, 2007), 324–27.

32 Paul Neurath, "Paul Lazarsfeld und die Institutionalisierung der empirischen Sozialforschung: Ausfuhr und Wiedereinfuhr einer Wiener Institution," in *Exil, Wissenschaft, Identität. Die Emigration deutscher Sozialwissenschaftler 1933–1945*, edited by Ilja Srubar (Frankfurt: Suhrkamp, 1988), 67–105, see esp. 77–78.

33 Max Horkheimer to Raymond Aron, 30 Dec. 1936, fol. 281, Na 1, 2-Korrespondenzen A, fol. 281–83, UBA Ffm; Max Horkheimer to Paul Lazarsfeld, 16 May 1935, Na 1, 31-Korrespondenzen L, fol. 235, UBA Ffm. See also Clemens Albrecht et al., *Die intellektuelle Gründung der Bundesrepublik. Eine Wirkungsgeschichte der Frankfurter Schule* (Frankfurt: Campus, 1999), 279–80.

34 Wheatland, *Frankfurt School*, 37–60. See Anthony R. Oberschall, "Paul F. Lazarsfeld und die Geschichte der empirischen Sozialforschung," in *Geschichte der Soziologie. Studien zur kognitiven, sozialen und historischen Identität einer Disziplin*, vol. 3, edited by Wolf Lepenies (Frankfurt: Suhrkamp, 1981), 15. See also Fleck, *Transatlantische Bereicherungen*, 110, 250.

scientific autonomy and the Frankfurt scholars continued projects they had begun in Europe. Horkheimer continued to publish the institute's *Journal of Social Science* (*Zeitschrift für Sozialforschung*) in German,[35] but in 1939 changed its title to *Studies in Philosophy and Social Science*. He began to accept articles by American scholars such as Margaret Mead, Charles Beard, and Harold Lasswell. In contrast to Horkheimer, Fromm was much more interested in establishing scientific contacts with American scholars, especially with Robert Lynd who became a close friend.[36] Horkheimer, Löwenthal, Marcuse, and Neumann were asked to teach at Columbia University (Adorno did not move to New York City until 1938 and then worked for Lazarsfeld.[37]) In 1937, Horkheimer gave a lecture on "Authoritarian Doctrines and Modern European Institutions" and Neumann in 1936/37 offered a class on the totalitarian state. But lecturing proved to be a problem. In 1941, Horkheimer wrote to Harold Laski in London, that "it will take us some time before we are able to express our ideas adequately in English— that is to say in such a way that language per se conveys some of the meaning we hope to attach to it."[38]

While the institute for some time continued on in the course it had assumed in Frankfurt, major epistemic transformations began to alter its research program after 1939. By that time, Weil and Pollock had lost a large amount of the institute's capital on the American stock market and Horkheimer was forced to reduce the institute's staff.[39] Marcuse and Neumann found office jobs in the American secret service. Fromm's dismissal by Horkheimer, however, was prompted by intellectual as well as economic dissonances as Horkheimer and Fromm disagreed, in particular, on how to approach Freudian theory. During the early years of exile, Fromm had distanced himself from Freud's theory on human instincts and had turned to a more existentialist philosophical anthropology, which for Horkheimer was related to vitalistic metaphysics. From this point on,

---

35 Jay, *Dialectical Imagination*, 114; Gunzelin Schmid Noerr, "Die Emigration Max Horkheimers und seines Kreises im Spiegel des Briefwechsels," in *Exil, Wissenschaft, Identität. Die Emigration deutscher Sozialwissenschaftler 1933–1945*, edited by Ilja Srubar (Frankfurt: Suhrkamp, 1988), 252–80, see esp. 255–56.

36 Wheatland, *Frankfurt School*, 65–78.

37 Ziege, *Antisemitismus*, 45.

38 Max Horkheimer to Harold Laski, 10 March 1941. Cited in *Max Horkheimer. Gesammelte Schriften*, vol. 17, *1941–1948*, edited by Gunzelin Schmid Noerr (Frankfurt: Fischer, 1996), 17–18.

39 Wiggershaus, *Frankfurter Schule*, 257.

Horkheimer cooperated more intensively with Adorno in order to advance his critical theory of society.[40] At the same time, Columbia University evaluated the IfS and came to the conclusion that the institute should finally publish the research results of its empirical projects in English.[41] In order to increase output, university representatives suggested that Horkheimer should fuse his institute at Columbia with Lazarsfeld's Bureau of Applied Social Research. For Horkheimer, this was not an option. Although he had collaborated with Lazarsfeld on research projects, Horkheimer considered Lazarsfeld's work and thinking "positivistic".[42]

The loss of the economic independence of the IfS and the pressure to increase its number of publications in English resulted in the need to mobilize funding, which meant adapting to the structures of American social sciences. These were characterized as follows: 1) social science predominantly meant empirical social research, using quantitative statistical methods; 2) research results had to be published in journals on a regular basis; 3) research projects in the social sciences were often funded by philanthropic organizations such as the Rockefeller and the Ford Foundation or the Carnegie Corporation.[43] In order to receive funding, social scientists had to submit applications;[44] 4) research was usually embedded in collaborative projects, which meant that researchers and institutes had to cooperate;[45] and 5) social scientists in the US usually claimed to provide applied knowledge for the progressive improvement of society.

With the dismissal of Fromm, Horkheimer had lost his major empirical researcher. Horkheimer himself was first and foremost a philosopher who advised the empirical projects in Frankfurt, but did not practice empirical research.[46] Pollock directed empirical projects, but was not really familiar with quantitative methods, and Adorno, who was a philosopher and musicologist, had no experience and no interest in empirical social research. Since 1938 Adorno had worked on Lazarsfeld's "Princeton Radio Research Project," but did not adapt his philosophical thinking to the empirical and

---

40 Abromeit, *Max Horkheimer*, 336–41.

41 Wheatland, *Frankfurt School*, 72, 81–85; Ziege, *Antisemitismus*, 25, 43–44.

42 Wheatland, *Frankfurt School*, 86; Wiggershaus, *Max Horkheimer*, 165–73.

43 See Volker Berghahn, *Transatlantische Kulturkriege. Shepard Stone, die Ford-Stiftung und der europäische Antiamerikanismus* (Transatlantische Historische Studien 21) (Stuttgart: Steiner, 2004), 183–85.

44 See Albrecht et al., *Die intellektuelle Gründung*, 73–74.

45 Fleck, *Transatlantische Bereicherungen*, 41–42, 229, 237–55.

46 Abromeit, *Max Horkheimer*, 211.

more positivistic research program of Lazarsfeld and his team. Lazarsfeld, who had studied mathematics,[47] applied quantitative and fact-oriented methods of empirical research. The aim of the Princeton project was to gain knowledge about the various personalities of radio listeners through the analysis of their listening habits, and Lazarsfeld was not as interested as Adorno in a critical reflection on radio listening and listeners. Instead he wanted to create knowledge that the radio industry could use.[48] Lazarsfeld's and Adorno's different epistemic goals were closely linked to profound methodological differences between the two scholars. According to Adorno, theory should inform the techniques of exploring the radio listeners' habits. For him, empirical facts without any theoretical precondition were useless. Contrary to this view, Lazarsfeld created his types of analytical categories from the quantitative exploration of the empirical material. Lazarsfeld preferred the quantitative-inductive method, which opposed the methodological approaches decisive for the IfS, as Horkheimer had formulated in the early 1930s, namely that analytical categories should be created through philosophical investigations.[49] After two and a half years, the Rockefeller Foundation decided not to prolong Adorno's employment.[50]

Horkheimer and his fellow scholars could only translate some elements of their research program to the American social sciences. One of these elements was empirical social research, which was elastic enough to be adapted to the methodology of American social sciences. American social scientists or émigrés such as Lazarsfeld, however, refused to mix empirical social research and philosophical-critical knowledge. They preferred "administrative research" to create useful knowledge for politics and industry. Robert Lynd was a leftist liberal, heavily influenced by New Deal politics, who had conducted critical social research in his 1929 study on Mid-

47 Neurath 1988, "Paul Lazarsfeld," 70.

48 Wolfgang Bonß, "Kritische Theorie und empirische Sozialforschung—ein Spannungsverhältnis," in *Adorno-Handbuch. Leben—Werk—Wirkung*, edited by Richard Klein, Johann Kreuzer, and Stefan Müller-Dohm (Stuttgart: Metzler, 2011), 238–39. See David Jenemann, *Adorno in America* (Minneapolis: Univ. of Minnesota Press, 2007), 16–18.

49 Fleck, *Transatlantische Bereicherungen*, 357.

50 Detlev Claussen, "Die amerikanische Erfahrung der kritischen Theoretiker," in *Keine Kritische Theorie ohne Amerika*, edited by Detlev Claussen, Oskar Negt and Michael Werz (Hannoversche Schriften 1) (Frankfurt: Verlag Neue Kritik, 1999), 27–45, 32; Fleck, *Transatlantische Bereicherungen*, 276–77, 284–96, 330; Wheatland, *Frankfurt School*, 86.

dletown. But a critical and philosophical perspective akin to Horkheimer's was not part of his scientific program.[51]

Despite his aversion to "administrative research," Horkheimer applied for funding from the Rockefeller Foundation and at organizations such as the American Council for Learned societies.[52] One of these projects was "Cultural Aspects of National Socialism," which was to be carried out in cooperation with Eugene Anderson and which included a committee of well-known American scholars.[53] Yet none of these applications was successful.[54] Two private institutions, however, were willing to finance the empirical research of the IfS: the American Jewish Committee (AJC) and the Jewish Labor Committee (JLC). Horkheimer suggested a topic that employees of the IfS had investigated in European contexts only, namely anti-Semitism. In the early 1940s, this was a highly debated topic in Anglophone social sciences and in social psychology.[55] American and English scholars investigated fascism, authoritarianism, and anti-Semitism in order to provide scientific instruments for the prevention of the establishment of "totalitarian" radicalism in their own countries. Horkheimer thus adopted topics which American social scientists considered relevant and important. In 1941, he presented a research project on a typology of an anti-Semitic personality type in the *Studies in Philosophy and Social Science*.[56] Another idea for a project was a prize competition published in the New York émigré-journal *Aufbau* that called for articles depicting personal experiences of Nazi anti-Semitism.[57] Finally, as a cooperative effort between the IfS and the AJC between 1943 and 1944, the *Studies in Anti-Semitism* served as a pilot for a larger research project on the same topic.[58]

---

51 Robert S. Lynd and Helen M. Lynd, *Middletown: A Study in Contemporary American Culture* (New York: Harcourt, Brace, and Company, 1929).

52 Fleck, *Transatlantische Bereicherungen*, 50–64, 307–09.

53 Frederick Pollock to the American Council for Learned Societies, 27 June 1941, Na 1, 1-Korrespondenzen A, fol. 83–88, UBA Ffm.

54 Walter-Busch, *Geschichte*, 126. See Jay, *Dialectical Imagination*, 169.

55 Ziege, *Antisemitismus*, 177.

56 Fleck, *Transatlantische Bereicherungen*, 358.

57 Institute of Social Research, B. "Attitudes of the German People: The Institute's Contest on the German People and Antisemitism under Hitler," in *Studies in Anti-Semitism*, edited by Robert McIver and Frederick Pollock (New York: Institute of Social Research, 1944), 240–76.

58 Institute of Social Research, *Studies in Antisemitism: A Report to the American Jewish Committee*, vol. I–IV (New York: Institute of Social Research, 1944).

The new alliances with American-Jewish organizations provided the possibility of loose connections between the IfS and Columbia University in 1944/45. At the same time, the cooperation with the AJC and the JLC meant that Horkheimer was forced to adapt his research to the intentions of his clients.[59] Horkheimer and his colleagues sought to balance contradictory epistemic demands: They were supposed to conduct "administrative research" like Lazarsfeld, while they also wished to analyze a politically charged problem for which they needed a critical perspective. In addition, the AJC was a conservative organization and this prompted Horkheimer to downplay or hide his (and his staff's) early Marxist philosophical works.[60] Horkheimer stated that the "bourgeois position of the Jews in America [...] admittedly could only be asserted if one gets to grips with anti-Semitism with the help of every theoretical and practical means available."[61] This was a very different view compared to the one he held in the essay "The Jews and Europe," published in 1939. In this essay, Horkheimer heavily criticized liberalism and the bourgeoisie, because he thought that liberal society must end in fascism. Connected to this was a critique of Jews, particularly of Jewish intellectuals in exile.[62] Horkheimer criticized Jewish faith in reason. According to him, Jews had not seen that it was exactly this reason that had turned to fascism and, therefore, that it had now turned against Jews.[63] Thus, the new alliances between the IfS and the American-Jewish organizations resulted in a transformation of Horkheimer's critical view of liberalism towards a more positive image of liberal society. By the mid-1940s, his own Jewish background became more important for Horkheimer.[64]

---

59 Wiggershaus, *Max Horkheimer*, 165–73.

60 Ziege, *Antisemitismus*, 28, 80–81, 234, 69. Another reason for hiding his Marxist attitude was the investigations by the FBI of Horkheimer and the IfS, because the FBI and the police suspected the German émigrés of being communists. See Jenemann, *Adorno in America*, xii–xiv. See also Wheatland, *Frankfurt School*, 73.

61 Draft, Projektbeschreibung, Beilage zum Memorandum zum Antisemitismus-Projekt, not dated [probably spring 1940], fol. 29, 30, Na 1, 5-Korrespondenzen B, fol. 29–34, UBA Ffm.

62 Max Horkheimer, "Die Juden und Europa" [1939], in *Max Horkheimer. Gesammelte Schriften*, vol. 4, *1936–1941*, edited by Alfred Schmidt (Frankfurt: Fischer, 1988), 308–31, 323.

63 Horkheimer, "Die Juden," 324.

64 See Manfred George, Editor der Zeitschrift "Der Aufbau" to Max Horkheimer, 5 Dec. 1940, Na 1, 2-Korrespondenzen A, fol. 327, UBA Ffm; Draft, Projektbeschreibung, Beilage zum Memorandum zum Antisemitismus-Projekt, not dated [probably spring

In 1944, two big conferences on anti-Semitism took place, one in New York, organized by the AJC, the other in San Francisco by the Psychoanalytic Society. Horkheimer was among the many American and émigré social scientists who attended. Discussions at these conferences laid the foundation for the cooperation between the IfS and researchers such as Nevitt R. Sanford and Else Frenkel-Brunswik, who presented their pilot study on the "anti-Semitic Personality" in San Francisco. That same year, Horkheimer was promoted to research consultant of the AJC. The final product of this large project was the five volume series *Studies in Prejudice*, published in 1949/50.[65]

Yet Horkheimer did not limit his scientific work to empirical research. Since the early 1930s, he had planned a comprehensive study on dialectical logic, but his duties as director of the IfS, the need to get funding for research projects and to connect with American colleagues prevented him from working on this study. But when Horkheimer reduced the number of employees of the institute and intensified the cooperation with Adorno in their new domicile in the Pacific Palisades, California, he finally carried out his theoretical investigations.[66] The results were *Dialectic of Enlightenment* (*Dialektik der Aufklärung*) by Horkheimer and Adorno and *Eclipse of Reason* by Horkheimer. Both books were published in 1947. *Eclipse of Reason*, in which Horkheimer showed why the Nazis labelled their politics as "reasonable," was addressed to American scholars because it provided tools to prevent such a development in capitalist-democratic America. *Dialectic of Enlightenment*, in contrast, was written in German and, therefore, appeared as a work of genuine critical thinking. Horkheimer and Adorno knew that the theoretical outlook and the writing style of this book would differ heavily from established social theory in the US.[67]

In American social sciences during the 1930s and 1940s, various theoretical approaches were common: positivistic and law-oriented behaviorism, structural functionalism (prominently represented by Talcott Parsons who merged Max Weber's approach of ideal types with rational choice-

---

1940], Na 1, 5-Korrespondenzen B, fol. 29–34, UBA Ffm. See also Ulrich Oevermann, "Denn du Ewiger bist meine Zuversicht," *Frankfurter Allgemeine Zeitung*, Sept. 9, 2013, www.faz.net/aktuell/finanzen/2.3017/denn-du-ewiger-bist-meine-zuversicht-12577324.html.

65 Ziege, *Antisemitismus*, 175, 70.

66 Max Horkheimer to Franz Alexander, 22 September 1938, Na 1, 1-Korrespondenzen, folder 33, UBA Ffm. See Abromeit, *Max Horkheimer*, 302, 394–95.

67 See Schmid Noerr, "Die Emigration," 253.

theory and with American pragmatism[68]), the interpretative sociology of Charles Horton Cooley or Howard P. Becker, cultural anthropological studies such as those by George Herbert Mead, and Bronislaw Malinowski's widely endorsed ethnographical and structural-functionalist analyses.[69] While phenomenological social psychology and clinical psychology offered methodological and also theoretical points of juncture for Horkheimer, he fervently criticized the epistemological approach of logical empiricism even though it was strongly represented by émigrés, particularly by the Vienna circle.[70] Horkheimer published four critical essays, of which "Traditional and Critical Theory" (1937) is the best-known. In this article, Horkheimer rejected epistemological developments including logical empiricism, attacking its historical ignorance and static viewpoint, and argued that the development of a social theory that focused on actual problems of society required a historical consciousness that took dynamic social processes into account. Horkheimer suggested that a belief in a mathematical and logical structure of social phenomena led logical empiricists into a new metaphysics.[71]

In contrast to these approaches, Horkheimer and Adorno wanted to demonstrate that the Enlightenment unleashed both the emancipation and the self-destruction of the bourgeoisie.[72] They thought that the self-destruction of bourgeois liberalism was caused by the development of classical liberalism into fascism, a development which Horkheimer and Adorno observed both in Europe and in the US. *Dialectic of Enlightenment* was a fundamental critique of modern society as such, and addressed

---

68 Robert C. Bannister, "Sociology," in *The Modern Social Sciences*, edited by Theodore M. Porter and Dorothy Ross, vol. 7 of *The Cambridge History of Science* (Cambridge: Cambridge Univ. Press, 2003), 344–48; David Paul Haney, *The Americanization of Social Science: Intellectuals and Public Responsibility in the Postwar United States* (Philadelphia: Temple Univ. Press, 2008), 71–74; Helmut R. Wagner, "Der Einfluss der deutschen Phänomenologie auf die amerikanische Soziologie," in *Geschichte der Soziologie*, edited by Lepenies, 4:205–06.

69 Steinmetz, "American Sociology," 316–39. See Jürgen Hartmann, *Geschichte der Politikwissenschaft. Grundzüge der Fachentwicklung in den USA und in Europa* (Opladen: UTB, 2003), 49–99, 57.

70 Max Horkheimer to Friedrich Pollock, 9 June 1943, cited in Wiggershaus, *Frankfurter Schule*, 383.

71 Max Horkheimer, "Traditionelle und kritische Theorie" [1937], in *Gesammelte Schriften*, vol. 4, 162–225. See. Abromeit, *Max Horkheimer*, 301–21; Jay, *Dialectical Imagination*, 82–83. See also Dahms, *Positivismusstreit*, 57.

72 Max Horkheimer and Theodor W. Adorno, *Dialektik der Aufklärung. Philosophische Fragmente.* (Frankfurt: Fischer Taschenbuch Verlag, 2008 [1947]), 7.

Western liberal societies, fascist regimes, and socialist states as well. While there were critical books on modern American society, one well-known example being David Riesman's *The Lonely Crowd* (1950), the highly speculative philosophy of history proposed by Horkheimer and Adorno was hardly compatible with theoretical studies in American social sciences and social philosophy.[73] The thrust of their *Dialectic of Enlightenment* was highly pessimistic and American readers would have perceived this as an attack on their political and cultural values. Horkheimer and Adorno, however, did not intend to expose their theoretical investigations to American readers as they consciously wrote *Dialectic of Enlightenment* in German.[74] But Horkheimer's and Adorno's philosophical assumptions formed the IfS research program's persistent "hard core."[75]

The theoretical and philosophical works of Horkheimer and Adorno mutually influenced empirical projects they carried out at the time, and a deep gap opened up between the empirical and the theoretical-philosophical levels of social knowledge.[76] Theory meant for Horkheimer and Adorno a sphere of intellectual independence in which they developed ideas without the need to seek funding or with the stipulation that they convey such ideas to American readers.[77] Their empirical research, however, had to generate knowledge that was applicable to the public, the economy, and to politics and education. This gap was caused by the successful translation of the empirical approaches already developed in Frankfurt into the empirical research used by American social scientists, and by the fact that neither Horkheimer nor Adorno intended to convert their theoretical assumptions into theories common among American social scientists. This does not mean that the empirical research carried out by the IfS was completely uncritical and should be regarded as merely "adminis-

---

73 Ziege, *Antisemitismus*, 99–100.

74 See Martin Jay, "Adorno in America," in *Keine Kritische Theorie*, edited by Claussen, Negt, and Werz, 46–76, 46–49. See Ziege, *Antisemitismus*, 50, 180. For Adorno's strong commitment to German language see Theodor W. Adorno, "Fragen an die intellektuelle Emigration" [1954], in Theodor W. Adorno, *Vermischte Schriften I: Theorien und Theoretiker. Gesellschaft, Unterricht, Politik* (Gesammelte Schriften 20.1) (Frankfurt: Suhrkamp, 2003), 352–59.

75 Theodor W. Adorno to Heinz Norden, 23 June 1950, fol. 1–2, fol. 1, folder USA, M-Z, Korr. H. Marcuse, Archiv des Instituts für Sozialforschung an der Johann Wolfgang Goethe-Universität (Archive of the Institute of Social Research at the Johann Wolfgang Goethe-University, hereafter cited as Archive IfS).

76 See Ziege, *Antisemitismus*, 101–02.

77 See Jay, *Dialectical Imagination*, 197.

trative". Instead, their empirical research and philosophical-theoretical thinking was no longer part of the concise research program Horkheimer had developed in Frankfurt.

This gap caused several epistemic contradictions. The correspondence between Horkheimer and Adorno shows that they severely criticized Lazarsfeld because of his "affirmative" position, which, according to them, resulted in the reproduction and confirmation of the status quo. At the same time, Horkheimer cooperated with émigré scholars such as Käthe Leichter, Maria Jahoda, or Herta Herzog whom he accused of being "affirmative" towards society or of being "positivistic". Similar contradictions arose in the relationship between Horkheimer and American social psychologists, colleagues such as R. Nevitt Sanford or Daniel J. Levinson, who applied behavioristic approaches that Horkheimer had fervently criticized.[78] Another example is that the basic assumptions of the volume *Dynamics of Prejudice* by Bruno Bettelheim and Morris Janowitz in the series *Studies in Prejudice*, contradicted critical theory because the authors maintained an affirmative position towards existing power relations in family structures, and this undermined the results of Fromm's earlier empirical study on the family. Horkheimer was aware of these contradictions but he did not ask Bettelheim and Janowitz to change their book.[79]

Horkheimer, Pollock, and Adorno directed the field research for two empirical projects about anti-Semitism and authoritarianism, "Anti-Semitism among American Labor" (unpublished) and *Studies in Prejudice*.[80] The field research of the "Labor Study" was carried out between spring and November 1944, encompassing surveys in New York, Philadelphia and areas in New Jersey including Camden and Newark, also in Pittsburgh, Los Angeles, Detroit, San Francisco as well as areas in the states of Massachusetts, Maryland, and Wisconsin. The method applied in this project was that of the "attitude surveys" developed by American social scientists around 1920. "Attitude" here meant the "state of mind of the individual toward a value." Researchers used questionnaires and coded answers with indexes and scales. According to Lazarsfeld, an index had to establish a correlation between certain dimensions of measurable characteristics of behavior to a certain number that valued these characteristics. The total

---

78 Fleck, *Transatlantische Bereicherungen*, 381–82; Ziege, *Antisemitismus*, 48–49. See Jay, *Dialectical Imagination*, 239.

79 Ziege, *Antisemitismus*, 239–41.

80 Ibid., 200, 224.

score was the index value.[81] In order to investigate anti-Semitic attitudes in an indirect way, researchers used a "group method"[82] that sought to facilitate casual remarks and prompt effective and honest statements.[83] Lazarsfeld, Jahoda, and Zeisel had created the basic idea for such interviews in the early 1930s.[84] Randomly chosen workers were to converse with colleagues and results were entered in standardized protocols. A group of four research assistants, two secretaries, and fourteen research collaborators coordinated interviewers. The research team sent out 4,500 questionnaires and recruited 1,000 workers as possible interviewers, of which 500 agreed to take part in discussions with their colleagues. 270 of the interviewers sent 613 protocols back to the research team. For the *Studies in Prejudice* project, the methods of American psychologists and social researchers were decisive as well. Based on the technique of scaling, which had been standard in American social science since around 1940, Sanford and Levinson developed a scale for measuring anti-Semitic attitudes. The well-known scale on fascism, the F-Scale, and other scales for measuring authoritarianism, would later be derived from Sanford's and Levinson's methodological approaches.[85]

Adorno's original contribution in the volume on the *The Authoritarian Personality* of the series *Studies in Prejudice* was the qualitative and psychological-critical interpretation of the empirical facts. The goal of the analysis was the development of a typology of the "authoritarian personality."[86] This method was, in Adorno's words, "a phenomenology based on theoretical formulations and illustrated by quotations from the interviews" with the goal "to exploit the richness and concreteness of 'live' interviews to a degree otherwise hardly attainable."[87] Adorno tried to combine empirical typology based on indexes and scales with the critical-theoretical typecast. This combination remained unsystematic, however, as Adorno mixed speculative thoughts about the psychodynamic roots of anti-Semitism with typological patterns of empirical indicators, merging speculative explanations of authoritarian personality types with empirically proven real types. In addition, Adorno constantly changed the methodological criteria of his

81 Fleck, *Transatlantische Bereicherungen*, 400.
82 Jay, *Dialectical Imagination*, 226.
83 Ziege, *Antisemitismus*, 188–89.
84 Ibid., 189–90, 196–97.
85 Ibid., 174.
86 Fleck, *Transatlantische Bereicherungen*, 402–04.
87 Cited in Wiggershaus, *Frankfurter Schule*, 463.

typology.[88] It was, therefore, unclear how an "authoritarian personality" could be distinguished from a "fascist personality."[89] Adorno's interpretation of empirical facts generated from field research took the form of an empirical confirmation of a priori assumptions developed in Horkheimer's and Adorno's theoretical work.

From their empirical and theoretical studies on anti-Semitism and authoritarianism in the US, Horkheimer and Adorno concluded that fascism was part and parcel of modernity. This implied that every modern society had the potential sooner or later to become fascist.[90] Fascist thinking and anti-Semitic attitudes were not confined to individual nations and societies such as Germany but resulted from a psychological pathology that arose from problematic developments in modern society. According to this view, in modern society anyone could adopt a fascist mindset but anyone could also become liberal and democratic. For Horkheimer and Adorno, "fascist behavior" became visible only when the general situation of a society changed dramatically, as was the case in Germany after the Nazis took power.[91] As a consequence of these observations, Horkheimer and his colleagues concluded that they could be based anywhere in the modern world, in the US or in Germany, because the development of fascism and democracy was possible everywhere. For them, the question was not whether fascism and anti-Semitism could be erased as such, but how an outbreak could be prevented. The key vaccination was provided by teaching Christian and democratic values. The "Labor-Study" demonstrated that workers' anti-Semitic and racist stereotypes correlated inversely to their level of education and exposure to Christian values.[92] In the late 1940s, the perspective of mediating democratic values so as to prevent anti-Semitic and fascist outbreaks provided Horkheimer, Pollock, and Adorno with a key motive to return to Frankfurt.

---

88 See ibid., 408, 414–15.

89 See the critique by Edward Shils. See Jay, *Dialectical Imagination*, 247.

90 Ziege, *Antisemitismus*, 170–71. See Horkheimer, "Die Juden." See also Abromeit, *Max Horkheimer*, 426–28; Jay, *Dialectical Imagination*, 269–70.

91 Theodor W. Adorno, *Studien zum autoritären Charakter* (Frankfurt: Suhrkamp, 1995 [1973]), 2–5, 9–10.

92 Ziege, *Antisemitismus*, 211, 214–15.

4. Empirical Social Research and Democratization Politics in Frankfurt

In October 1945, representatives of the Johann Wolfgang Goethe-University, Frankfurt am Main, the Hessian Ministry of Culture and Education, and the city of Frankfurt asked Max Horkheimer to return to Frankfurt and to reestablish the *Institut für Sozialforschung*. With Frederick Pollock and Adorno, he accepted the opportunity to successfully reestablish the IfS at the Goethe-University. This was a rare achievement. Not many Jewish academic émigrés were as successful as this group upon their return to Germany. Returnees were fiercely criticized by international and American Jewish organizations that refused to support them. Anti-Semitic prejudices had hardly been eliminated from post-war West German society.[93] But Horkheimer, Pollock, and Adorno returned to Frankfurt for three reasons.

First, in the late 1940s, the relationship between the IfS and its American partner institutions became increasingly difficult. The cooperation between the IfS and the American Jewish Congress in the project about anti-Semitism became challenging because the representatives of the AJC were pushing the IfS to publish the results of its empirical investigations. It was unclear, furthermore, in what way Horkheimer and his colleagues should continue their work in the US.[94] They welcomed the invitation to Frankfurt as an opportunity to reestablish their academic careers in philosophy and sociology at a German university and in a revived and well-supported IfS.

Second, with their theoretical work, neither Horkheimer nor Adorno were able to penetrate American social science and philosophy. Their American colleagues dismissively referred to their work as unscientific "metaphysical speculations" of little interest.[95] With the exception of some

93 Werner Bergmann, "'Wir haben Sie nicht gerufen'. Reaktionen auf jüdische Remigranten in der Bevölkerung und Öffentlichkeit der frühen Bundesrepublik," in *'Auch in Deutschland waren wir nicht wirklich zu Hause'. Jüdische Remigration nach 1945*, edited by Irmela von der Lühe, Axel Schildt, and Stefanie Schüler-Springorum (Göttingen: Wallstein, 2008), 19–39, 24; Michael Brenner, "Jüdische Geistesgelehrte zwischen Exil und Heimkehr: Hans-Joachim Schoeps im Kontext der Wissenschaft des Judentums," in *Ich staune, dass Sie in dieser Luft atmen können'. Jüdische Intellektuelle in Deutschland nach 1945*, edited by Monika Boll and Raphael Gross (Frankfurt: Campus, 2013), 29–39, 27–28.

94 Wiggershaus, *Max Horkheimer*, 165–73.

95 Wheatland, *Frankfurt School*, 81–85.

emigrants who read German, few scholars during the late 1940s and 1950s took note of *Dialectic of Enlightenment*.[96] But the theory and philosophy of history was for Horkheimer and Adorno more important than empirical research, even though the holistic epistemology of Horkheimer intended to merge the two epistemic levels. In addition, Horkheimer and Adorno were steeped in German philosophical culture and neither could bend their language skills to American philosophical thinking.[97]

Third, the scholars' empirical projects on anti-Semitism showed that social knowledge was useful and applicable to actual social and political problems. After the end of World War II and the Nazi regime, West Germany and the US both dealt with the latency of anti-Semitism and fascist attitudes in democratic political structures.[98] Their explorations of anti-Semitic and anti-democratic thinking, and their work on authoritarian attitudes, provided Horkheimer, Pollock, and Adorno with empirical knowledge to help resolve problems in reconstituting West German society after 1945. The usefulness of their insights provided them with an important motive to return to Frankfurt.[99]

After his return, Horkheimer in July 1949 assumed a professorship at the University of Frankfurt's Faculty of Philosophy. Pollock was appointed adjunct professor (*apl. Prof.*) in 1951. Seven years later, he was offered a full chair at the Faculty of Economics and Social Sciences. Shortly thereafter, Horkheimer was elected dean of the Faculty of Philosophy, then president (*Rektor*) of the university.[100] With the financial support of the McCloy-Funds, other private foundations, the government of Hesse, and the city of Frankfurt, the IfS resumed work in May 1950 and officially reopened in November 1951.[101]

---

96 Detlev Claussen, "Malentendu? Theodor W. Adorno: Eine Geschichte von Missverständnissen," in *'Ich staune, dass Sie in dieser Luft atmen können'*, edited by Boll and Gross, 375–92, 375.

97 See Lorenz Jäger, *Adorno: Eine politische Biographie* (Munich: Pantheon, 2009), 230.

98 See Jay, *Dialectical Imagination*, 230–34.

99 Max Horkheimer to Fritz Karsen, 17 September 1948, Na 1, 26-Korrespondenzen, fol. 293, UBA Ffm.

100 Walter-Busch, *Geschichte*, 34–35; Wiggershaus, *Frankfurter Schule*, 479.

101 Memorandum Theodor W. Adorno to Prof. Dr. Benecke, 5 Oct. 1953, fol. 1–6, fol. 1–2, Adorno-Korrespondenz, B, 2, Archive IfS. See also Der Hessische Minister für Erziehung und Volksbildung (gez. Metzger) to the Universitäts-Kuratorium, Wiesbaden, March 1951, Abt. 134, no. 234, fol. 118, Universitätsarchiv Frankfurt (University Archive, Frankfurt, hereafter cited as UAF).

Fig. 3. Max Horkheimer (right) with German chancellor Konrad Adenauer (left) and Frankfurt mayor Walter Kolb during university festivities in 1952.

<small>(Courtesy of the Archivzentrum der Universitätsbibliothek Frankfurt am Main)</small>

The reestablishment of the IfS as an institution for democratic expertise was the result of Horkheimer's intelligent management as university president. He promoted the IfS as a research institution to apply the "highest developed empirical research methods of the modern American social sciences" and as a consultant for West Germany's most pressing political and social issues.[102] The directors of the IfS could demonstrate to German politicians and to German social scientists that they promoted American scientific modernity, thereby validating the institute's raison d'être. German scholars and politicians believed that the IfS perfectly facilitated the country's new alliance with the US (*Westbindung*). Politicians who advocated a pro-Western and "anti-totalitarian" attitude, both members of the Social Democratic Party such as Theodor Heuss and conservative Christian Democrats such as Konrad Adenauer, considered Horkheimer an im-

---

102 Wiggershaus, *Frankfurter Schule*, 480.

portant intellectual ally.[103] In turn Horkheimer regarded them as important partners for implementing his re-education policy and for making the IfS one of West Germany's key institutions for social research, "providing training at a high university level for those elements about to undertake responsible work in industry, administration and various other vocations."[104]

The institute's success was facilitated through alliances with academics, politicians, and industrialists. Hellmut Becker, son of the Prussian Education Minister in the Weimar Republic, a lawyer and later founder of the Max-Planck-Institute for Education Research who had been a member of the NSDAP and member of the George circle, assumed a key role in establishing this network. In the fledgling Federal Republic of Germany, Becker was well-connected, with an extensive network of friends in politics, academia, and among intellectuals and artists.[105] He proved to be completely loyal to the IfS, not only as a lawyer, but also as its general adviser. At the IfS, he attended almost every meeting with potential clients, connecting Horkheimer and Adorno to businessmen and politicians, especially with respect to education.[106] Teacher training was of particular relevance for the IfS because the State of Hesse, like other regions in Germany, suffered from a shortage of teachers at a time when they were considered essential for the country's successful democratization.[107]

In the 1950s, the IfS was predominantly concerned with empirical research projects. As in the US, empirical research for the institute was a tool to establish itself in West German social sciences. The best-known project was the "group experiment" (*Gruppenexperiment*) financed by the High Commissioner for Germany (HICOG), the goal of which was to examine the attitudes of Germans towards the Nazi regime, anti-Semitism, the occupation forces, and democracy. The project team consisted of fifteen to twenty researchers, students or doctoral candidates without training in empirical methods. Adorno acted as conceptual adviser for the empirical

---

103 Albrecht et al., *Die intellektuelle Gründung*, 113–14.

104 Institut für Sozialforschung an der Johann-Wolfgang-Goethe-Universität Frankfurt am Main, Professor Horkheimer, Professor Adorno, Dr. v. Friedeburg, undated [probably 1952/53], fol. 1–2, fol. 2, Adorno-Korrespondenzen C, 3, Archive IfS.

105 Ulrich Raulff, *Kreis ohne Meister. Stefan Georges Nachleben* (Munich: Beck, 2010), 25, 383, 385–400, 403–409, 457–58, 470–77, 481–96.

106 Friedrich von Weizsäcker to Hellmut Becker, 7 Oct. 1955, 1–3, Project A 10, Betriebsuntersuchung Mannesmann, folder 1.1, file "O. Vorarbeiten", Archive IfS.

107 See Albrecht et al., *Die intellektuelle Gründung*, 131.

work and he represented the IfS in public while Pollock served as the project's executive director.[108] Although Horkheimer, Pollock, and Adorno had done empirical social research in the US, they lacked training in statistics and demographic research. In 1951, the IfS recruited Diedrich Osmer for such expertise. In 1957, the institute hired Rudolf Gunzert as department head of quantitative studies, and in 1959, he became one of the institute's directors. Gunzert, a statistician and social scientist, had been a member of the NSDAP and had produced statistical studies about the city of Mannheim for Nazi organizations. He remained director of the empirical studies of the IfS until 1977. His tenure at the IfS reflects Horkheimer's intention to work with younger scholars and with students who, in his view, had had no choice but to join Nazi organizations if they had wished to pursue a career. To Horkheimer, the real Nazis were not among the younger colleagues but among cynical professorial mandarins who had welcomed the Nazi regime and joined its ranks without being forced to do so.[109]

Much like in his work on the *The Authoritarian Personality*, Adorno intended to interpret empirical facts by merging quantitative methodology and critical philosophical analysis.[110] The result of his work in Frankfurt was the same as in America. There was a deep discrepancy between Adorno's phenomenological-psychological typology and types of "authoritarian personalities" deduced in empirical studies. Furthermore, after the project had gotten underway, Adorno demanded changes and an expansion of scales and criteria for quantitative analysis that slowed completion of research in the field.[111] Methodological confusion resulted from Adorno's critique of empirical research methods, a critique he articulated in a much more pronounced manner than in the US. When compared to the results of qualitative analysis, Adorno considered anti-Semitism un-

---

108 See Archive IfS, Project 2, Gruppenstudie, folder 1, Bericht über die Sitzung on 3 March 1952, 1–2.

109 Alex Demirović, *Der nonkomformistische Intellektuelle. Die Entwicklung der Kritischen Theorie zur Frankfurter Schule* (Frankfurt: Suhrkamp, 1999), 115.

110 Protokoll der Sitzung vom 1. März 1957: "Zum Verhältnis von Soziologie und empirischer Sozialforschung." Anwesende: Prof. Adorno, Dr. Dahrendorf, Dr. v. Friedeburg, Prof. Gunzert, Dr. Habermas, Prof. Lieber, Dr. Noelle-Neumann, Dr. Popitz, Prof. Stammer, fol. 1–29, fol. 24–25, S 1: Tagungen 1950–1961, folder 1 1950–1952, Archive IfS.

111 See Interview Schema für Einzel-Interviews, 4 Jan. 1950, fol. 1–2, Projekte 2 (6): Gruppenexperiment/Gruppenstudie, Archive IfS.

derrepresented in quantitative data because the practice of "coding" (the attribution of answers to categories of scales) divided up anti-Semitic remarks into a larger number of categories than pro-Semitic statements. Adorno also criticized that many remarks he considered anti-Semitic disappeared into the rubric "other matters." In his estimate, certain categories were too narrow and excluded a large quantity of answers from attribution to his typology. Quantitative methods thus distorted the empirical facts. Only the empirical researcher himself could adequately evaluate the study's results because he understood the methods.[112] But Adorno was unable to develop an alternative form of empirical research that would have included a critical-philosophical perspective. Adorno called the formal conception of questionnaires, the selection of samples, and quantitative analysis through punch cards "fetishized machinery" because the procedure remained for him a "black box." But he left unanswered the question of how ideologies could properly be recorded using research terminology, a matter that was central to his team and its "group experiment."[113]

Alliances struck by the IfS included institutes that focused on opinion and market research. These other institutes employed several researchers who had been diehard Nazis during the regime and others who had worked for Nazi organizations, including Elisabeth Noelle-Neumann, head of the Institute for Demoscopy in Allensbach. Helmut Schelsky, professor at the *Akademie für Gemeinwirtschaft* (Academy of Public Enterprise) in Hamburg since 1948, had been a member of the SA and department head of the National Socialist German Students' League.[114] When Adorno became advisor of the *Darmstadt-Studie*, a survey of that German city carried out between 1949 and 1952, he worked with Max Rolfes, a former scientific adviser to Heinrich Himmler, who was engaged in racial research (*völkische Forschung*) in Alsace and Lorraine for Himmler's settlement poli-

---

112 Bericht über die Sitzung vom 8.7.1952, fol. 1–2: Thema: Gruppenstudie, quantitativer Teil und Termine, anwesend: Adorno, Gretel Adorno, Osmer, Beier, Freedman, v. Hagen, Koehne, Dr. Maus, Sardemann, Schmidtchen, Dr. Sittenfeld, fol. 1, Projekte 2 (1), Gruppenexperiment/Gruppenstudie, Archive IfS.

113 Entwurf, 2 Oct. 1952, Staff Meeting Gruppenstudie, und Plessner, v. Schlauch, fol. 1–4, fol. 1–2, Projekte 2 (1), Gruppenexperiment/Gruppenstudie, Archive IfS.

114 Entwurf eines Auswahlplans für das Projekt "Auswahlstudie", 15 Febr. 1953, fol. 1–2, A 20, P 14, Bundeswehr, folder 1.1, Archive IfS. For Noelle-Neumann see Elisabeth Noelle, *Meinungs- und Massenforschung in U.S.A. Umfragen über Politik und Presse* (Frankfurt am Main: M. Diesterweg), 63, 67, 94. For Schelsky see NSDDB, Der Reichsamtsleiter, to Dr. W. Greite, 7 March 1936: Betr. Dr. Helmut Schelsky, Deutsche Forschungsgemeinschaft, box 5, file "Correspondence S", Hoover Institution Archives.

tics during the early 1940s.[115] We do not know exactly, of course, what Adorno knew about the Nazi past of his colleagues but the peaceful cooperation of Nazi victims and former Nazi supporters perhaps comes as a surprise. Remarkably, "empirical methods of the modern American social sciences" were applied by German researchers who had not been trained in the US but in Germany. They had learned about American empirical methods in the social sciences during the Nazi regime.[116]

Ironically, there was a demand in post-war Germany for the expert knowledge social scientists had developed during the Nazi regime to meet two major challenges: urban planning to provide housing and the integration of more than forty million German refugees. In order to tackle these problems, the Hessian State Ministry in August 1946 set up a commission that included researchers at the state's universities, administration officials, and urban planners. The commission's projects carried well into the 1950s.[117] Except for the absence of references to Nazi organizations, its research programs were hardly distinguishable from their Nazi-era forerunners. In 1953, the IfS also conducted a project on the state's "social climate in the urban and rural districts" (*Gemeinden*), financed by the district's office of regional planning. Akin to the earlier *Gruppenexperiment*, the project sought to aid the development of settlement and economic policies through statistical and quantitative knowledge to help ameliorate the state's social problems.[118]

---

115 Carsten Klingemann, *Soziologie und Politik. Sozialwissenschaftliches Expertenwissen im Dritten Reich und in der frühen westdeutschen Nachkriegszeit* (Wiesbaden: VS Verlag für Sozialwissenschaften, 2009), 21.

116 Ibid., 19–22.

117 Großhessisches Staatsministerium, Der Minister für Wiederaufbau und politische Befreiung, an die Herren Rektoren der Technischen Hochschule in Darmstadt, Universität Frankfurt am Main, Hochschule für Naturwissenschaften Gießen, Universität Marburg, Statistische Landesamt Wiesbaden, Landesamt für Bodenforschung, die Herren Regierungspräsidenten Darmstadt, Wiesbaden, Kassel, 26 Aug. 1946, Abt. 1, no. 76, fol. 254–55, UAF; Der Rektor der Goethe-Universität an Heinz Sauermann: Bitte um Wahrnehmung des Sitzungstermins am 20. September 1946, Abt. 1, no. 76, fol. 257, UAF; Rundschreiben no. 1/47, Der Rektor der Goethe-Universität, 27 Jan. 1947, Abt. 1, no. 76, fol. 264, UAF.

118 Friedrich Pollock to Ministerialdirektor Wittrock—Landesplanung—beim Hessischen Ministerpräsidenten, 6 June 1953; Prof. Dr. Ludwig Preller to Ministerialdirektor Wittrock, Landesplanungsstelle b. Ministerpräsidenten des Landes Hessen, 15 July 1952, Adorno-Korrespondenz W, 24, Archive IfS; Tätigkeitsbericht: laufende Projekte des IfS, 28 Aug. 1953, fol. 1–5, fol. 3, Adorno-Korrespondenzen L (1950–1960), 12, Archive IfS.

Such policies and research strategies met with great success and bolstered the institute's standing. At the University of Frankfurt in 1954, the cooperation between the IfS and social scientists elsewhere in Germany resulted in the establishment of the country's first degree program (*Dipl.*) in sociology.[119] During the 1950s and early 1960s, the IfS expanded and the University of Frankfurt became known as one of the most innovative centers for social research in West Germany.

Horkheimer and Adorno had little time to advance their research program of a critical theory of society. Horkheimer was busy taking care of his duties as university president and Adorno and Pollock were in charge of research projects. But they made use of their critical philosophical thoughts in university lectures and public commentary such as radio broadcasts and newspaper articles on democracy, education, their experience abroad in the US, and Germany's Nazi past. Horkheimer and Adorno sought to mediate liberal values to the West German population by emphasizing the significance of individualism and freedom and by appreciating non-conformism, anti-authoritarianism, and self-esteem.[120] In doing so, Adorno drew on his American radio experience. US military officials in the HICOG controlled West Germany media but they cooperated with liberal and left-liberal German intellectuals including Adorno.[121] American support served to legitimize the IfS as a modern, democratic, and anti-totalitarian institution[122] and it provided the West German public with information about basic values in a liberal democracy.[123] For their new role as reeducators in Germany, it was essential that Horkheimer and Adorno had dissociated

---

119 Der Dekan der Philosophischen Fakultät an den Hessischen Kultusminister, 16 June 1965, Abt. 504, no. 12.411, fol. 7–8, Hessisches Hauptstaatsarchiv (Main State Archive of Hesse, hereafter cited as HHStAW). See Demirović, *Der nonkomformistische Intellektuelle*, 406.

120 Gunzelin Schmid Noerr, "Aufklärung und Mythos. Von der *Dialektik der Aufklärung* zur *Erziehung nach Auschwitz*," in *Das Feld der Frankfurter Kultur- und Sozialwissenschaften nach 1945,* edited by Richard Faber and Eva-Maria Ziege (Würzburg: Königshausen & Neumann, 2008), 30–31.

121 Monika Boll, *Nachtprogramm. Intellektuelle Gründungsdebatten in der frühen Bundesrepublik* (Münster: LIT, 2004), 6–7, 11.

122 Albrecht et al., *Die intellektuelle Gründung,* 206–08.

123 See Adorno's public lectures "*Was bedeutet: Aufarbeitung der Vergangenheit?*" ("What does dealing with the past mean?," lecture held at the occasion of the Coordination Council for Christian-Jewish cooperation, autumn 1959) and "*Erziehung nach Auschwitz*" ("Education after Auschwitz," lecture for the Hessian Radio, broadcast 18 April 1966). See Theodor W. Adorno, *Erziehung zur Mündigkeit. Vorträge und Gespräche mit Hellmut Becker 1959–1969,* edited by Gerd Kadelbach (Frankfurt: Suhrkamp, 1970), 10–28, 88–107.

themselves from their earlier dedication to political Marxism,[124] as anti-totalitarian ideology during the early Cold War shut Marxist scholars out of powerful academic positions.[125]

The transatlantic reconfiguration of alliances and the transfer of sociological knowledge, however, should not only be viewed in the context of the democratization of the Federal Republic of Germany. During the early Cold War, such shifts also helped create and stabilize cooperation between West Germany and the US.[126] During World War II, social knowledge had been used to investigate Germany as an enemy. After the war, such knowledge came to mediate democratic and liberal education and politics, and it was reissued to counter communist ideas. American military officials and politicians in charge of education policy in the State of Hesse welcomed reeducation ideas emanating from the IfS because these ideas helped legitimize the transatlantic liberal alliance, and because they helped bring German politicians and scholars out of the Nazi shadows.

5. Conclusion: Shifting Alliances, Social Knowledge, and the Stabilization of West German Democracy

The transatlantic cases of Horkheimer, Pollock, and Adorno show that unlike their critical philosophical approach, methods of empirical social research could be translated and transferred between Germany and the US. For German-Jewish émigrés, empirical methods offered an opportunity to integrate with the social sciences in the US. They also allowed researchers who had remained in the country during the Nazi regime to adapt to new American approaches represented by the IfS after the war. Empirical methods and practices, therefore, provided a language for mediating between the various groups of social scientists who cooperated in helping build a democratic West German state. In contrast to this role assumed by empirical methods, key assumptions of critical social theory and philoso-

---

124 Alex Demirović, *Der nonkonformistische Intellektuelle. Die Entwicklung der Kritischen zur Frankfurter Schule* (Frankfurt: Suhrkamp, 1999), 331, 430.

125 Albrecht et al., *Die intellektuelle Gründung*, 83.

126 Ibid., 171–72; Nicolas Guilhot, "Reforming the World: Georges Soros, Global Capitalism, and the Philanthropic Management of the Social Sciences," *Critical Sociology* 33 (2007): 453.

phy developed by Horkheimer and Adorno proved to be non-translatable. Anti-positivistic and anti-pragmatic critical theory constituted the "hard core" of a local research program that could not be translated into theories common in the American social sciences. A gap appeared between the core of critical philosophical thought and the "elastic and protected area" represented by empirical social research. Even though Adorno intended to reenchant critical theory and empirical research after the IfS returned to Frankfurt, the gap between these epistemic levels remained. In their radio broadcasts and public lectures, Horkheimer and Adorno reissued the core of critical theory as reeducation politics.

This transformation of critical theory buffered the legitimacy of West German democracy. Adorno's insistence on individualism, non-conformism, and freedom went hand in hand with democratic reeducation politics developed by the IfS. Critical intellectuals such as Adorno should be considered in the context of an intellectual and scientific culture of the Western hemisphere. Whether by chance or design, the IfS was part of a Cold War sphere set up against the intellectual and political culture of the communist "Eastern Bloc."

Legitimizing Islamic Studies after 9/11

Andreas Franzmann

1. Introduction

Legitimizing science in the nation-state is complex because it involves numerous aspects. A productive approach is to discuss these aspects in terms of examples taken from particular disciplines, something I do in a project in which I investigate Islamic studies in both the US and Germany pertaining to the impact of public debates on them.[1] The role of the public as an affirmative and even critical background to scientific enterprises has often been mentioned but the influence of the public on scientific fields is not easy to show precisely because of many labyrinthine and hidden effects within this correlation.[2] Nor, additionally, is it easy to compare national configurations in this relationship.

But Islamic studies are well suited for such an investigation, both because of the strong political impact that ongoing public debates in Western countries concerning Islam have had since the late seventies, and due to the many different branches and national traditions of Western Islamic studies that are concerned with a country's relationship to the Muslim world. So in what follows I will use the example of Islamic studies to explore some changes in this research field brought about by the ongoing transformation in public attitudes of Western countries relating to Islam, Muslims, and countries in the Middle East.

In starting out, it might be helpful to describe characteristics of national branches of Islamic studies that are associated with each country's political history. In order to trace the impact of political developments on the field

---

1 With Axel Jansen in a research project (at the Universität Tübingen and at the University of California, Los Angeles, funded by the Volkswagen Foundation) on the "Professionalization and Deprofessionalization in the Public Context of Science since 1970."

2 Peter Weingart, *Die Wissenschaft der Öffentlichkeit. Essays zum Verhältnis von Wissenschaft, Medien und Öffentlichkeit* (Weilerswist: Velbrück, 2005).

in recent decades, I have conducted interviews with Islamic studies experts and I will turn to a discussion of these interviews below. But my interpretation of these interviews will make more sense once I have outlined some important trajectories of the field's history in countries important for its development. Hence I will turn to the field's historical origins first (section 2), then to its development in selected countries, first in Europe (sections 3 and 4), then in the US (section 5). I will spell out the consequences of recent political developments for the field in all countries in sections 6 and 7, and then discuss passages from interviews in section 8 prior to my conclusion.

2. Historical Origins

Historically, the study of Islam as an academic field in Western countries emerged during the Reformation, when scholars opened their mind to religious texts in their original languages. In the sixteenth century, professorships for Arabic were established in Germany, France (*Collège de France*, 1539), and the Netherlands (Leiden, 1576), in the seventeenth century also in England (Cambridge, 1632; Oxford, 1634) to improve a better understanding of the Quran.[3] Up to the French Revolution these interests remained embedded within a theological context that, since medieval times, had struggled with religious rivals in European countries, especially in Spain, southern Italy and the Balkans. First steps towards a secular academic field were taken, when revolutionary France founded a national school for Oriental languages (*École spéciale des langues orientales*) in 1793 to encourage the study of Semitic languages. In this context, Silvestre de Sacy established the study of Arabic and other languages as a comparative, grammar-oriented philology that shook off theological interests in the origins of the bible and early Christian communities while it also left be-

---

3 Hartmut Bobzin, *Der Koran im Zeitalter der Reformation. Studien zur Frühgeschichte der Arabistik und Islamkunde in Europa* (Beirut and Stuttgart: Steiner Verlag, 1995); Albert Hourani, *Islam in European Thought* (Cambridge: Cambridge Univ. Press, 1991), 13. Mohammed Arkoun, "The study of Islam in French Scholarship," and Jacques Waardenburg, "The Study of Islam in German Scholarship," both in *Mapping Islamic Studies: Genealogy, Continuity, and Change*, edited by Azim Nanji (Berlin and New York: De Gruyter, 1997), 33–44 and 1–32, respectively.

hind a focus in the humanities on languages of the classical period, on Latin and Greek, as a heritage of ancient European culture.

Until the end of nineteenth century, the study of Islam was profoundly embedded within this philological approach and even early historian's accounts of Mohammad's life or early Caliphates or other episodes of Islamic History were originally motivated by the pursuit of finding and translating unknown texts from the Arab world.[4]

3. National Branches: France, UK, and the Netherlands

In European countries, this mode of studying Islamic culture was maintained for many decades before it began to change gradually with the growing interest of governments and companies of various countries with colonies to engage in the practical training and education of their personnel. Due to the needs of their colonial administration the UK or France founded central institutions in London or Paris to improve more practical agendas by instructing officers, administrators, merchants, and missioners not just in languages but in the culture and religion of contemporary Muslim societies with respect to local peculiarities and traditions.[5] The agenda of Islamic studies shifted from classical texts and periods to contemporary Muslim societies.

British, French, or Dutch Islamic studies, for instance, are profoundly embedded in the long colonial and postcolonial histories of their own countries with all the effects these have on their respective teaching activities and everyday working lives; this implies its practical involvement in diplomatic, commercial, military or missionary efforts, with a special focus

---

4 Regarding early histories, including accounts of the life of Mohammad, see Rudi Paret, *Arabistik und Islamkunde an deutschen Universitäten. Deutsche Orientalisten seit Theodor Nöldeke* (Wiesbaden: Franz Steiner Verlag, 1966), 9–12; Josef Horovitz, "The Earliest Biographies of the Prophet and their Authors," *Islamic Culture*, vol. 1 (1927): 535–559, vol. 2 (1928): 22–50; 164–182; 495–526.

5 The famous School of Oriental and African Studies (SOAS) in London was opened in 1916 as a part of the University of London, following the Reay Report, published in 1909, which underlined a necessity for "the training of persons who are going to the East or Africa, either for public service or private business." See "Report of the Committee Appointed by the Lords Commissioners of HM Treasury to consider the Organisation of Oriental Studies in London" (London: HM Treasury, 1909), quoted in Hourani, *Islam in European Thought*, 67.

on areas and regions their countries are connected with,[6] and at least career-paths with opportunities for scholars to explore Muslim countries attending colonial services. Many members of the colonial services exceled at collecting manuscripts, arts and crafts, coins, that were brought to western collections and museums. Scholars such as Christiaan Snouk Hurgronje, a lecturer at Leiden School for Colonial Civil Servants, had been visiting Mecca and later on Sumatra and Indonesia for years, half on behalf of his government, half on his own initiative, until he became a professor at Leiden University. We owe Hurgronje a collection of rare and outstanding late nineteenth-century photographs of Muslim pilgrims.

The overlap of academic and practical interests, moreover, led to a strong ethnographic tradition in the field, an enrichment of knowledge with biographical experiences and first-hand observations, on the one hand, personal judgements and interpretive patterns regarding contemporary Muslims civilization, on the other. The prominent standing of ethnological approaches and cultural anthropology in the UK and in France goes back to colonial times, when even officers or missionaries participated in reports on and inquiries into Muslim countries. France had even established centers for Arabic studies in Muslim cities such as at the University of Algier, though these advanced posts of oriental studies were dissolved after Algiers's independence 1962.[7] In the Netherlands, the Institute for Social Science, founded in The Hague in 1952, played a key role in establishing new approaches regarding developmental aspects in Muslim countries. This in turn confirms the traditional and contemporary relevance of the field with central institutions and comparatively tight connections between academic and non-academic elites.

In his widely discussed book *Orientalism*, Edward Said argued in 1979 that the French and British academic fields were not even collaborating in

---

6 Whereas the UK had expanded its Empire to cover India, Egypt, Sudan, and South East Asia, France was prevailing in Tunisia, Morocco, Algeria, and Lebanon, and the Netherlands were oriented towards South East Asia ("East India," Indonesia). Both had to rule territories with respective Muslim populations. Additionally, this influence in the Muslim world again grew immensely after World War I, when the UK and France received League of Nations mandates to govern several countries in the Levant such as Jordan, Syria, Iraq, and Palestine, that had formerly been ruled by the Ottoman Empire.

7 Although an ethnographic tradition in the French tradition of Islamic studies was strong, the first Institute for Islamic studies was founded only after World War II, when Sorbonne opened his *Institute d'Études Islamiques*. Further institutions in Aix-en-Provence, Boudeaux, Lyon and Strasbourg were established in or after 1962, as centers for research on Mediterranean regions.

parts of the colonial system in practical terms, but were confirming an inaccurate view of Muslims through fictional depictions of Muslim people and cultures as "exotic," "mysterious," "feminized," and "threatening"— always using binary concepts to describe the Orient as the non-European Other.[8] Using Foucault's approach to analyzing discourses, Said showed how orientalists' perceptions of the Orient were derived from fictional images of places or practices such as harems, bazaars, or Saracens as wild horsemen. As such they were adopting their concepts to what they were experiencing in the colonial services and evaluating Muslim realities from an administrator's perspective instead of reporting the realities of Muslim life in a neutral way, which might have been seen as the duty of the profession. In Said's view, authors such as Lane, Renan, or Massignon tended to legitimize Western colonialism by assuming an "inferiority" in Oriental civilizations and explaining their state of development by concepts that reduced the reality of Muslim people to a theoretically overestimated "essence," which in their view described the "Orient." Concepts of inferiority regulated Western perceptions and discourses, be they cultural or religious (as in the case of Massignon) or even racial (as in Renan's case).

Said's thesis of Orientalism has often been criticized for its controversial and accusatory style, for offering many historical interpretations in detail,[9] for not considering the many attainments of early Orientalists,[10] and for the relativistic consequences of its criticism, that attacks "ethnocentrism" but refuses to reveal its own point of view.[11] But certainly the debate on Orientalism opened eyes regarding hidden agendas in Islamic

---

8 Edward Said, *Orientalism* (New York: Knopf Doubleday, 1979).

9 Urs App, *The Birth of Orientalism* (Philadelphia: University of Pennsylvania Press, 2010).

10 The controversy went beyond a historical question because Said and his supporters suggested that a prolonged effect of Orientalism could be felt up to the present day and even in the United States, despite the US (except for its claims on the Philippines and other semi-colonial control over territories abroad) never having been a true colonial power. Among his critics were historians such as Bernard Lewis, "The Question of Orientalism," in: *New York Review of Books*, Vol. 29, No. 11, June 24th, 1982, 1–20; Bernard Lewis: *Islam and the West* (New York: Oxford University Press, 1993), 99, 118, and Islamic studies scholars such as Albert Hourani, *A History of the Arab Peoples* (Cambridge: Harvard University Press, 1991), Robert Irwin, *For Lust of Knowing: The Orientalists and their Enemies* (London: Allen Lane, 2006), and Ernest Gellner, "The Mightier Pen? Edward Said and the Double Standards of Inside-out Colonialism," *Times Literary Supplement*, February 19, 1993, 3–4 (review of Edward Said, *Culture and Imperialism*).

11 In Said's view, even the economic and political practices of contemporary Arab elites indicate that they have internalized Western concepts of the Oriental created by British, French and American orientalists.

studies, thus marking a point of no return because debate in the field could no longer ignore such aspects. The debate coincided with Said, who grew up in the US and was born in Palestine, personally incorporating the matter of how "insiders" can participate in Islamic studies and avoiding an outsider's view.

4. *Islamwissenschaft* in Germany

Islamic studies in Germany, by contrast, emerged and developed in a different way for almost a century. Since 1884 and 1918, Germany has had smaller colonial territories in Islamic areas, but lost these colonies after World War I. Similar to institutions in London or Paris, growing interests in colonial affairs had led in 1887 to the establishment of the Berlin *Seminar für Orientalische Sprachen* (Seminar for Oriental Languages, SOS) to improve the language skills of colonial servants. In Hamburg, the *Kolonial-Institut* in 1919 became part of the university there. In that city, Carl Heinrich Becker (later an influential Prussian minister of cultural affairs, 1925–30) was appointed professor of history and culture of the Orient in 1908, which was the first facility dedicated to the study of modern Muslim societies in Germany at that time.[12] Alongside Martin Hartmann, a former student of Fleischer and member of the Foreign Service and interpreter in Beirut with interests in current political and sociological issues, Becker tried to open German Islamic studies to sociological and socio-cultural approaches to study contemporary Muslim societies. Due to the breakup of colonialist interests and post-war difficulties in funding science in Germany, these approaches remained unsuccessful. With a few exceptions during the wars and the Third Reich, German *Islamwissenschaften* had no noteworthy political use or importance for postcolonial services or any involvement in military actions before the end of the twentieth century. The absence of any practical function in this field meant the persistence of philologically based study programs. Up to today it remains necessary to learn the Arab language thoroughly, as well as Persian, Turkish or Urdu in addition, to pursue a degree in *Islamwissenschaften*. This development prevented an early opening

---

12 Alexander Haridi, *Das Paradigma der "islamischen Zivilisation"—oder die Begründung der deutschen Islamwissenschaft durch Carl Heinrich Becker (1876–1933)* (Würzburg: Ergon Verlag, 2005).

of the field to questions regarding contemporary Muslim societies and slowed a shift from orientalist topics to social science approaches. It otherwise reflected and confirmed its deep roots in the German tradition of the humanities. The German concept of *Geisteswissenschaften* emphasized the peculiarities of a foreign world as a unique culture that followed its own rules that were unrecognizable without a "deep" understanding of that culture's language, history, and its key religious texts.[13] Analogous to other fields in the humanities, Islamic studies scholars at this time emphasized the need to understand the Islamic world on its own terms. This methodological concept forced candidates to work hard on classical texts and historical details in order to gain permission to work on larger projects, which, of course, remained the ambition of young aspiring scholars.

The methodological concept of *Geisteswissenschaften* strengthened a strong philological and historian approach, which for a long time had normally been focused on topics such as editorial projects, the many highlights of Arabic and Persian literature, Arabian science, arts and architecture and the political history of Islamic empires, from their beginnings after Mohammad's life up to the late Osman Empire, whose decline began only in the early modern age and was not complete with the early beginnings of Islamic studies. This configuration maintained a conception of the Muslim world which was mostly oriented towards the past and which usually refrained from looking at current societal developments. Therefore, within the field, emerging questions concerning the many historical and contemporary differences between oriental and occidental societies were rarely addressed. German Orientalism didn't have a theoretical and methodological framework at its disposal for comparing societies and cultures with respect to their economic, political, jurisdictional, and cultural/religious developments. Nor did it take much note of discussions about methodo-

---

13 The program of *Geisteswissenschaften* has been elaborated upon significantly by philosophers such as Wilhelm Dilthey or Heinrich Rickert, who pronounced the need to find adequate methods for the set of human sciences relating textual and language-based materials such as documents, books and cultural utterances, on which History, Philology and many other fields are usually based on. These methods are considered to be comprehensive, whether they are hermeneutical or psychological, in contrast to explanatory methods that are applicable in Sciences ("*Naturwissenschaften*"). The strong distinction between "*Naturwissenschaften*" and "*Geisteswissenschaften*" in the German tradition has often been criticized as inadequate, and the adequate classification of disciplines such as Psychology or the Social Sciences has always been a contentious issue. For an overview, see Julian Hamann, *Die Bildung der Geisteswissenschaften. Zur Genese einer sozialen Konstruktion zwischen Diskurs und Feld* (Konstanz: UVK, 2014).

logical approaches in other fields such as sociology, where Max Weber at least had suggested a framework based on his concepts of universal processes of "rationalization," asking systematically about its origins and its constraints in an analytical manner. Hence, while early German sociologists such as Weber avoided advanced statements due to their lack of knowledge, German orientalists for a long time did not address such questions at all due to their academic seclusion, which nonetheless was very productive due to their positivistic diligence.[14]

Edward Said had excluded German Islamic studies from what he called Orientalism due to its lack of political relevance.[15] He saw in the German tradition only a local branch of the French philological tradition, partly due to the prosopographic lineages of German professors who had studied Arab philology in Paris at de Sacy; its most successful German scholar, Heinrich Leberecht Fleischer, who dominated the field for over a half of a century, was professor at Leibzig University and in 1845 was a co-founder of the German Oriental Association.[16]

Historians of the discipline, such as Ursula Wokoeck, argue in another way. In her view, the German tradition of Islamic studies differs from others insofar as Germany's federal university system, despite its competition sourcing a relatively well equipped and nation-wide establishment of thirty-four professorships and institutes, did not allow Islamic studies to

---

14 On Weber's view of Islam, see Walter Sprondel "Max Weber's Protestant Ethic. The Universality of Social Science and the Uniqueness of the East," in *Recent Research on Max Weber's Studies of Hinduism*, edited by Detlef Kantowsky (München: Weltforum Verlag, 1986), 59–72; John Love, "Max Weber's Orient," in *The Cambridge Companion to Weber*, edited by Stephen Turner (Cambridge: Cambridge Univ. Press, 2000), 172–99; Benjamin Nelson, "On Orient and Occident in Max Weber," *Social Research* 43, no. 1 (1976): 114–29.

15 Said, *Orientalism*, 17–19. Meanwhile, many authors tried to show how the principles of Orientalism in Germany also were prevalent. Bernd Adam, *Saids Orientalismus und die Historiographie der Moderne: Der "ewige Orient" als Konstrukt westlicher Geschichtsschreibung* (Hamburg: Diplomica. 2013); Suzanne L. Marchand, *German Orientalism in the Age of Empire: Religion, Race, and Scholarship* (Cambridge: Cambridge Univ. Press, 2009); Andrea Polaschegg, *Der andere Orientalismus. Regeln deutsch-morgenländischer Imagination im 19. Jahrhundert* (Berlin: de Gruyter, 2004); Todd Curtis Kontje, *German Orientalisms* (Ann Arbor: Univ. of Michigan Press, 2004).

16 Concerning Ursula Wokoeck, Said could not recognize the German tradition as an authentic approach, because he would otherwise have had to acknowledge the branch to be an acceptable type of Western Islamic studies outside his understanding of Orientalism as always involved in imperialistic power. See Ursula Wokoeck, *German Orientalism: The Study of the Middle East and Islam from 1800 to 1945* (New York: Routledge, 2009), 9.

emancipate itself form its philological tradition to expand their approaches.[17] With reference to sociologist Rudolf Stichweh's conception of scientific disciplines, Wokoeck considers Islamic studies a minor discipline within the field of classical philology, to which it belonged from its very beginnings. The field could not achieve a status independent of linguistic agendas, an institutional and intellectual autonomy in driving an agenda in its own right, because of the needs of language training.[18] Unlike other main disciplines such as English studies (*Anglistik*) or the natural sciences including chemistry or physics, Islamic studies was not based on a secondary school curriculum, which would prepare students in the basic language skills and help them find the profession reflecting their interests.[19] It has to start from scratch, it must train all languages from the very beginning, attracting only gifted students familiar with classical languages; and therefore it does not have the opportunities to follow a self-stimulating research agenda that exceeds and is able to include any other approach than the philological.[20] This does not mean that Islamic studies in Germany would not have made steps toward embracing cultural and sociological studies. But they could not succeed. Due to these "limitations" in the discipline, authors have noticed a long-standing resistance in German Islamic studies to integrate social science methods to adequately approach questions regarding contemporary Muslim society.[21] But this exceptional position changed a little during the 1960s.

---

17 Ibid., 27 ff.

18 Until the seventies, Chairs for Islamic studies were usually called Chairs for Semitic or Arab languages.

19 Rudolf Stichweh, *Zur Entstehung des modernen Systems wissenschaftlicher Disziplinen: Physik in Deutschland, 1740–1890* (Frankfurt: Suhrkamp, 1984).

20 "Although the modern German research tradition in Middle East studies began in the first half of the nineteenth century, an investigation of university positions reveals that not even a single chair for Middle East Studies, defined as Arabic and Islamic studies, existed at any German university throughout the nineteenth and the first half of the twentieth centuries. In other words, Middle East studies were not a discipline. [...] In institutional terms, Middle East studies were part of university positions which were initially defined as morgenländische/orientalische Sprachen (Oriental languages), potentially including the entire "Orient", i.e. Asia and Africa, in practice mostly languages from a region stretching from the Middle East to India." Wokoeck, *German Orientalism*, 2.

21 Ursula Wokoeck, German Orientalism, 10; Jacques Waardenbourgh, *Muslims as Actors: Islamic Meanings and Muslim Interpretations in the Perspective of the Study of Religions* (Berlin and New York: de Gruyter, 2007).

At that time, Islamic studies had to respond to pervasive transformations in Muslim countries following decolonization, the rise of Arab and Turkish nationalism, and the seeming loss of significance of Islamic religious traditions. Some Islamic studies scholars such as Fritz Steppat, Jörg Krämer, and Gustave E. Grunebaum in Los Angeles, opened their work up towards contemporary Muslim societies.[22] In the 1970s the field came under the influence of the social sciences. A few left-wing students went into the field to pursue their interests in the Israel-Palestinian conflict, and initial projects included investigations into contemporary Muslim migrant milieus in Berlin, while some Scholars such as Baber Johansen opened up their agenda towards contemporary and sociological subjects such as Islamic law (*Fiqh*). This did not transform the field as a whole, but encouraged new approaches only as subsidiaries. The efforts of Steppats and others led to the foundation of the Center for Modern Oriental Studies in Berlin in 1980, which since 1996 was established as a research institute outside the university.[23] Relating to the religious and cultural aspects of Islam, the field remained oriented towards the past. In the aftermath of the secular zeitgeist of 1968 only a minority of scholars were considering the political role of Islam and or even religion in the present. After the oil crisis brought about by the Organization of the Petroleum Exporting Countries (OPEC) in 1973, debates on post-colonial Muslim countries were related to the question of whether or not post-colonial countries were driven by "neo-patrimonial" interests of rent-seeking economies.[24]

Hence German Islamic studies more or less remained shielded from the need to address political controversies until the nineties. Thus when the renowned scholar Annemarie Schimmel received flak for her com-

---

22 Fritz Steppat, "Die arabische Welt in der Epoche des Nationalismus," in *Geschichte der arabischen Welt*, edited by Franz Taeschner (Stuttgart, Kröner 1964), 178–236. Jörg Krämer, "Der islamische Modernismus und das griechische Erbe," *Der Islam* 38, no. 1 (1962): 1–26. Gustave von Grunebaum, "Das geistige Problem der Verwestlichung in der Selbstsicht der arabischen Welt," *Saeculum* 10 (1959): 289–328.

23 Meanwhile the Centre organizes international research and exchange programs. For further information see "Zentrum Moderner Orient," https://www.zmo.de/.

24 Primarily discussed by Hossein Mahdavy, "The Pattern and Problems of Economic Development in Rentier States: The Case of Iran," in *Studies in the Economic History of the Middle East. From the Rise of Islam to the Present Day*, edited by Michael Cook (London: Oxford Univ. Press 1970), 428–67; see also Shmuel N. Eisenstadt, *Traditional Patrimonialism and Modern Neopatrimonialism* (Beverly Hills: Sage, 1973); Peter Pawelka, *Herrschaft und Entwicklung im Nahen Osten: Ägypten* (Heidelberg: Müller Juristischer Verlag, 1985), 22–97.

mentaries on Ayatollah Khomeini's fatwa against Salman Rushdie in 1995, in which she expressed her sympathy for Muslim's indignation about the *Satanic Verses*, she argued that she was non-political as a scholar as well as as a person, and thus excused herself for some of her mistaken comments to get out of the line of fire.[25] But according to some interviews I have conducted with younger scholars and to which I will turn in a moment, this incident marked a turning point in the field as academics began to realize the impossibility of an empathic non-political statement.

5. Middle East Studies in the US

In the US we find a third model for Islamic studies. Here, the study of Islam is embedded in different institutional contexts. Originally established as Orientalist studies at the same time as in Europe, with the foundation of the American Oriental Society in 1842, the study of Islam developed more and more towards area studies after World War II. So-called "Islamic studies" programs do not exist, with new exceptions at the University of California Los Angeles, Hartford, Michigan or Georgetown. Study programs based on classical philological training can be found in departments of Near Eastern languages and civilizations at older Universities such as Harvard, Princeton, Yale, Columbia, Chicago or Pennsylvania.[26] And, finally, studies of Islam are practiced at departments of religious studies, where Islam is taught alongside other religions on the basis of classical texts (*Qur'an, Hadith*), departments of the history of arts and architecture, or at faith-based study programs in divinity schools. But the main approach is based on area studies.

25 Arno Widmann, "Ein Friedenspreis für die Zensur?," *Zeit*, Sept. 15, 1995; Ludger Lütkerhaus, "Zum Weinen. Eine Antwort auf Annemarie Schimmels Rushdie-Schelte," *Süddeutsche Zeitung*, May 11, 1995; Wolfgang Frühwald. "Es riecht nach Hexenjagd," *Rheinischer Merkur*, June 9, 1995; Ariane Müller, "Unglücklich über dieses Buch – Friedenspreisträgerin Schimmel präzisiert im Fall Rushdie," *Frankfurter Rundschau*, May 11, 1995; Gernot Rotter, "Das Weinen der Muslime," *Zeit*, May 11, 1995; "Der Friedenspreis: Ein verdeckter Angriff. Annemarie Schimmel und Gernot Rotter über den Friedenspreis, Salman Rushdie und Taslima Nasrin," *Spiegel*, May 22, 1995.

26 Higher Education Funding Council for England (HEFCE), "International Approaches to Islamic Studies in Higher Education. A report to HEFCE" (Subject Centre for Languages, Linguistics and Area Studies, June 2008), 39, http://www.hefce.ac.uk/pubs/rereports/year/2008/islamicstudiesinheintl/.

According to the interdisciplinary concept of area studies, academic debates in the US are much more oriented towards geographical regions, monitoring them with respect to contemporary societal and political aspects.[27] In general, area studies bring together multiple perspectives from different disciplines such as history, the social sciences, economics, and anthropology. In the US, academics traditionally are more used to being confronted with high expectations from politicians and funders of think tanks that they work with on specific topics relevant to the public because of ongoing political developments. While the United States are geographically distant from Arab homelands and have never had colonies in the way European countries have, the US has been involved in many conflicts in the Middle East since 1945.

Due to the important role of the Israeli-Palestinian conflict for US foreign policy, Middle and Near Eastern studies tend to assume either a more pro-Palestinian or a more pro-Israel perspective. Broadly speaking, academic autonomy in the US is the result of an ongoing political struggle for balance within the academic field to represent the subject adequately, whereas in Germany it is the result of claims of a non-political perspective by the academy at large.

Another difference derives from the importance of funding. In contrast to many European countries, most leading universities draw on private contributions and enrollment fees. With respect to Islamic studies, it might be important to note that the US has never seen a large immigration of Muslim workers' families as has occurred in Europe. Muslim families in the US are strongly connected with former Iranian or Arab elites and often wealthy and well-educated. Therefore these milieus play an important role as donors to university programs in funding chairs and stipends in Middle East studies. Increased student demand has led to a variety of programs. Subsequently the academic system in the US seems to be more flexible with regards to arrangements to integrate new approaches in Islamic

---

27 As is generally known, Area Studies are strongly rooted in the US history and came up in the post WW II context when the US policy realized a lack of academic knowledge and expertise concerning areas in Asia, Africa or the Middle East, the US had to deal with due to the upcoming Cold war. Private Companies such as the Ford foundation, the Rockefeller Foundation and the Carnegie Corporation pushed programs (FAFP), in order to support areas studies trainings. See David L. Szanton, "The Origin, Nature and Challenges of Area Studies in the United States," in: *The Politics of Knowledge: Area Studies and the Disciplines*, edited by David L. Szanton (Berkeley: Univ. of California Press, 2004), 10–11.

studies such as trans-regional subjects, Islam in South-East Asia or the study of Islam in the West.[28] It also offers more opportunities for academics with a Muslim background, either domestically or coming from abroad, to pursue their careers within the field.

6. Transformations in the Political Context for Islamic Studies

Islamic studies are well suited for a comparison of national branches regarding legitimizing sciences due to the strong political impact on the field which emerged through the ongoing contemporary debates about Islam and "Islamism." This political impact differs from nation to nation with respect to a western country's policy and preliminary relationship to the Muslim world, but in its entirety it is quite similar, and we can see how nation-specific responses follow a common sequence of internationally meaningful events and developments.

A brief historical mapping of incidents and current Islam-related debates will help recall the significance of their political impact. The Iranian Revolution in 1979 at the very least raised the west's public awareness of the power of new Islamic movements that emerged during the seventies as a result of the failure of postcolonial regimes in Iran, Egypt, and several Arab countries. Like the Egyptian Muslim Brotherhood or the Taliban or Salafi movements later, Ayatollah Khomeini and his supporters placed a strong emphasis on Islam as a source for political reform in Muslim societies, whether the movement is Sunni or Shiite. Islamist's movements differ from country to country but address unsolved problems of tribal inequality, corruption in the elite, and social injustice quite similar to the reinforcement of Islamic jurisdictions in order to reestablish a legitimized authority based on one's personal behavior, in keeping with the *Koran* and Mohammad's advice.[29] Many events during the eighties and nineties offered examples of smoldering conflicts within Muslim countries as well as

---

28 Examples are the Abbasi Program at Stanford and the Prince Alwaleed Bin Talal Programs at Georgetown University and at Harvard University.

29 Islamistic movements draw upon moderate intellectuals such as Dschamal ad-Din al-Afghani (1837–97), Muhammad Abduh (1849–1905), Raschid Rida (1865–1935), Sayyid Abul Ala Maududi (1903–1979) or even more radical Intellectuals such as Sayyid Qutb (1906–1966).

between Islamic movements, Western countries, and Israel. The assassination of Egypt's Anwar El Sadat in 1981, suicide bombings after the 1982 Lebanon War, Khomeini's Fatwa against Salman Rushdie in 1989, the role of Hamas in Palestinian's intifadas against Israel after 1993, the Taliban's destruction of the Buddha monuments of Bamiyan in 2001, and even the Danish cartoon disputes over the depiction of prophet Mohammad's image in print, all prompted anxious public debates about the power and meaning of Islamic movements.

Conflicts were growing more and more acute, and suicide bombings in New York and Yemen pointed to a change of strategy by terrorist groups within Islamic movements as they began attacking western facilities and personnel because of their support for the Arab regimes.[30] This development culminated in the terrorist attacks on September 11, 2001, followed by a western coalition's decision to invade Afghanistan and Iraq in 2001, and subsequent terrorist attacks in Madrid (in April 2011) and in London (in July 2005). While military conflicts dominated public debates for over a decade, the so-called Arab Spring and revolutions in Tunisia, Egypt, Libya, and elsewhere after 2010 suggested to western audiences that these countries may be turning a corner. Debates shifted form a strong focus on the Islamic menace towards democratic developments within Muslim societies. But after years of civil war in Syria and roll-backs in Egypt (2013) or Tunisia (2015), the focus shifted once again. Today, the Iraq and the Syria conflicts have taken center stage and military successes by the so-called "Islamic State" (ISIS) in Iraq and Syria seem to have compounded all these different conflicts into a greater and indeed more dangerous conflict zone that connects Turkey, Iran, and Jordan to Israel and Lebanon, and European heartlands too due to new terrorist attacks such as in Paris in 2015 (Charlie Hebdo) or through the mass movement of waves of refugees.

This course of events was accompanied by several domestic political debates about the aftermath and attendant circumstances of the long-standing migration of Muslim families into western societies since the late sixties. Here, the US differs from European countries due to the amount, and social stratification, of Muslim immigrants.[31] Such debates, including

---

30 For this interpretation see Guido Steinberg, *Der nahe und der ferne Feind. Die Netzwerke des islamistischen Terrorismus* (München: Beck, 2005).

31 In the US, the Census Bureau does not track religious affiliation, so estimated numbers of Muslims population differ from 3 to 7 million. According to the Pew Research Center (2011), in 2010 5.25 million Muslims lived in the Americas, with a high percentage

controversies about Islamic headscarves in French or German schools in July 2004,[32] Mosques in urban neighborhoods, forced marriages or honor killings of Muslim women in 2005,[33] as well as the emergence of political parties and right-wing populism critical of Islam or even openly anti-Islamic,[34] all made clear that Islam was no longer a topic of foreign affairs but also affected western ways of living. The debates made clear the right of Muslim citizens to practice their religion as well as their duty to accommodate these practices to the law of secular states. And yet domestic political debates and debates on Islamic terrorism often became combined and confused, damaging an open-minded atmosphere of discourse.

Governments and local authorities had to reconsider their immigration policy to address the many problems of everyday life. These include language problems in schools, inheritance law, funerals, legal assistance, Muslim holidays, observance of dietary requirements, and citizenship. Local civil services or police units employ staff and social workers with a Muslim background. States such as Germany invited universities to create new departments of Islamic theology so as to provide academic training programs for imams and high-school teachers.[35] Therefore governments have to come to terms with representatives of Muslim communities.[36]

(45%) of American born converts, most of them African-American, and an increasing immigration rate of Muslims from South East Asia, Turkey, and the Arab world. In contrast, 44.1 million Muslims (or 6 percent of the total population), most of them in Russia and the Balkans, lived in Europe. France counts a Muslim population of 5–6 million (8.2 percent), many of them from Maghreb and Africa, Germany places a number of 3–4 million, 2.3 of them of Turkish and Kurdish origin. See http://www.pewforum.org /2011/01/27/the-future-of-the-global-muslim-population/. According to the United Kingdom Census 2001, 2.7 million Muslims lived in the UK (or 4.5 percent of the population) and most of them were from Pakistan and Bangladesh.

32 Clémence Delmas, *Das Kopftuchverbot in Frankreich. Ein Streit um die Definition von Laizität, Republik und Frauenemanzipation* (Frankfurt: Lang, 2006); Katharina Haupt, *Verfassungsfragen zum muslimischen Kopftuch von Erzieherinnen in öffentlichen Kindergärten* (Frankfurt: Lang, 2010).

33 Selen A. Ercan, "Same Problem, Different Solutions: The Case of 'Honour Killing' in Germany and Britain," in *"Honour" Killing and Violence: Theory, Policy and Practice*, edited by Aisha K. Gill, Carolyn Strange, and Karl Roberts (London: Palgrave Macmillan, 2014), 199–218.

34 Wodak, Ruth, Brigitte Mral and Majid Khosravinik, editors. *Right wing populism in Europe: politics and discourse.* (London: Bloomsbury Academic, 2013).

35 In German public schools, faith-based instruction in separate classes for Catholic and Protestant students had been offered all along, and these classes were now to be complemented by optional instruction in the Islamic faith. Centers for faith-based Islamic studies in Germany were established at the Universities of Frankfurt,

To reduce all these debates to a common denominator one can say that Islam-related topics gradually became a "hot issue" in western public debates since 2001. This had significant consequences not only for the relationship between Muslims and non-Muslims but also for Islamic studies.

7. Consequences for the Field

Some of these consequences are obvious and easy to describe. (1) The field had to respond to growing public interest, as academics were asked by governments and the media to provide their expertise to the public. The more journalists or politicians were asked for advice the more the field had to respond with analyses. Therefore, the field had to become acquainted with specific issues such as, for instance, strategy debates of Islamist groups, their recruitment via the internet or new political and secular movements upcoming with the Arab Spring.[37] (2) As a result of shifting debates, the field had to adapt its agenda more and more towards contemporary topics. Therefore it began to merge with the field of Near East and Middle Eastern studies in the political sciences, on the one hand, and with social sciences interested in Muslim migration or new modes of religious faith, on the other. Both disciplines began to contribute to Islam related topics significantly from the 1980s. A consequence for Islamic studies was the need to compete with and to legitimize its traditional approach includ-

---

Münster/Osnabrück, Erlangen, and Tübingen. The Government promotes the process with an amount of twenty million euros, following recommendations by the German Research Council. See *Wissenschaftsrat* (German Research Council), "Empfehlungen zur Weiterentwicklung von Theologien und religionsbezogenen Wissenschaften an deutschen Hochschulen" (Drucksache 9678-10, Berlin, Aug. 2010), http://www.wissenschaftsrat.de/download/archiv/9678-10.pdf.

36 In Germany, the federal government in 2010 founded an *Islam-Konferenz* to provide a platform for Muslim representatives. See Reinhard Busch and Gabriel Goltz, "Die Deutsche Islam Konferenz – Ein Übergangsformat für die Kommunikation zwischen Staat und Muslimen in Deutschland," in *Politik und Islam*, edited by Hendrik Meyer and Klaus Schubert (Wiesbaden: VS Verlag, 2011), 29–46.

37 For an overview, see Rüdiger Lohlker, ed., *New Approaches to the Analysis of Jihadism* (Wien: Vienna Univ. Press, 2012); Rüdiger Lohlker, ed., *Jihadism: Online Discourses and Representations* (Vienna: Vienna Univ. Press, 2013); Tilman Seidensticker and Hans G. Kippenberg, *The 9/11 Handbook: Annotated Translation and Interpretation of the Attackers' Spiritual Manual* (London: Equinox, 2006).

ing its strong philological training and knowledge of history and culture. Meanwhile, academics such as Juan Cole, John Esposito, James Galvin, as well as their German colleagues Gudrun Krämer, Navid Kermani, or Guido Steinberg became prominent spokesmen of Islamic studies, because they represented a type of scholar who contributed publicly on both the original working fields of Islamic studies, and political debates on current incidents that are controversial and pushed by the media. But considering the academic field as a whole, tensions and restrictions in the development of the field's agenda were inevitable. Against this background, scholars in the US on average have tended to adjust their working agenda to upcoming political debates, while the German *Islamwissenschaften* try to continue to work on their own academic agenda, which is not necessarily, but often is, without obvious relevance for actual processes of political decision making.[38] (3) But shifting debates towards the contemporary also challenged the institutional settings. While in the US, Middle East studies encourage interdisciplinary approaches, the German *Islamwissenschaften* have not had the facilities or the framework to thoroughly integrate interdisciplinary approaches in university research and teaching due to its philological teaching requirements. So contributions to contemporary topics remain a special service. Similarly, the field had to adapt its professional association and communication to the need to discuss new and an increasing quantity of topics. While the Middle East Studies Association (MESA) founded in 1966 provided a framework for interdisciplinary debates in the US, many German researchers from outside the *Islamwissenschaften* felt a lack of adequate opportunities to discuss contemporary issues in the field's long-standing *Deutsche Morgenländische Gesellschaft*, which had been founded in 1845. They decided to create the *Deutsche Arbeitsgemeinschaft Vorderer Orient* (DAVO) in 1993, an association for the investigation of contemporary Muslim societies. Although grievances about deficient language skills and a lack of historical knowledge are prevalent, both groups began to cooperate. Meanwhile each conference of the DMG holds panels on contemporary topics too. (4) The field had to continue to work on its own

---

38 In recent years, prominent research projects in German *Islamwissenschaft* have been DFG-Research Kolleg, "History and Society during the Mamluk Era (1250–1517)," University of Bonn; "Corpus Coranicum," a research project at the Berlin-Brandenburg Academy of Sciences and Humanities, working towards a critical edition of the Quran under the direction of Prof. Dr. Angelika Neuwirth at the Free University of Berlin. The project is currently funded through 2025.

agenda to maintain a better understanding of Islamic cultures by delving into the specific history and diversity of its cultures and of Islam's denominations. But on the other hand, and against the background of political controversies, the field had to explain and communicate this diversity to the public to maintain an open-minded atmosphere of discourse. This had to be done to preserve its own intellectual integrity so as to publically claim the academic's role of working vicariously on such subjects in a professional scientific manner.

But with respect to this role of a vicariously active field of Islamic studies it also had to reconsider the adequacy of some of its previously formed history-related concepts such as the "Jihad", which had mostly been interpreted as elements of the early beginnings of Islam but not as applicable or transferable to contemporary political movements.[39] In order to make sense of the new Islamic and militant movements, the field had to revise its interpretative patterns and recognize the return of religion as a creative power in contemporary political and economic affairs. Hence political controversies activated scientific debates and had repercussions in the manner in which the field interpreted contemporary concepts.

(5) In the context of debates about Jihad or Sharia, scholars were frequently asked to participate in panel discussions, talk shows or to write popular articles for newspapers and cater for a flourishing book market. Scholars thus exposed themselves to non-academic forms of criticism and had to deal with political suspicions of being partisan or personally non-engaged. Their role as professional researchers demands a strong attitude of academic neutrality and freedom from value judgments, but this does not mean avoiding "hot issues" because of the fear of becoming involved in struggles or being misunderstood or even alienating themselves from academic circles.

The field had no option but to answer these requests. Otherwise it would have risked being called into question at all, and losing its standing within the already hard-pressed field of the humanities, whose branches were all competing for jobs, third-party funds and the public's approval at not being some irrelevant ivory tower.

---

39 Albrecht Noth, *Heiliger Krieg und Heiliger Kampf in Islam und Christentum* (Bonn: Röhrscheid, 1966); Patrick Franke, "Rückkehr des Heiligen Krieges? Dschihad-Theorien im modernen Islam," in *Religion und Gewalt. Der Islam nach dem 11. September*, edited by André Stanisavljevic and Ralf Zwengel (Potsdam: Mostar Friedensprojekt, 2002), 47–68.

(6) The field had to recognize the increase in numbers of students, in particular among those with a Muslim migration background, as a consequence of both that migration itself and the persistent focus on Islam-related issues in the public debates. This led to new modes of teaching as well as partial shifts in students' motives. To put it bluntly: While non-Muslim western Orientalist studies scholars had in the past only studied Oriental arts, languages and history as part of an entirely strange hemisphere, and were driven by an intellectual curiosity and gift for languages, Muslim students and scholars of Islamic studies today have personal roots and relatives in Muslim countries. As a consequence, the Islamic religion is no longer merely an intellectual subject of enquiry and curiosity. This new generation of students brings in new perspectives and interests based on or informed by their personal experiences of and intimacy with two cultural hemispheres. Due to their careers, Muslim scholars of Islamic studies are no longer part of the subject-matter but are colleagues working actively on it.

(7) Islamic studies also had to cope with changes in its academic environment due to the emergence of the discipline of faith-based Islamic studies at universities. In recent years, western countries felt the need for new modes of collaboration with Muslim communities to integrate the religious education of Muslims into the public education system in order to encourage academic standards within theological debates and teaching. The Siddiqui-Report in the UK mentioned that "there is a need for the community and the universities to find ways to co-operate and collaborate in order to widen the influence of higher education among Muslims."[40] While US universities have had a long tradition of teaching Islam-related topics within its divinity departments, while avoiding teaching imams, some UK universities started collaborations with Muslim institutions to offer a range of graduate development courses.[41] Recently, some German universities began to establish departments of Islamic Theology and faith-based Islamic studies at the Universities of Tübingen, Münster, Erlangen, Frankfurt/Gießen, Osnabrück, partly in cooperation with the Turkish Presidium of Religious Affairs. The process were assisted by the Federal Government

---

40 Ataullah Siddiqui, "Islam at Universities in England: Meeting the Needs and Investing in the Future" (DFES, London, 2007), 35, http://dera.ioe.ac.uk/6500/1/Updated%20Dr %20Siddiqui%20Report.pdf.

41 For instance the Universities of London, Gloucestershire, Middlesex. See HEFCE Report, "International Approaches to Islamic Studies in Higher Education" (2008), 16.

and accompanied by the Council of Science and Humanities (*Wissenschafts-rat*).[42] The process was disputed both in Muslim Communities due to a lack of confidence in the newly appointed professors, who were seemingly teaching their own version of "Euro-Islam" aside from traditional doctrines,[43] and in the academic *Islamwissenschaft*. Some Orientalists contributed critically, for instance, to the question of who might have the legitimacy to decide what a true Islam is, or about the questions of whether Islamic theology should have its own faculty or be integrated into the philosophical faculties, as previously with the departments of the humanities, in order to guarantee academic standards in the new field. And some Academics feared that *Islamwissenschaften* could have difficulties distinguishing between academic secular *Islamwissenschaften*, on the one hand, and faith-based Islamic theology, on the other, because both were called "Islamic studies." For the first time in history, professors published an open letter including a statement on behalf of the German Orientalist Association (DMG), in which all these criticisms were summarized.[44]

These consequences are more or less obvious and easy to describe, but they make it clear that Islamic studies is in the middle of a profound change, which challenges and in some respects overburdens the field due to its traditional limits and institutional possibilities. Islamic studies have to persist with their own scientific agenda in continuing the study of Islam's history and languages. But due to recent developments the field has had to expand its working fields to include new developments in Arab, Turkish or Persian Muslim countries as well as new Muslim migrant milieus in western

---

42 The process is well documented and comprehensible. See the Council's webpage: http://www.wissenschaftsrat.de/presse/veranstaltungen/islamische_studien_in_deutschland_tagung.html.

43 Conflicts arose for instance in Münster on Prof. Dr. Sven Kalisch and on his successor Prof. Dr. Mouhanad Khorchide, who were not accepted by several Muslim organizations. See Koordinationsrat der Muslime (coordinating committee of Muslims), "Stellungnahme mit Gutachten des KRM zum Münsteraner Islamlehrstuhlinhaber Mouhanad Khorchide," http://www.koordinationsrat.eu/detail1.php?id=138&lang=de; see also Mouhanad Khorchide and Klaus von Stosch, *Herausforderungen an die islamische Theologie in Europa/Challenges for Islamic Theology in Europe* (Freiburg im Breisgau: Herder, 2012); Mouhanad Khorchide and Marco Schöller, *Das Verhältnis zwischen Islamwissenschaft und islamischer Theologie. Beiträge der Konferenz Münster, 1.–2. Juli 2011* (Münster: Agenda, 2012).

44 Almost all colleagues signed the letter. See "Stellungnahme von Fachvertreterinnen und -vertretern der Islamwissenschaft und benachbarter akademischer Disziplinen zur Einrichtung des Faches 'Islamische Studien' an deutschen Universitäten." http://www.dmg-web.de/pdf/Stellungnahme_Islamstudien.pdf.

capitals or the cultures of Indonesia, Malaysia, India or Pakistan, which in terms of population and economic impact are no longer peripheral. At the same time, Islamic studies' scholars have to contribute to public debates on Islam-related topics, whether these debates are initially triggered by foreign policy, war and terror incidents or domestic controversies. Since the public and the media expect to receive expertise, comments and analysis from Islamic studies' scholars, the field has had to respond to it. Therefore, Islamic studies are at the forefront of a permanent struggle, in which academic disciplines are challenged with claiming scientific standards in publically dealing with a subject, which refers to societies' value judgments, beliefs, anxieties and controversies.

But there are other elements of this transformation which are not so easy to describe and which remain far more hidden. These consequences concern the attitudes and feelings of academics in dealing with the political transformation of their own field. My own interest in these habitual transformations led me to conduct interviews with Islamic studies scholars in the US, in Switzerland, and in Germany, which I will focus on in what follows.

8. Habitus Transformation and Professional Behavior

One of my interviewees, a European Islamic studies scholar currently lecturing at an American university, described his response to the terrorist attacks in the US on September 11, 2001. I will focus on what he has to say after this long excerpt from our interview:

I'm Irish as you can gather (I: mhm) and and I grew up in [...] (I: mhm) so I spent part of my Ph.D. time in Berlin in fact and in kind of among kind of Turkish Afghan Berlin and part of it in London and very much multicultural London and the time of my most of my Ph.D. in the late nineties felt sort of a wonderful time of kind of multicultural jollying and then suddenly with nine-eleven things (I: mhm) really that day then for the several years afterward really did feel like a very big turning point based in in sort of in the general . communities I was moving among you know (I: mhm) and also for my with my own academic life as well I mean suddenly everything I did # I mean throughout the nineties what I had worked on seemed to be kind of irrelevant I think would think well why would you bother doing that well maybe when I was doing my Masters I thought oh you might become a diplomat that was the only (I: mhm) possibility but wanting the Ph.D.

people thought was just indulgent and it kind of was you know there's no reason for it (I: mhm) but then you know as soon as nine-eleven happened it almost flipped the other in a negative sense that it I seemed to be suddenly asked about to explain and and justify and you know it was almost as if can you justify this can you you know ah as if I were somehow meant to be (.) sort of explaining the Muslim perspective as a scholar (I: mhm) and you know ahm for someone you know just coming out of the Ph.D. you know I mean technically I finished in 2002 but really you know my first academic job kind of my first research position was actually September you know kind of 2001 (I: mhm) so going straight into that at that kinda point (I: mhm) was I think was actually kind of difficult psychologically there's sort of a lot of burden to to to explain (I: mhm) I mean it was a massive turning point I mean you can see even now ten years later I don't quite know how to make sense of it.[45]

I would like to highlight two points. (1) The interviewee expresses his uncertainty in dealing with the biographical coincidence of his career's beginnings as an academic and 9/11 and its aftermath. Against his relatives' skepticism he decided to take the risk of an academic career and so follow his passion for contemporary Islamic cultures. At this point, the public shifted seamlessly from indifference concerning Islam-related topics to overwhelming Islamic studies' scholars in pursuit of explanations. Friends and colleagues outside the subject area were constantly looking for the chance to address their own needs by discussing the political background. Due to his passion for Islam-related topics, the interviewee was not only asked to explain things as an expert but to justify them. This meant that he had to cope with the subliminal expectations of people who suspected him of being sympathetic to Muslims or Islam or opposing a standpoint critical of Islam. He found himself being asked for his loyalty on this issue. At this critical phase of his career he didn't yet have the standing to be able to respond to this demand adequately. The only way to continue was to work and remain secluded from public debates. As he mentioned, this tenuous situation went on for about ten years. Obviously he eventually made landfall by getting a permanent post in his academic field.

(2) The second point is that, as the interviewee mentions, he needed ten years "to make sense of it," i.e. the aftermath of 9/11 and 7/7. So what he is saying here is that he had difficulties arriving at an adequate interpretation of the event. 9/11 had confronted him with aspects of Islam that he

---

45 Interview with Irish Islamic studies scholar, no. 1, Sept. 2012. Dots indicate a pause, the number sign a break, and "I:" stands for interviewer.

couldn't integrate into his own conception and former understanding of his subject-matter. He was attuned to working on urban Muslim milieus in Asian cities as well as in western capitals, and he was interested in culturally mixed neighborhoods in which new modes of cultural life emerge in the form of cooperation, cuisine or religious exercises and thus as an expression of mutual assimilation. 9/11 and its aftermath called into question the conditions for the opportunity for such a cultural coexistence and brought up the crucial issue of terrorists' authenticity in claiming Islam's concept of Jihad. So the interviewee had to come to terms with some destructive aspects of his subject-matter that he had not confronted before.

Until you know nowadays I think I feel more . secure (I: mhm) (I: mhm) and I think I'm confident (I: mhm) but I think actually to be honest I think it's taken myself kinda ten years to really I think probably clarify my position I think kinda nine-eleven I think for many people was was and then the all the other things that came afterwards it's not as if it was one event . ah I think takes kind of a . a lot of processing really because as I say on the one hand there's been a position of people wanting to explain it but you know you could well how far do you go towards not say like justify an event but kind of let's say in giving a positive spin on it (uv.) saying yeah you can't condemn a whole civilization because (I: mhm) (I: mhm) but you know one goes down that point now I'm I think I'm in a position when you know I don't necessarily feel the need to do that I actually feel that my position is much more to say well I want to explain the dark side rather than say (I: mhm) probably like many Middle Eastern studies people were saying well hang on you know my job is to go against Fox News or something (I: mhm) my job is to give the other side and ah you know and I I I think probably now I feel kind of .. I feel you know enough professional stability and perhaps confidence to say well you know I I don't wanna # I don't feel the need to be among the defenders of Islamic civilization (I: mhm).[46]

In this passage the interviewee's thoughts take an interesting turn. At present, he has found a stable position due to the success of his career; he has a social position, which gives him stability, and on the basis of this he is able to clarify his professional attitude. He doesn't condemn those Middle East studies colleagues who are willing to oppose the perceived wrongs and unfairness of FOX News' public agitation against Muslims. But he feels he is not in a position to be able to fight primarily for a balance in political statements on these issues. As a professional scientist he doesn't want to lose himself in such educative projects and in some respects he feels that defending Muslims on behalf of universalistic values such as

---

46 Ibid.

humanity, fairness and truthfulness might be taxing scientific standards and autonomy if this leads to ignorance of issues or aspects which might be part of the big picture and cause complexity to be omitted and ignored in public controversies. So if he mentions that he "wants to explain the dark side", he is looking for the bigger picture and doesn't want to ignore aspects that nobody can identify with. However, this doesn't mean that he has changed his allegiance. In my view he just intended to approach the difficult questions that arose after 9/11 and does not want to omit or ignore them. Therefore he exemplifies a scientist's role, which is to be dedicated to the ongoing task of understanding the issues in question that are hard to deal with and even critical for a society facing divisive controversies. The interviewee thus maintains a third position in order to sustain a regulatory idea of truth.

Let me now turn to passages from an interview with a German Islamic studies scholar; a university professor in northern Germany, the same age as the previous interviewee but working in a history-related field. I have selected passages in which the interviewee describes his experiences after 9/11.

I felt this as a challenge and how can I say I experienced situations to whom I felt entirely displeased with my own responses. After 9/11, I felt really terrible about Islamic studies, about the lack of an Islamic studies community, in which one could exchange one's positions, but that didn't happen. And this didn't happen even in our institute, and somehow my director said, "Well, you know there is much need for discussion of the issue in the public, let's organize a panel discussion. And then, we appeared on that Panel but we didn't have compared notes with each other on how to respond on it. And everybody was actually alone with his/her thoughts concerning 9/11 and what had happened. And I was totally suppressed by that at that time. I must say for days I couldn't do anything with it, because in my feeling it had to do something with me. And what did I have to do with this Islam? So, after 9/11 I was really at a loss with it. And then suddenly, I was supposed to speak on a panel discussion eight days later (…) and in this situation I felt enormous difficulties to quickly get a grasp of it and it wasn't clear to me how I should deal with it, because the only register that was available to me were somehow like reflexes in how to respond as an Islamic studies Islamic studies scholar to specific questions, they were simply so routinized and were not in line with what had happened. So as a start I had to realign myself completely.[47]

---

47 Interview with German Islamic studies scholar, no. 2, May 2012, my translation. German original: "Also ich habe das als eine Herausforderung empfunden und äh hab da auch sehr viele ähm wie sagt man ja ich bin da einfach ich habe da durchaus Situationen

At first sight one notices the concordance between both cases. 9/11 was a shock for the German Islamic studies scholar too. He worked under similar conditions as a postdoc; public requirements came suddenly and found him unprepared. He wanted to answer but couldn't find an adequate interpretation. It was a real crisis, if by crisis we mean the collapse of routine, here it constituted the collapse of interpretative patterns, as expressed in the words "register" and "reflex". And in his view the crisis affected not only him, but the entire discipline. And finally he felt the need to address his person feelings. It was a crisis of his biographical concept of an Islamic studies scholar's professional life as well as of his own concept of Islam. As a matter of course, he knew as a scientist that violence was part of Mohammad's life as well as of subsequent Islamic conquerors. But he obviously didn't expect to get roped into contemporary conflicts personally and be confronted with the ugly, "dark side" of his subject-matter. In contrast to his colleague above, the German scholar seems to be more horrified by Islamism than by the public's hysterical response to it.

---

erlebt wo ich überhaupt nicht zufrieden war hinterher mit dem wie ich darauf reagiert habe. (I: mhm) ja also gerade nach dem 11. September (I: mhm) das war ganz schrecklich also ich hab das wirklich als schrecklich empfunden und das habe ich auch als sehr unangenehm empfunden, daß die Islamwissenschaft daß es keine islamwissenschaftliche Community gibt innerhalb derer man sich darüber ausgetauscht hat. (I: mhm) das hat überhaupt nicht stattgefunden das hat auch nicht bei uns am Institut stattgefunden (klappern) (I: mhm) sondern irgendwie unser Institutschef also mein Chef der sagte (I: mhm) ja komm es gibt ja jetzt so großen Bedarf in der Öffentlichkeit äh über das Thema zu sprechen äh: lass uns mal ne Podiumsdiskussion machen und dann sind wir auf dieser Podiumsdiskussion äh erschienen (I: mhm) wir haben uns vorher aber gar nicht ausgetauscht wie man überhaupt darauf reagieren solle und jeder war eigentlich mit seinen Gedanken darüber über den 11. September und was da geschehen ist eigentlich alleine] (I: mhm) und mich hat das damals total fertiggemacht. (I: mhm) ich muß sagen ich war ich war ich hab wirklich über Tage hinweg wußte ich überhaupt nichts damit anzufangen weil ich irgendwie das Gefühl hatte das hat was mit mir zu tun. und was habe ich eigentlich überhaupt mit diesem Islam zu schaffen? also ich war nach dem 11. September wirklich sehr ratlos und dann sollte ich plötzlich acht Tage später in so ner Podiumsdiskussion sprechen und das waren auch viele Leute die da kamen und in [Universitätsstadt] das ist jetzt nicht das Publikum wie in Frankfurt oder so aber ich würd mal sagen so 120 130 (I: ja) Leute und dann sollte ich plötzlich dazu sprechen und ich empfand das als enorm schwierige Situation plötzlich die Begrifflichkeit zu finden dafür und ich war mir selbst noch nicht im Klaren darüber äh wie ich damit umgehen soll weil die ganzen Register die bei mir vorhanden waren irgendwie auch Reflexe wie man als Islamwissenschaftler auf bestimmte Fragen reagiert die waren einfach so eingeschliffen und paßten überhaupt nicht zu dem was da passiert ist daß ich mich auch erst einmal völlig neu sortieren mußte (I: mhm)."

What are the differences between these cases in terms of a national discipline's relationship to the public? There are a few interesting points to highlight. In contrast to the Irish interviewee, the German scholar emphasizes a general crisis in his field due to his experiences after 9/11. German Islamic studies didn't offer an adequate response to the event, but in addition the field wasn't able to communicate options within his inner academic circle. The young scholar experienced the lack of an intellectual infrastructure in the field, the absence of meetings, places where the field could have exchanged views. In dealing with his director, he missed a countenance facing the crisis, an awareness of the turning point, but he had to adopt a more or less detached attitude. The field was adjusting itself to cope with political requirements strategically rather than dealing with the issues. Compared with the British or even the American public sphere, this is a wide difference between the two. German scholars were not used to looking for a balance in political statements and therefore didn't have a routine that would have allowed them to deal with current incidents. This began to change slightly in the wake of 9/11 when a new generation of Islamic studies scholars such as the interviewee began to establish forums dedicated to discussing political concerns within the field.

Therefore, in Germany a new type of professor might have arisen with a new generation, which has been affected by 9/11 and its aftermath. But to what extent these incidents changed the field's dominant philological tradition is uncertain. A piecemeal look at recently filled professorships of German universities shows both, that the subject-matter of projects as well as of publications and student courses are predominantly still not dedicated to political or contemporary issues, but to literature, history and cultural studies as before, and it shows, that in almost all relevant publication lists only a few articles referring to issues of political relevance could be found. Although the need to contribute to contemporary and political issues is apparently widely accepted, the philological and historical tradition (in Germany) remains predominant. The shift might be explicable by a real change in the academic interests of Islamic studies scholars but also by an adaption of academic publication strategies: Today, Islamic studies scholars are widely expected to come up with an agenda inclusively reflecting the contemporary moment.

It is clear that this kind of consideration only scratches the surface of what is actually really happening in the field with respect to methodological considerations and new approaches. A third interviewee gives some interesting inside views. He holds a professorship in southern Germany and was educated in the seventies, so he is a representative of a former generation of academics.

After all, Islamic studies [in Germany] Islamic Studies has been established three times. First, around the seventies, eighties of the nineteenth century, then, second, in 1910, 1920, and third, in the seventies [in the twentieth century, AF]. Not till the seventies did it become possible to define Islamic studies as an autonomous discipline. I think the first *Habilitation* [i.e. postdoctoral lecture qualification] in *Islamwissenschaft* without a dedication to philology took place in Berlin, it was Fritz Steppat, if I'm right, who got the first *Habilitation* in *Islamwissenschaft*. But in the seventies, it wasn't clear whether the field should become an autonomous discipline or not, and what does this mean. Then, when the field began to prevail strategically, this was a question of competition. Will it still define itself within the field of classical philology? Or will it define itself in a wider spectrum, scope of social sciences, history. And sure, the focus of early *Islamwissenschaft* [after 1970] was first to history, and merely in addition to social sciences. But henceforward, there were efforts in *Islamwissenschaft* not to cross over towards, say, a nomothetic position but to develop ability for diagnosing trends, oriented towards theories, that is an ability to develop theoretical generalizations with which older philology always has had difficulties. Just as in the seventies, discourses regarding generalizations came up, one could hear accusations, like, "Jesus, that's a relapse into these old big theories of culture by Troeltsch or Weber. It won't work. We have to stick to small details." But nevertheless it became a trend in the seventies and *Islamwissenschaft*, therefore, partly were able to respond to seventy-nine [i.e. the Iranian Revolution in 1979] and it could offer answers. [...] And one read Bourdieu or Foucault and how they saw the Iranian Revolution and in some way one liked it. But *Islamwissenschaft* as a whole, except for some authors such as Heinz Halm or others, didn't respond to it.[48]

---

48 Interview with Swiss Islamic studies scholar, no. 3, March 2012, my translation. German original: "Die Islamwissenschaft ist eine Disziplin, die im Grunde drei Mal gegründet worden ist. Einmal . so siebziger achtziger Jahre des 19. Jahrhunderts. [I: mhm] Dann das zweite Mal 1910 1920 und das dritte Mal dann in den siebziger Jahren. . Und erst in den siebziger Jahren wurde es möglich, die Islamwissenschaften als eine autonome Disziplin zu definieren. Also ich glaube die erste Habilitation in der Islamwissenschaft nur Islamwissenschaft ohne Verweis auf Philologie äh ist glaub ich in Berlin erfolgt wenn ich nicht mich recht entsinne war's Fritz Steppat, der die erste Habilitation für Islamwissenschaft bekommen hat. Und dann in den siebziger Jahren war es noch nicht ganz klar soll das ne autonome Disziplin sein oder nicht und wenn ja was ist sie dann? Als autonome Disziplin. Und als sich in den siebziger Jahren diese Disziplin dann doch strategisch durchzusetzen begann [...] war im Grunde das die Konkurrenzfrage. Definiert sich jetzt die Islamwissenschaft weiter im Spektrum der klassischen Philologie? Oder definiert sie sich in nem modernisierten Spektrum von mhm ich möchte jetzt sa-

As above I'd like to highlight some aspects. The interviewee tells the story of the transformation's starting point, which he participated in as a young scholar in the seventies, and which he obviously identified with. *Islamwissenschaft*, and by this we mean here a discipline not dedicated to language teaching and Semitic philology, was going to be emancipated as a field with its own institution and professorships.[49] Therefore, the field had to find answers regarding to what extent it had to conceptualize and absorb new approaches and reduce language-teaching related requirements. As in other disciplines in the seventies, younger scholars in the field were influenced by sociological authors and debates, which emphasized the need for theoretical approaches to analyze entities and developments using methods of comparison and generalization of common grounds or "structures." The interviewee uses the phrase "ability for diagnosing trends," which he explicitly distinguishes from "finding laws," which means forecasting and extrapolating given occurrences by interpreting them from a theoretical point of view. And "theoretical" in this respect doesn't mean "abstract," "hypothetical" or "normative," but ruled by an accumulated knowledge and scientifically evaluated experience with a prognosis.

---

gen im Sozialwissenschaften ganz im breiten Feld nicht? historische sozialw . historische und Sozialwissenschaft Und sicherlich war der Hauptaugenmerk der frühen Islamwissenschaft der siebziger Jahre im Bereich der Geschichte und erst im Nachgang dann im Bezug auf die Sozialwissenschaften hin ausgerichtet. Aber in dem Moment wo das in den siebziger Jahren passierte gab es dann auch immer mehr Bemühungen in der Islamwissenschaft hm eben solche eher also nicht in die nomothetische Haltung überzugehen, aber so ne Art [I: ja] Trendbestimmungsfähigkeit zu entwickeln oder theorieorientiert zu werden also Theorien . Verallgemeinerungen zu bringen, die . ähm mit denen sich die alten Philologen sehr schwer getan haben. In dem Moment, wo so Verallgemeinerungsdiskurse entstanden in den siebziger Jahren, war dann auch vielfach die Anklage da Jeses des is en Rückfall in die alten großen Kulturtheorien der zwanziger Jahre Troeltsch und Weber oder wat weiß ich und det ging nicht wir müssen eben bei dem Kleinteiligen bleiben aber es hat sich dann doch in den siebziger Jahren dieser Trend ergeben so dass ein Teil der Islamwissenschaft dann schon auf 79 reagieren konnte also auf die iranische Revolution und dann gewissermaßen schon Antworten anbieten konnte die . [...] und man las dann Bourdieu und ähm oder man las dann Foucault und wie der dann die iranische Revolution gesehen hat und das fand man irgendwie gut aber die Islamwissenschaft selbst bis auf Autoren wie Heinz Halm oder andere haben das relativ wenig bedient."

49 After World War II, Fritz Steppat (1923–2006) had been working for years as a journalist, when he became director of the DMG's Orient-Institute in Beirut in 1963. He held the first professorship of "Islamwissenschaft" at Berlin University from 1969 until 1988 and was a representative of an interdisciplinary, presence-based approach.

Representatives of the traditional field were arguing against this ambition, and it is clear that from the interviewee's point of view these arguments were nothing more than strategies to prevent an opening of the field to new questions and methods, in order to preserve the traditional routines of philology. In contrast, the interviewee wanted to see the field merging with social sciences and history, which in his view partly occurred since the seventies.

In addition, the interviewee was impressed by authors such as Bourdieu and Foucault, probably less regarding their Islam-related content, but because these authors were not inhibited in dealing with incidents such as the Iranian Revolution. It is a matter of habit. Sociologists were more used to act as intellectuals, and the interviewee was impressed by this habitus. So his story is clear. Although Islamic studies at that time couldn't provide answers to questions about what was happening in Iran, Professor Heinz Halm was one of a few colleagues who could respond adequately. As such he represents a kind of *Islamwissenschaft* which became more and more oriented towards the present.[50] This was achieved within the historical approach, as historians, such as Gudrun Krämer later, and others like the interviewee himself, began to draw up an agenda and research program dealing with theoretical concepts such as "secularization," "capitalism" or "rationalization," and "discourse," which were widely discussed in the social sciences. But, as the interviewee expresses elsewhere, it needed decades to implement this approach in Islamic studies. And the process is not finished yet.

Only now Islamic studies are going to operate with methods and strategies of theory formation, which exist in social and political sciences, but within its own empirical field. We are no longer confined to saying 'Now we have all the theoretical proposals of social sciences. And they please might make aware all information, we are working out. We want to participate actively on theory formation. We are

---

50 Prof. Dr. Heinz Halm, born in 1942 and a former student of Annemarie Schimmel in Bonn, held a professorship in *Islamwissenschaft* at Tübingen University from 1980 until his retirement. He focused in his work on the history of the Middle East, Egypt, Maghreb, and Syria. He is a specialist in the history of *Shia* and well known for his books about Shiites, Arabs, and on Islam. As he had worked as journalist before his appointment, he tended to address the public and was a protagonist of bringing together Islamwissenschaft and public debates. See Heinz Halm, *Shi'ism*, transl. by Janet Watson (Edinburgh: Edinburgh Univ. Press, 1991); Halm, *The Empire of the Mahdi. The Rise of the Fatimids*, transl. by Michael Bonner (Leiden: Brill, 1996); Halm *Shi'a Islam: From Religion to Revolution*, transl. by Allison Brown (Princeton, NJ: Wiener, 1997).

using our own empirical background to operate with these theories in a new way. In former times, Islamic studies offered their information, delivered it to social sciences and ceded systematical conclusions to social sciences. Now there is a process of integration or virtually fusion, in which Islamic studies gets more and more self-assurance in saying like: 'Ok. Max Weber, we can, too. We don't have problems with that.' And we are not afraid of saying 'Well, we see deficits in European social sciences due to the fact, that they formatted their theories only against a very limited empirical background. Now we have a global knowledge community.[51]

Although the interviewee doesn't give examples for his claim here, there are no reasons to doubt his optimistic views. In recent years, lots of books have been published in which authors use methodical and theoretical frameworks such as discourse analysis to deal with Muslim or western perceptions of Islam or Jihad, women's rights, violence and secularism, etc.[52] Political developments such as 9/11, the Iraq war, ISIS, or the many chapters of the Arab rebellion have driven this diversification of projects. However, the question to what extent this new kind of extended *Islamwissenschaft* has succeeded cannot yet be answered here.

---

51 Interview with Swiss Islamic studies scholar, no. 3, March 2012, my translation. German original: "Das was sie eigentlich jetzt erst macht nämlich selbst die Methoden und Theoriebildungsstrategien, die in den systematischen Wissenschaften existieren, . zu bedienen, aber aus nem eigenen empirischen Feld heraus. Also sich nicht einfach darauf zu beschränken und zu sagen ja jetzt haben wir die Theorie und Methodenangebote der systematischen Disziplinen und die mögen doch bitte dann unsere Informationen einfach nur zur Kenntnis nehmen. Nein wir wollen jetzt sozusagen an der Theorie- und Methodenarbeit selbst aktiv teilnehmen. Wir benutzen unseren empirischen Hintergrund, um das Theoriefeld . neu zu bedienen und . neu . zu gestalten. Und das wurde früher nicht gemacht, ne. Früher war das sozusagen allenfalls sozusagen der Informationshintergrund der durch die Islamwissenschaft aufgearbeitet wurde und dann die systematische Arbeit den Disziplinen jenseits der Islamwissenschaft überlassen. Jetzt gibt's diesen Integrations- und schon fast schon Fusionsprozeß wo die Islamwissenschaft immer mehr Selbstbewußtsein gewinnt zu sagen 'Ok . Weber können wir auch' [I: mhm] da haben wir kein Problem mit mehr, also wir scheuen uns dann nicht und wir sehen dann die Defizite einfach in der europäischen oder in der westeuropäischen Sozialwissenschaft einfach darin daß sie ihr Theoriefeld allein vor einem sehr engen empirischen Hintergrund aufgebaut hat. Und jetzt haben wir globale Wissenslandschaft."

52 See for instance, Thomas Bauer, *Die Kultur der Ambiguität. Eine andere Geschichte des Islams.* (Berlin: Insel Verlag, 2011).

8. Conclusion

While I have not provided a comprehensive map or model of national differences in the relationship between a research field and the public, I have provided an example of how an interview-based comparison helps identify characteristics that are otherwise not easily accessible. In this case, responses to 9/11 not only show how Islamic studies scholars have dealt with this incident but how their reaction is part of a larger and unfinished transformation of their field. The third featured interview allows for a closer look at this transformation.

These extracts reveal a deeper crisis in the field in terms of, on the one hand, the need to adjust professional behavior and infrastructure to the requirements of politics and the media; and, on the other hand, a mental change in dealing with concepts of Islam that are established in historical configurations that have been bypassed due to the return of political Islam since the seventies. In addition, a long-term amplification of the field with respect to its methodological approaches abounded, although reasons for structural limitations of the field caused by the need to teach languages were preserved.

Considering the terms of legitimizing Islamic studies, this overview shows how difficult it is to carve out all the important dimensions. While the return of religion, political Islam, Muslim migration, and the many conflicts in the Middle East seem to legitimize Islamic studies self-evidently in terms of academic needs, the same developments raised expectations in public and politics, which forces the field to deal with all these issues promptly and by exposing itself in a more controversial field. The potential for conflict in all political issues made it more crucial to answer public expectations adequately. Therefore, the field had to face these critical debates, especially in the United States, where Middle East studies programs have been accused of being unbalanced, partisan and ignorant of Islamic threats and furthermore insufficiently independent due to the many donors from Arab countries.[53] On the other hand, the field

---

53 For instance, see Martin Kramer, *Ivory Towers on Sand: The Failure of Middle Eastern Studies in America* (Washington Institute for Near East Policy, policy paper no. 58, 2001); Mitchell Bard, *The Arab Lobby: The Invisible Alliance that Undermines America's Interests in the Middle East* (New York: Harper Broadside Books, 2010); also see the website "Campus Watch: Monitoring Middle East Studies on Campus," http://www.campus-watch.org/. In Germany, Clemens Heni suspected Islamic studies of being driven by anti-Israel and

seems in the meantime to have achieved a critical mass of scholars and projects, which are not only able to deal with all these issues sufficiently but to incorporate controversies within the field. The field is frankly debating its challenges at conferences and within books.[54] So Islamic studies has to face a more sensitive and demanding situation which threatens to delegitimize Islamic studies in cases of a failure or imbalance in its responses. But it now has more possibilities of achieving this.

---

hence anti-Semitic feelings. See Clemens Heni, *Schadenfreude. Islamforschung und Antisemitismus in Deutschland nach 9/11* (Berlin: Edition Critic, 2011).

54 See, for instance, Abbas Poya and Maurus Reinkowski, eds., *Das Unbehagen in der Islamwissenschaft. Ein klassische Fach im Scheinwerferlicht der Politik und der Medien.* (Bielefeld: transcript, 2008); Stefan Weidner, *Aufbruch in die Vernunft. Islamdebatten und islamische Welt zwischen 9/11 und den arabischen Revolutionen* (Bonn: Verlag Dietz, 2001).

Stem Cell Debates in an Age of Fracture

Axel Jansen

1. Introduction

Historians have recently begun to revisit the 1970s in order to identify historical trajectories that lead into our own time. In Germany and in the US, they have moved away from a sometimes self-absorbed focus on the sixties to reframing the seventies as an age introducing the intellectual outlook and market-oriented language that continues to hold today. In doing so, they have tried to make sense of an incipient economic decline in Western countries and the unraveling of the Soviet bloc, replacing a binary and state-focused perspective dominant during the Cold War.[1] For the case of the US, Daniel T. Rodgers has offered a particularly subtle framework for conceptualizing the four decades after 1970. His central tenet concerns the overall direction taken by discourses on the left and right, both converging towards a narrow conception of rights and obligations on a federal level. "The domain of citizenship, which had expanded in the post-World War II years ... began to shrink," Rogers suggests, "into smaller, more

---

1 The historiographical impetus to revisit the 1970s has been stronger in Germany than in the US. See, for example, Geoff Eley, "End of the Post-War? The 1970s as a Key Watershed in European History," *Journal of Modern European History*, 9:1 (2011), 12–17; Niall Ferguson et al., eds., *The Shock of the Global. The 1970s in Perspective* (Cambridge: Harvard Univ. Press, 2010; Andreas Wirsching, ed., "The 1970s and 1980s as a Turning Point in European History?," *Journal of Modern European History* 9 (2011), 8–26; Ariane Leendertz and Wencke Meteling, eds., *Die neue Wirklichkeit. Bedeutungsverschiebungen, Bezeichnungsrevolutionen und Politik seit den 1970er Jahren* (Frankfurt and New York: Campus, 2015); Anselm Doering-Manteuffel and Lutz Raphael, *Nach dem Boom: Perspektiven auf die Zeitgeschichte seit 1970* (Göttingen: Vandenhoeck & Ruprecht, 2008). In his recent comprehensive overview, historian of science Jon Agar has considered key developments to be part of a "Sea Change in the Long 1960s." See his *Science in the Twentieth Century and Beyond* (London: Polity Press, 2012), chap. 17.

partial contracts: visions of smaller communities of virtue and engagement."[2]

I would like to complement and refine Rogers' assessment. With the
announcement of the first successful in-vitro cultivation of human embryonic stem cells (hES cells) by James Thomson and John Gearhart in 1998,
the field of biology was on the front pages across the globe. Stem cells
were associated with groundbreaking research and radical medical promise,
but there were strong ethical reservations because they were taken from
fertilized human eggs, i.e. embryos. The matter touched on the role of
industry, as investors stood to profit from their use. Few other public debates about science during the past fifty years have been so heated, so
international, and yet so national in their results. A close look at the
American debate and its prehistory complicates Rodgers' perspective.
Within a federal framework that precluded a decisive stance on core ethical
issues, biomedicine in the US remained free to mobilize support for controversial research from sources other than the federal government. In the
case of stem cell research, public endorsement of such a field has not only
devolved from the national to the private sector but also to the state level.
New coalitions have preserved scientific impetus in the US but they have
accompanied a reduction of its symbolic significance for national achievement.

In order to contextualize later stem cell debates, I will begin by charting
the sometimes complicated regulatory history of biology and biomedicine
in the seventies and eighties. In the US, biomedicine had replaced physics
by the late 1970s as the one field representing the larger scientific enterprise in public. At that time, important regulatory decisions were made in
the context of research on recombinant DNA (rDNA) and in connection
with evolving research on human fetal tissue and in vitro fertilization.
These decisions set the stage for subsequent discussions about biomedicine. I will end with a discussion of the post-1998 confrontation over hES
cells in order to show how debates about this rather limited area of biomedical research eventually transformed the role and legitimacy of science
in the American federation.

---

2 Daniel T. Rodgers, *Age of Fracture* (Cambridge: Harvard Univ. Press, 2010), 198. See also
  Sean Wilentz, *The Age of Reagan: A History, 1974–2008* (New York: Harper Collins,
  2008).

2. Trust in Biology: The Asilomar Legacy

The history of stem cell research is a strong reminder of the old observation that science transgresses national boundaries. Despite its transnational characteristics, however, science has fared quite differently in the various national states in which it took hold. In negotiating the role of science in society, individual fields of research have at times come to represent the broader scientific enterprise. This has occurred when a token issue has evolved from research to prompt a reevaluation of the role of science as such. In the US, such an issue arose in the early 1970s when molecular biologists stood poised to alter a living organism's DNA. The ensuing debate over rDNA turned the field of molecular biology (and biology at large) into a cultural symbol for renegotiating the terms of science's public responsibility.

In 1972, most people in the US probably had never heard of rDNA, but researchers had been interested in the general idea of altering DNA for some time. Since 1953, when James Watson and Francis Crick had published their work on the structure of the DNA, the emerging field of molecular biology had made significant progress in understanding mechanisms such as DNA replication and protein synthesis and had developed tools to alter and replicate DNA. Some argued that it might be possible to use genetic engineering to create bacteria that express (produce) expensive proteins such as insulin and interferon.[3] Such work seemed feasible when Paul Berg's group at Stanford University Medical School developed techniques in 1971 to cut open and join together DNA from a bacterial lambda virus and a monkey tumor virus.[4] The following year, Berg planned to introduce rDNA into *Escherichia coli*, a bacterium commonly found in the human gut. While he continued with other work, Berg postponed this particular experiment because he was concerned about creating a tumor virus carrier.[5]

---

3 The idea was proposed by Joshua Lederberg. Robert Bud, "History of Biotechnology," *The Modern Biological and Earth Sciences*, edited by Peter J. Bowler and John V. Pickstone, vol. 6 of *The Cambridge History of Science* (Cambridge: Cambridge Univ. Press, 2009), 535.

4 In creating rDNA, his group used enzymes isolated in Herbert Boyer's laboratory at the University of California, San Francisco.

5 Susan Wright, *Molecular Politics: Developing American and British Regulatory Policy for Genetic Engineering, 1972–1982* (Chicago and London: Univ. of Chicago Press, 1994), 72–73.

*Fig. 4. Maxine Singer, Norton Zinder, Sydney Brenner, and Paul Berg at the 1975
Asilomar Conference.*

(Courtesy of the National Library of Medicine, www.nlm.nih.gov/hmd/ihm)

In considering Berg's situation and subsequent developments, it is important to note that no regulation for potentially hazardous biological experiments existed at that time. In 1973, after some of Berg's colleagues
decided to share their concerns over the safety of the new technology in a
letter that was published in *Science*, Berg and others around him were concerned that the field could lose control over how they could do their work.
What if politicians chose to pick up on a supposed danger associated with
their experiments and shut it down or restrict it? Harvard microbiologist
Bernard Davis pointed to the "danger of public over-reaction."[6] In
responding to such fears, Berg and his colleagues chose to take the initiative in order to maintain control.

In charge of a committee appointed by the National Academy of Sciences to resolve the matter, Berg gathered leading molecular biologists,
including David Baltimore and James Watson. In a letter published in the
National Academy's *Proceedings* in July 1974, the scientists framed the issue
as a matter of safety for laboratory workers rather than the public at large.

---

6 Quoted in Wright, *Molecular Politics*, 135.

They called for a moratorium on certain experiments until an international conference had assessed the risks. By the time approximately 140 biologists and selected journalists from the US and abroad gathered for the famous conference on rDNA-technology in Asilomar, California in February 1975, Berg had lined up a regulatory process within the National Institutes of Health. The NIH was (and remains) the largest funder of biomedicine in the US and provides the field with a home base within the federal funding structure.

Asilomar took place under the auspices of the NIH which stood ready to create a committee to establish guidelines for research on rDNA.[7] Berg sought to avoid a tedious legislative process and more stringent NIH regulations by preparing the ground for such guidelines.[8] At one point, Berg warned his colleagues that if

our recommendations look self-serving, we will run the risk of having standards imposed. We must start high and work down. We can't say that 150 scientists spent four days at Asilomar and all of them agreed that there was a hazard—and they still couldn't come up with a single suggestion. That's telling the government to do it for us.[9]

Historian Susan Wright has suggested that their strategy was shaped, in part, by the economic prospects associated with the research. She has also criticized Berg and his colleagues for excluding issues such as the potential use of rDNA technology by the military. But there is little evidence that Berg and his group were not simply trying to keep restrictions under their control in order to preserve their field's autonomy. One could argue that they should have broadened the issue to include the risk of rDNA-research to the general public, but they refused to become public intellectuals.[10]

Twenty-five years later, the debate on human embryonic stem cells would once again put science in the limelight through a contentious issue arising from biological research. By that time, the legacy of Asilomar had long shaped American regulation of biology and biomedicine. According

---

7 Ibid., 138.

8 Donald S. Fredrickson, *The Recombinant DNA Controversy, a Memoir: Science, Politics, and the Public Interest 1974–1981* (Washington, D.C.: ASM Press, 2001), 50–51.

9 Quoted in Wright, *Molecular Politics*, 153.

10 For Wright's critique, see *Molecular Politics*, chap. 3, and explicit references on pp. 140, 158. For general accounts of Asilomar, see Richard Hindmarsh and Herbert Gottweis, "Recombinant Regulation: The Asilomar Legacy 30 Years on," *Science as Culture* 14, no. 4 (2005): 299–307; Wright, *Molecular Politics*, chap. 3; Sheldon Krimsky, *Genetic Alchemy: The Social History of the Recombinant DNA Controversy* (Cambridge: MIT Press, 1984).

to Wright, the conference "contributed powerfully to defining and rein-
forcing the central role of the biomedical research community in policy-
making."[11] The NIH took charge of the safety rules that researchers had
decided on, and it turned them into mandatory rules for all grant recipients. The 1976 NIH *Guidelines for Research Involving Recombinant DNA Mole-
cules* marks the beginning of biotechnology policy in the US.[12] Because the
NIH was the largest funder of biological research, these rules applied to
most laboratory work on recombinant DNA in the US and abroad as
well.[13] By taking charge of the regulatory process, Berg and his colleagues
were able to deflect criticism from those who argued that NIH regulations
were not sufficient and that Congress should turn such regulations into
law. Microbiologist Maxine Singer later recalled that the scientific community "succeeded in … hav[ing] reasonable regulations but not legislation."[14]
The NIH, of course, had been created by law and both the legislative and
the US government retained ultimate control. Hence the responsibility for
the safety of basic research devolved to the NIH as the biomedical community's representative, and the organization's guidelines represented
standards for its work at universities and other institutions where NIH-
sponsored research was ongoing.

3. Defining the Issue: Embryo and Fetal Tissue Research

From the late 1970s, however, an increasing amount of work in biomedicine was not sponsored by the NIH. While the Asilomar-approach had
assumed that NIH guidelines would regulate most (or all) experimental
research in the US, the rise of biotech in the 1980s and 1990s created a

---

11 Wright, *Molecular Politics*, 159.
12 Adam D Sheingate, "Promotion Versus Precaution: The Evolution of Biotechnology
   Policy in the United States," *British Journal of Political Science* 36, no. 2 (April 2006): 246.
13 In the US, the NIH remained in charge of regulating laboratory research while emerging
   products ("green" biotechnology such as modified crops and "red" biotechnology)
   would eventually be regulated by other agencies, including the United States Department
   of Agriculture (USDA) and the Federal Drug Administration (FDA). Ibid., 248–55.
   British and German researchers who were present at the Asilomar conference quietly
   implemented these rules in their own countries.
14 Maxine Singer, Interview Maxine Singer, July 24, 1998, NIH, http://history.nih.gov/
   archives/downloads/singer.pdf.

large area of research that remained unregulated. The evolving situation is best described by focusing on issues in genetics and developmental biology.

By the 1970s, livestock breeders had long been interested in reproduction. Cattle embryo transfer (i.e. fertilizing an egg in vitro and then placing the embryo in the uterus) had evolved into an international business as the procedure became routine. While in vitro fertilization in humans had been reported since the 1950s, no such claim could be substantiated until physiologist Robert Edwards, at the University of Cambridge, reported the birth of Louise Brown in 1978.[15] Public attention soared and countries looked for ways to deal with new ethical questions arising from the medical procedures and in vitro fertilization (IVF) research such as the procedure's safety, licensing, and the fate of left-over fertilized eggs. These included whether left-over eggs should be made available to other couples or to research and if so, under what conditions, and at what point a a fertilized egg should be considered a human being, hence requiring the protection of the law. The biomedical community could not answer such questions because they touched on matters well beyond the scope of assessing the risk for lab workers. The particularity of American policy stands out when compared to its counterparts abroad.

In the UK during the eighties, IVF became routine, but continued to be unregulated. In 1982, the British government established a committee chaired by Dame Mary Warnock to consider the ethical status of left-over eggs and their possible use in research. After two years of deliberation, the committee in 1984 found that

in practical terms, a collection of four to sixteen cells was so different from a full human being, from a new human baby or a fully formed human foetus, that it might quite legitimately be treated differently.[16]

Rather than cutting off all research involving human eggs, therefore, the committee proposed that such research should be allowed for up to twenty days after fertilization. It recommended that it "shall be *a criminal offence* to handle or to use as a research subject any live human embryo derived from

15 Nick Hopwood, "Embryology," in *Modern Biological and Earth Sciences*, edited by Peter J. Bowler and John V. Pickstone, vol. 6 of *The Cambridge History of Science* (Cambridge: Cambridge Univ. Press, 2009), 314.

16 Warnock Report, 1985, p. xv, quoted in Herbert Gottweis, Brian Salter, and Cathy Waldby, *The Global Politics of Human Embryonic Stem Cell Science: Regenerative Medicine in Transition* (Basingstoke: Palgrave Macmillan, 2009), 62.

in vitro fertilisation beyond that limit."[17] In 1990, after eight years of public debate, the British government translated these recommendations not into regulation but into law. In Germany, a legal solution was found in 1990 as well. But the *Embryonenschutzgesetz* placed the country at the other end of the legislative scale. While the UK had passed a law that allowed research to go forward under well-defined rules, much of this work became illegal in Germany.[18]

In the US, in vitro fertilization came to be discussed under the highly contested rubric of abortion. After the US Supreme Court legalized abortion in its 1973 decision in *Roe v. Wade*, pro-life groups mobilized against the liberalization of abortion policy. In turning to the federal level rather than the states, they also directed their campaign at research that involved human embryos, including research on in vitro fertilization.[19] By the time Louis Brown was born in 1978, the US government had stopped paying for research that involved fetal tissue. While the NIH proposed that fetal research be allowed to proceed with oversight by a new Ethics Advisory Board, in 1980 Congress effectively blocked NIH-sponsored fetal tissue research (including research on in vitro fertilization) by allowing the Board's charter to lapse. In response to research in biology and biomedicine during subsequent decades, several attempts were made to reintegrate such work with the NIH. The first significant opportunity for a revision of federal funding policies arose at a time when the concept of the stem cell had resulted in promising perspectives for biomedicine.

---

17 My emphasis. "The Warnock Committee," *British Medical Journal (Clinical Research Ed.)* 289, no. 6439 (July 28, 1984): 238–39.

18 Thomas F. Banchoff, *Embryo Politics: Ethics and Policy in Atlantic Democracies* (Cornell Univ. Press, 2011); Simon Fink, "Ein deutscher Sonderweg? Die deutsche Embryonenforschungspolitik im Licht international vergleichender Daten," *Leviathan* 35, no. 1 (March 1, 2007): 107–28.

19 James Mohr, *Abortion in America: The Origins and Evolution of National Policy* (Oxford and New York: Oxford Univ. Press, 1978). Herbert Gottweis, "The Endless hESC Controversy in the United States: History, Context, and Prospects," *Cell Stem Cell* 7, no. 5 (November 5, 2010): 555. In addition to the issue of rDNA, the contemporary issue of research on human fetal tissue in the wake of the Supreme Court 1973 ruling on abortion in *Roe v. Wade* also shaped this tradition. Congress in 1974 created the National Commission for the Protection of Human Subjects of Biomedical and Behavioral Research to investigate and to make recommendation what research should be supported by the Department of Health, Education, and Welfare (via its National Institutes of Health). Owen C.B. Hughes, Alan L. Jakimo, and Michael J. Malinowski, "United States Regulation of Stem Cell Research: Recasting Government's Role and Questions to Be Resolved," *Hofstra Law Review* 37, no. 02 (2008): 399–411.

4. Emergence of the Stem Cell Concept

The idea of stem cells had emerged since World War II in the context of work using radiation. In 1949, Leon Jacobson had found that protecting the spleen during otherwise lethal radiation allowed a mouse to live. Joan Main and Richmond Prehn in 1955 were able to show that the source for the reconstitution of the blood system was cellular rather than humoral.[20] At the Ontario Cancer Institute in Toronto, Ernest McCulloch and James Till in 1963 inferred the existence of cells that could both self-renew and generate most if not all cells in the blood.[21] But isolating these cells required the knowledge of surface markers to pick them out. In mouse bone marrow, for example, only about one in ten thousand cells is a blood-forming stem cell and in normal blood the number is ten-fold lower.[22] There were obvious medical uses for such cells. In 1969, E. Donnall Thomas had performed the first bone marrow transplant in a leukemia patient who had previously been exposed to radiation to kill cancer cells. The intrusive use of bone marrow to restart the blood system could perhaps be avoided by using blood (and hematopoietic stem cells) taken from the vein. In 1981, both Martin Evans at the University of Cambridge and Gail Martin at the University of California, San Francisco isolated and cultured what Martin called "embryonic stem cells" in mice, cells that were "pluripotent" because they could create the blood-forming system as well as all other types of cells in the body. Evans and Martin also found a way to culture such cells in vitro and to preserve their ability to differentiate. "The availability of such cell lines," Martin pointed out, "should make possible new approaches to the study of early mammalian development." This included the feasibility of isolating pluripotent stem cells carrying mutant genes to grow tissue.[23] A few years later, Martin Evans and

20 E. Donnall Thomas, "A History of Haemopoietic Cell Transplantation," *British Journal of Haematology* 105, no. 2 (May 1, 1999): 330–39.

21 A. J. Becker, E. A. McCulloch, and J. E. Till, "Cytological Demonstration of the Clonal Nature of Spleen Colonies Derived from Transplanted Mouse Marrow Cells," *Nature* 197, no. 4866 (February 2, 1963): 452–54.

22 Jos Domen, Amy J. Wagers, and Irving L. Weissman, "Regenerative Medicine," National Institutes of Health, *Regenerative Medicine*, (2006), 19, http://stemcells.nih.gov/info/Regenerative_Medicine/Pages/2006Chapter2.aspx.

23 G. R. Martin, "Isolation of a Pluripotent Cell Line from Early Mouse Embryos Cultured in Medium Conditioned by Teratocarcinoma Stem Cells.," *Proceedings of the National Academy of Sciences of the United States of America* 78, no. 12 (December 1981): 7634–38. Similar results by M. J. Evans and M. H. Kaufman, "Establishment in Culture of Pluripotential

Matthew Kaufman showed that genetically altered embryonic stem cells could transmit traits to offspring in mouse.[24] This raised the prospect of creating mice with certain characteristics that could be studied in the lab. By 1988, Charles Baum in Irving Weissman's lab at Stanford isolated blood-forming stem cells in mice, and by 1991, Weissman isolated candidate human hematopoietic stem cells.[25] Biologists anticipated using such work done in the mouse model to guide research on human cells. Even though NIH funding for such work with human cells was blocked, changed seemed imminent when William J. Clinton was elected president in 1992.

Biologists looked forward to support from a president who had campaigned on health issues and who picked Nobel laureate Harold Varmus to lead the NIH. Things looked promising when Clinton lifted the longstanding ban on federal funding for fetal tissue research in 1993 but by December 1994, the opportunity for liberalizing NIH funding policy was lost when an expert panel consisting of bioethicists and biologists overplayed its hand. The NIH had formed the Human Embryo Research Panel (HERP) to recommend guidelines for research on pre-implementation embryos, as such guidelines were required for any research funded by the NIH.[26] The panel sought to identify ways to deal with fetal research expected to evolve during subsequent years, and the panel's assessments proved to be remarkably lucid. It discussed potential research such as growing human stem cells in vitro or the prospect of using somatic cell nuclear transfer ("therapeutic cloning") to create donor-specific cells. In December 1994, the panel recommended that the NIH allow researchers to use fertilized eggs from IVF clinics that were no longer needed, either because the donor couple had been treated successfully, or because the eggs were too old to be used and bound to be discarded anyway. But in a

---

Cells from Mouse Embryos," *Nature* 292, no. 5819 (July 9, 1981): 154–56. For an overview, see Davor Solter, "From Teratocarcinomas to Embryonic Stem Cells and Beyond: A History of Embryonic Stem Cell Research," *Nature Reviews. Genetics* 7, no. 4 (April 2006): 319–27.

24 Kenneth Paigen, "One Hundred Years of Mouse Genetics: An Intellectual History. II. The Molecular Revolution (1981–2002)," *Genetics* 163, no. 4 (April 1, 2003): 1228–29.

25 C. M. Baum et al., "Isolation of a Candidate Human Hematopoietic Stem-Cell Population," *Proceedings of the National Academy of Sciences* 89, no. 7 (April 1, 1992): 2804–8. "Scientists Claim to Have Isolated Stem Cells," *New York Times*, Nov. 5, 1991, C12.

26 William A. Galston, *Public Matters: Politics, Policy, and Religion in the 21st Century* (Lanham, MD: Rowman & Littlefield Publishers, 2005), 75–76.

conclusion that turned out to be much more controversial, the panel also proposed that researchers be allowed to use federal funds to *create* embryos: "genetically diverse embryos from which stem cells could be derived, and reprogramming by somatic cell nuclear transfer [therapeutic cloning] to make pluripotent [stem] cells immunologically compatible with prospective patients."[27] Such embryos were not meant to be implanted in a uterus, of course, but in taking this significant step, the panel had thrown away an important opportunity to update NIH policy.

Anti-abortion groups had closely followed the panel's work. Activists put pressure on panel members such as Ronald Green, a bioethicist at Dartmouth, who received postcards of mutilated embryos.[28] Some members feared for their safety. The panel's recommendations were published right after midterm elections that put Republicans in charge of Congress.[29] Hours after the NIH officially endorsed the panel's report on December 2, 1994, Clinton wavered on his commitment to progressive science policy. The White House, by executive order, blocked any NIH funding of studies that involved the creation of embryos. Not to be outdone, Congress then passed its own measure. The Dickey-Wicker Amendment, as it came to be known, barred from funding any "research in which a human embryo or embryos are destroyed, discarded, or knowingly subjected to risk of injury or death." Congress, in addition to creating embryos for research, blocked federal funding for work on fertilized eggs discarded by IVF clinics.[30]

Asilomar had left an ambivalent legacy. Herbert Gottweis has argued that American policy was "characterized by a failure of government to comprehensively regulate biomedical and assisted reproduction research in the private sector, leading to a regulatory vacuum that undermines efforts to build social trust in new and sensitive scientific-medical developments."[31] Had the US chosen to regulate all research by law, this would have forced the issue in public. But as long as biologists feared that their research options would be curtailed, they preferred the leeway given them by restricting the issue to NIH funding. Dorothy Wertz would argue in

---

27 Varmus, *The Art and Politics of Science*, 190–205, quotation from p. 205. See also Stephen S. Hall, *Merchants of Immortality: Chasing the Dream of Human Life Extension* (Boston and New York: Hougton Mifflin, 2003), 105–113.

28 "Outside Laboratory, Moral Objections," *Washington Post*, Nov. 6, 1998, A14.

29 For Varmus' perspective on the panel's work, the scientific prospects and the ethical issues at the time, see Varmus, *The Art and Politics of Science*, 199–205.

30 Hughes and Malinowski, "United States Regulation of Stem Cell Research," 406.

31 Gottweis, "The Endless hESC Controversy in the United States," 557.

2002 that the "private sector remains unregulated and perhaps should remain so, as long as government remains hamstrung by conservative interests."[32]

5. Biotech as Research Shelter

For biologists interested in human stem cells, private industry offered viable alternatives. From the mid-1990s, private industry became a research shelter for human embryonic stem cell research.

The biotech industry had emerged along with rDNA technology. After Stanford applied for the first rDNA-related patent in 1974, the industry expanded at a time when the US government, following the oil crisis and stiff competition from Japan in several markets, looked for ways to stimulate growth through new industries. Revised tax laws facilitated investments in small companies by promising investors a return on rising stocks. From 1980, the Bayh-Dole Act granted universities, not-for-profit corporations and small businesses in the US the right to patent findings even if such findings resulted from research that the federal government had paid for.[33] Wall Street helped translate enthusiasm for investment into enthusiasm for biomedical engineering. In the 1980s, the term "Biotechnology" gained currency to help convey to investors the enthusiasm for developing applications in molecular biology.[34] Investments in biotechnology rose

---

32 Dorothy C. Wertz, "Embryo and Stem Cell Research in the United States: History and Politics," *Gene Therapy* 9, no. 11 (May 16, 2002): 678.

33 Sally Smith Hughes, "Making Dollars out of DNA: The First Major Patent in Biotechnology and the Commercialization of Molecular Biology, 1974–1980," *Isis* 92, no. 3 (September 1, 2001): 541–75. In 1983, President Ronald Reagan by Presidential Memorandum expanded such patenting opportunities to large corporations. Hughes, Jakimo, and Malinowski, "United States Regulation of Stem Cell Research," 392–99.

34 Robert Bud, "History of Biotechnology," in *The Modern Biological and Earth Sciences*, edited by Peter J. Bowler and John V. Pickstone, vol. 6 of *The Cambridge History of Science* (Cambridge: Cambridge Univ. Press, 2009), 524–38. Herbert Boyer and Stanley Cohen at Stanford were among the first to develop tools on the basis of their work on rDNA. When they took out patents, they dedicated royalties to their university. Nicholas Wade, "Biotechnology and Its Public," *Technology in Society* 6, no. 1 (1984): 17–21. Between 1980 and 2001, Stanford would earn 255 million dollars from the Cohen-Boyer patents. M. P. Feldman, A. Colaianni, and C. Liu, "Lessons from the Commercialization of the Cohen-Boyer Patents: The Stanford University Licensing Program," in *Intellectual Property Management in Health and Agricultural Innovation: A Handbook of Best Practices*, edited by A.

significantly as pharmaceutical, chemical, and oil companies sought to help develop products (or prevent others from developing products that might erode their traditional business).[35] At a crucial juncture in the development of that field, private industry thus came to play a key role in stem cell research.

In 1992, Mike West, founder of the biotech company Geron, approached developmental biologist Roger Pederson in San Francisco with the offer to fund efforts to isolate hES cells. Geron aimed to get a headstart on patenting such work. But Pederson declined. Clinton had just been elected president and, as mentioned, there was hope that NIH funding for such research would be unshackled. "'I think this area of investigation is something that is so at the headwaters,'" Pederson told West, "'that it's not appropriate for private investors to control the headwaters of the river. It's something that could benefit all the people,'" he added, "'and therefore it should be developed by the people, by the federal government.'" Two years later, however, efforts to implement new NIH guidelines had failed and Congress had shot down prospects for a liberal policy by passing the Dickey-Wicker Amendment. The political impasse in Washington changed Pederson's mind.[36]

Pederson now pointed West towards James Thomson at the University of Wisconsin who was working on isolating and culturing rhesus monkey embryonic stem cells. Thomson accepted West's offer to fund such research on human cells and work got underway in facilities separated from all his other NIH-funded work. Thomson used leftover fertilized eggs from IVF clinics in Madison, Wisconsin and Haifa, Israel.[37] Thomson was ultimately able to isolate human embryonic stem cells from the inner cell mass of the blastocyot, the ball of cells that evolves from an egg five to nine days after fertilization. He was also able to develop ways to preserve in vitro the ability of hES cells to grow into different types of human tissue. Because hES cells preserved the ability to grow into all the different tissues in the human body, they were considered pluripotent. Thomson's sponsor Geron had another egg in the stem cell basket through John

Krattiger, R.T. Mahoney, and L. Nelsen (Oxford and Davis: MIHR and PIPRA, 2007), 1797.

35 Agar, *Science in the Twentieth Century*, 439–41. Martin Kenney, *Biotechnology: The University-Industrial Complex* (New Haven and London: Yale Univ. Press, 1986).

36 Hall, *Merchants of Immortality*, quotations from pp. 98, 122.

37 Ibid., 122–125, 158–164.

Gearhart, the director of the IVF clinic at Johns Hopkins University. West learned of Gearhart's plans to isolate hES cells from aborted fetuses and proposed to Gearhart that Geron fund his work as well, which Gearhart accepted. The *Washington Post* reported in 1996 that Gearhart was "using private grants to isolate from aborted fetuses a class of cells with the potential to grow into various human tissues" and that this was considered "so sensitive that security officials have been apprised of the routes he takes between home and work, lest antiabortion activists or others try to harm him."[38]

Both Thomson and Gearhart were aware of the political implications of their work and they knew that if they were able to isolate and preserve hES cells in a petri dish, this would create a media frenzy. Two years before he would publish his work, John Gearhart in 1996 pointed out that there was "'going to have to be a real educational process on how to present this material to the public." The *Washington Post* observed that Gearhart's "interest is not in cloning people but in understanding human development and creating human tissues for transplantation." Gearhart feared, however, that "'[o]ur work ... has all the elements that are going to feed the public's worst imaginations.'"[39]

6. The Debate on Stem Cell Science

As news of human embryonic stem cells spread around the globe in November 1998, countries in which such research was ongoing had to respond. Depending on how they had come to discuss and define the "embryo", adjustment to the new findings was facilitated or slowed down.

The UK was prepared for the work published by Thomson and Gearhart, because eight years earlier the country had made a national decision on the status of the embryo. Legislation following the Warnock Committee's report had created the Human Fertilization and Embryology Authority (HFEA). The organization was charged with "comprehensive regulatory oversight of public and private embryo research and in vitro fertilization."[40] Gearhart had derived pluripotent stem cells from primordial germ

---

38 *Washington Post*, July 26, 1996, A03.
39 Ibid.
40 Gottweis, "The Endless hESC Controversy in the United States," 555.

cells in therapeutically aborted fetuses. The contentious issue concerning hES cells, however, was that Thomson's more promising human embryonic stem cell lines were derived from fertilized human eggs received from fertilization clinics. In Britain, the HFEA was now granted additional authority to regulate research on hES cells. The UK thus had official oversight for all embryo research regardless of funding source.

In the US, the ensuing controversy about stem cells was able to gain ground because the country had not made a binding legal decision on the core issue of using human eggs for research. While conservative forces seemed to have scored a victory by restricting NIH funding for embryo research, Thomson's privately funded work had remained perfectly legal. After 1998, news about such work kept fueling conservative rage, but this rage was sometimes falsely directed at the NIH, while the research profession avoided raising the issue of legislation because it sought to retain the freedom (and the private funding) for its work.[41] At the height of the stem cell controversy, Paul Berg explained that another Asilomar conference would not resolve the matter. "The issues that challenge us today," he wrote in 2001, "are often entwined with economic self-interest and increasingly beset by nearly irreconcilable ethical, religious, and legal conflicts, as well as by challenges to deeply held social values."[42]

In the context of rDNA research in the 1970s, researchers had limited the issue to the safety of experiments in laboratories. They had sought to retain the option of pursing research according to guidelines controlled by the NIH, an organization they considered as representing the biomedical research profession. Their aim had been to keep under their wing a matter they had brought to public attention. In the context of research on hES cells, the field once again challenged the public to wrap their head around new insights, this time with respect to a more nuanced understanding of the early stages of human development. Unlike rDNA-techniques during the early seventies, however, hES cells touched on a subject that the American public had discussed for some time.

41 Ibid.

42 Paul Berg, "Reflections on Asilomar 2 at Asilomar 3: Twenty-Five Years Later," *Perspectives in Biology and Medicine* 44, no. 2 (2001): 185.

Fig. 5. Harold Varmus presenting research by John Gearhart and James Thomson to the U.S. Senate Labor, Health, and Human Services, and Education Subcommittee on December 2, 1998. The meeting was broadcast by C-SPAN.

(www.c-span.org, accessed Sept. 2, 2015)

Debates about abortion had facilitated strict opposition to any reconsideration of the definition of the term "embryo." When they went public with their work, Gearhart and Thomson knew exactly that research on fertilized human eggs would prompt violent rejection. What ultimately turned the tide for stem cell research in the US, however, was that developmental biology could side with biomedicine to propose that the new understanding of human development would result in revolutionary therapies. Utilitarian arguments went hand in hand with a more general public learning process that began in 1998 and stretched out into the early 2000s, a process that made it increasingly difficult (but not impossible) for opponents of stem cell research to refer to fertilized eggs as "microscopic Americans" or "mini-children" to be "adopted," not "murdered."[43]

Such developments could hardly be foreseen in 1998. News of the isolation and cultivation of hES cells came at a time when biology had been on the news for some time. The previous year, Ian Wilmut at the Roslin Institute in Scotland had published his work on cloning. News of Dolly, a

---

43 Lynn Harris, "Clump of Cells or 'microscopic American'?," http://www.salon.com/2005/02/05/embryos/. A "certain critical mass of [public] attention," Hall observes, frequently facilitates a public learning process from repugnance to informed evaluation. Hall, *Merchants of Immortality*, 265.

cloned ewe, had framed the public debate on biomedicine and elicited dark fears, one the one hand, of a loss of individuality through cloning, and fantastic hopes for somehow prolonging life or copying individuals, on the other. Researchers struggled to explain the difference between what they called "reproductive" and "therapeutic" cloning, rejecting the former while defending the latter. Meanwhile, an IVF specialist and a reproductive physiologist announced that they would try to clone humans.[44] Researchers who had worked on cloning and stem cells for some time struggled to confront the idea that human reproductive cloning was safe while trying to prevent Congress from outlawing cloning altogether.[45] In combination with hES cells, cloning by somatic cell nuclear transfer could perhaps be used to create tissue with particular characteristics (such as Parkinson's), tissue that could then be used as a model for investigating diseases and therapies. Rudolf Jaenisch at MIT's Whitehead Institute pointed out that reproductive cloning did not work anyway as it led to abnormalities. Because no one in their right mind would try to clone humans, therefore, he considered the ethical issue to be more limited than both gloom-mongers and renegade physicians supposed.[46] By attacking reproductive cloning, researchers doubtless raised their public standing in the contemporary debate on stem cells.

While the issue of cloning evolved side by side with the new issue of stem cells, however, only the latter was associated with the abortion debate because of the controversial use of fertilized human eggs. The political setting established in the seventies, therefore, structured the cultural debate on stem cells after 1998. Within this setting, research on hES cells came to be evaluated along the familiar political lines of the abortion debate. When news of Thomson's and Gearhart's work broke in 1998, everyone connected it to the Dickey-Wicker regulatory context. Lawyers at the NIH scrambled "to settle that issue, even as Representative Jay Dickey (R-Ark.), one of the authors of the congressional ban, reasserted his intention to

---

44 Rudolf Jaenisch and Ian Wilmut, "Don't Clone Humans!," *Science* 291, no. 5513 (March 30, 2001): 2552.

45 National Academy of Sciences, "U.S. Policy Makers Should Ban Human Reproductive Cloning," January 18, 2002, www.nationalacademies.org (accessed July 1, 2015). The panel was chaired by Irving Weissman, Stanford University.

46 Rudolf Jaenisch, "The biology of nuclear cloning and the potential of cloned embryonic stem cells for transplantation therapy," in *Klonen in biomedizinischer Forschung und Reproduktion*, edited by Ludger Honnefelder and Dirk Lanzerath (Bonn: Bonn Univ. Press, 2003), 573–598.

keep the funding probation in place."[47] The political debate evolved in three distinct phases.

The first phase lasted from the 1998 publication of Gearhart's and Thomson's work to President George W. Bush's 2001 decision on federal funding for stem cell research. During this initial phase, the NIH assumed leadership of the debate in public because it stood to gain from a Congressional change of heart. The organization considered news of human embryonic stem cells to be an opportunity to sway public opinion on long-standing issues by highlighting "The Promise of Stem Cell Research." In January 1999, the NIH took the stance, certified by its legal counsel Harriet Rabb, that hES cells were not "organisms" and thus did not meet the criterion for exclusion from NIH funding in the Dickey-Wicker Amendment.[48] "Are these stem cells organisms?," asked Senator Tom Harkin who endorsed the NIH perspective.[49] The issue, of course, was that the derivation of hES cells from the blastocyst led to the latter's demise, even if the IVF clinic that had provided the fertilized egg, would have discarded it anyway. Whether the blastocyst should be considered an embryo, therefore, remained more pertinent than another issue, namely whether stem cells were "organisms." The NIH also emphasized the potential medical benefits of stem cell research. Director Harold Varmus called Gearhart's and Thomson's work "an unprecedented scientific breakthrough" with the "potential to revolutionize the practice of medicine."[50]

In April 1999, the NIH decided to act on the recommendation of its own legal counsel and begin drafting guidelines so that researchers who wanted to work with Thomson's hES lines could apply for funding. Considering the controversy's national significance, this was a momentous step for the NIH because it made an assessment and began to act on it without eliciting further political endorsement. Publication of the guidelines on August 25, 2000 prompted thousands of responses from the public as stem cells and cloning came to define national news. During this early phase of

---

47 "Crucial Human Cell Is Isolated, Multiplied," *Washington Post*, Nov. 6, 1998, A01.

48 *Washington Post*, Jan. 20, 1999, A02.

49 United States, *Stem Cell Research: Hearings before a Subcommittee of the Committee on Appropriations, United States Senate, One Hundred Fifth Congress, Second Session, Special Hearing, December 2, 1998, Washington DC, January 12, 1999, Washington DC, January 26, 1999, Washington DC*, Senate Hearing 105-939 (Washington: GPO, 1999), 29.

50 National Institutes of Health, "NIH Director's Statement on Research Using Stem Cells—12/02/98 [Stem Cell Information]," December 2, 1998, http://stemcells.nih.gov/policy/statements/pages/120298.aspx.

the stem cell debate, however, there was significant confusion as other biomedical topics interfered with a clear view on the stem-cell question.

Interference was caused by issues such as cloning, which continued to be debated in Congress, and they were also prompted by biomedical free-riders seeking the public limelight. Former Geron founder Michael West sought to tap public attention for his new employer, Advanced Cell Technology (ACT), by claiming that he had used a cow egg to clone human DNA (creating "embryos that are part human and part cow").[51] The Jones Institute, a private clinic, also sought the public's attention by announcing that it had fertilized human eggs with the specific aim of making them available for research. John Gearhart commented that he was "a bit perplexed by this," as you "will hear none of the scientists who are involved in this work talk about making embryos to destroy them in any way. We don't think it's necessary."[52] Despite this hodgepodge, however, one issue ultimately came to stand out in 2001, and that was the question of how the incoming US president would respond to the NIH decision to fund research on hES cells, a decision that would mark the beginning of the second phase.

In August 2001, George W. Bush went on national television for the first time in his presidency. In an announcement that commentators expected to shape his presidency, Bush explained that he would not allow the new NIH guidelines to stand and that he would restrict NIH-funded research on hES cells to such lines as already existed on the day of his speech. In keeping with a well-established pattern, the president wanted to rule out that federal funds were used for work that many of his supporters considered objectionable. But Bush's decision was actually in line with his predecessor's cautious approach and was not intended to shut down research.[53] Bush, after all, allowed research to go forward with some hES cell lines. But the limitations he imposed polarized public debate, with many Democrats gratefully taking up the cause of science and enlightenment. "I have concluded that we should allow federal funds to be used for research on these existing stem-cell lines, where the life-and-death decision has already been made," Bush argued in his address.[54] His critics responded

51 *Washington Post*, June 14, 1999, A01.
52 *New York Times*, July 11, 2001.
53 Gottweis, "The Endless hESC Controversy in the United States," 557.
54 "Bush's Address on Federal Financing for Research With Embryonic Stem Cells," *New York Times*, August 10, 2001.

that the life-and-death decision had been made before hES cells were derived from blastocysts because IVF clinics were bound to discard
thousands of fertilized eggs anyway.

The stem cell debate continued to be structured by decisions taken
during the seventies and eighties. Within that framework, however, the
debate during its second phase witnessed significant realignments. In Congress, some abortion opponents came out in favor of funding policies
more liberal than Bush allowed for. Pro-life forces were put on the defen-
sive by stem cell advocates who argued that they were also "pro-life," but
this time with reference to patients saved by future stem-cell cures. By
2004, stem cell research had become a rallying cry of "science vs. social
conservatism."[55] That year, John Kerry sought to unseat Bush in the White
House, and the Democratic Party turned the cause for stem cells into a
cause for science and the party. At their national convention, Democrats
were able to present Ronald Reagan's son Ron Reagan who explained in
his televised speech that Americans had to choose between "reason and
ignorance, between true compassion and mere ideology" and that they
should cast a vote for embryonic stem cell research.[56] Bush won the presidential election but the cultural polarization continued to help mobilize
stem cell supporters. And in keeping with policy characteristics developed
in the 1970s, the matter was no longer pursued in the federal arena, but on
subnational planes, as efforts shifted to the various American states.[57]

In response to Bush's 2001 decision, the *New York Times* had editorial-
ized that the "twin forces of science and capitalism will undoubtedly un-
leash their magic."[58] Because no federal law existed that would have
banned human embryonic stem cell research, private support remained
possible. In the absence of federal restrictions, however, states could also
decide to make up their mind about stem cell research. Some American

---

55 "Stem-Cell Debate another Division between Bush, Kerry," usatoday.com,
   http://usatoday30.usatoday.com/news/politicselections/nation/issues/2004-10-26-
   stem-cell-research_x.htm.
56 "Ron Reagan, Remarks to the 2004 Democratic National Convention," July 27, 2004,
   http://www.presidentialrhetoric.com/campaign/dncspeeches/reagan.html.
57 Aaron D. Levine, T. Austin Lacy, and James C. Hearn, "The Origins of Human Embryonic Stem Cell Research Policies in the US States," *Science and Public Policy* 40, no. 4 (August 1, 2013): 544–58.
58 Alan Wolfe, "Bush's Gift to America's Extremists," *New York Times*, Aug. 19, 2001.

states restricted work on such cells by criminalizing it.[59] But the prospects of stem-cell-based therapies developed so much momentum that states would eventually outdo the federal government in supporting such work.

From 2004, during the third phase of the stem cell debate, California turned itself into the poster child of stem cell research. "It's the first time I know of that a state has said that the federal government has neglected its opportunity to lead in this area of research," research veteran Irving Weissman at Stanford argued, "so it is the right of the states to take over where the federal government left off."[60] Real estate developer Robert Klein, father to a son with diabetes, initiated (and helped fund) a multimillion dollar campaign for Proposition 71. This state referendum created CIRM, the California Institute of Regenerative Medicine. Funded by a three billion dollar bond issue over ten years and shielded from state lawmakers through elaborate constitutional anchors, CIRM outflanked the NIH to become the largest funder of stem cell research in the world. Since it began work in 2007, CIRM has poured three hundred million dollars into stem cell research each year. At the time of the referendum in 2004, the US was bogged down in costly and unsuccessful wars in Iraq and Afghanistan. Klein and his supporters, the medical profession and patient-advocacy groups, used biomedicine as a means to reinvigorate ambitions for preserving California's leadership in science and technology, and to mobilize the state's progressivism against George W. Bush in Washington and against the state's political establishment in Sacramento. The pro-stem-cell line-up included well-known actors Michael J. Fox, who suffered from Parkinson's, and Christopher Reeves, who had suffered a severe spinal cord injury. Supporters envisioned a new Silicon Valley around stem cell research. Campaign managers overplayed the promise of therapies for major diseases even if some work has entered clinical testing.[61] But while key advisers such as Weissman helped design CIRM as an agency that would oversee therapies through clinical testing, the costly "valley of death" companies avoided crossing even with promising therapies. CIRM

59 Geoffrey Lomax, "Rejuvenated Federalism: State-Based Stem Cell Research Policy," in Benjamin J. Capps and Alastair V. Campbell, *Contested Cells: Global Perspectives on the Stem Cell Debate* (London: Imperial College Press, 2010), 369.

60 Daniel S. Levine, "Irv Weissman: A Researcher Becomes a Warrior," *San Francisco Business Times* 18, no. 43 (May 28, 2004): S12.

61 Eileen Burgin, "Human Embryonic Stem Cell Research and Proposition 71: Reflections on California's Response to Federal Policy," *Politics and the Life Sciences* 29, no. 2 (September 1, 2010): 73–95.

grants and loans provided to researchers, furthermore, are subject to agreements that provide the State of California with income from patents, royalties, and licenses that arise from their work.[62]

Other states followed the Californian example. John Gearhart in a 2006 interview expected as much when he pointed out that "places like California, with financial backing and legislative support, are going to be the real centres [sic] that lead the research in the USA and where you are going to see all the progress."[63] Gearhart's assessment played on rivalries between research-oriented states, and MIT-based researcher Rudolf Jaenisch added that Californians "have had a windfall with which we can't compete."[64] But science supporters in Missouri successfully initiated a state constitutional amendment to protect stem cell research in Kansas City that was funded with one billion dollars from cancer-survivors James and Virginia Stowers.[65] Research-oriented states such as New York and Massachusetts upped their stem cell research budgets. By 2007, all American states together anticipated spending over 500 million dollars a year.[66] Because it assumed a role in California akin to the NIH for the US at large, CIRM implemented structures such as funding guidelines and ethical review boards to provide oversight. The organization is aware that it effectively took the lead on dealing with ethical issues arising from work on hES cells,[67] and CIRM attests to major shifts in the relationship between the

---

62 Full Text of Proposition 71, voterguide.sos.ca.gov/past/2004/general/propositions/prop71text.pdf (accessed Sept. 1, 2015).

63 Howard Wolinsky, "A Decade of Stem-Cell Research. An Interview with John Gearhart," *EMBO Reports* 10, no. 1 (January 2009): 14.

64 Author's interview with Rudolf Jaenisch, 2012.

65 Kant Patel, "The Politics of Stem Cell Policy: Ballot Initiative in Missouri," *Social Work in Public Health* 26, no. 2 (March 2011): 158–75.

66 James W. Fossett, "Federalism by Necessity: State and Private Support for Human Embryonic Stem Cell Research," Rockefeller Institute Policy Brief (Albany, New York: Nelson A. Rockefeller Institute of Government, August 9, 2007), http://www.rockinst.org/pdf/health_care/2007-08-09-federalism_by_necessity_state_and__private_support_for_human_embryonic_stem_cell_research.pdf. For the case of Texas, see Joel B. Finkelstein, "Texas Prepares To Invest Billions In Cancer Research," *Journal of the National Cancer Institute* 100, no. 10 (May 21, 2008): 696–97.

67 CIRM modeled its regulations on earlier ones developed by the National Academy of Sciences. In a recent meeting, its Scientific and Medical Standards Working Group discussed changes to its MES regulations. During the meeting, Geoffrey Lomax, Senior Officer for Medical and Ethical Standards pointed out that CIRM regulations had evolved on the basis of National Academy guidelines but that the academy's committee had disbanded in 2010. "We are in a unique position," he said, "of being one of the only

states and the federal government since 1980, and in the shifting role of science in America.[68]

7. Conclusion

At the outset, I had suggested that science occasionally turns into a cultural icon for the public and I had also claimed that the use of this icon was transformed during the stem cell debates of the early 2000s. Because no federal decision had been made on the core issue of the embryo's ethical status during the seventies and eighties, this left open the door in the US for renewed debates after 1998. But when advocates of stem cell research mobilized the relevance of therapy against the pro-life camp, their success was relegated to the state level. The national issue remains unresolved and perhaps it will remain so even after researchers in 2007 discovered a way to create a stem cell (induced pluripotent stem cells) similar to hES cells, but without the need to derive them from a human blastocyst.[69] The new funding patterns for stem cell research have countered slackened support from the NIH, but federal funding remains the gold standard. "[No] matter how generous the private sector is," John Gearhart pointed out in 2006, "we need access to federal money; the NIH spends billions of dollars each year and funds 85 percent of all biomedical research in this country—private donations simply cannot compete."[70]

groups that's continuing to reevaluate that document … that has national implications. Everyone in the stem cell space more or less adopts those guidelines…. You're the only group that I'm aware of that's really having these kinds of deliberations." California Institute for Regenerative Medicine, "Minutes of the Regular Meeting, Scientific and Medical Accountability Standards Working Group," April 2, 2015, 64. See also Geoffrey P. Lomax, Zach W. Hall, and Bernard Lo, "Responsible Oversight of Human Stem Cell Research: The California Institute for Regenerative Medicine's Medical and Ethical Standards," PLoS Medicine 4, no. 5 (May 2007): e114.

68 Michael Mintrom, "Competitive Federalism and the Governance of Controversial Science," Publius 39, no. 4 (October 2009): 606–31; John Aubrey Douglass, "The Entrepreneurial State and Research Universities in the United States: Policy and New State-Based Initiatives," Higher Education Management and Policy 19, no. 1 (2007): 84–120; Council of State Governments, "State Responses to Federalism," Nov. 2007, www.csg.org/pubs/Documents/TIA_StateResponsestoFederalism.pdf.

69 Shinya Yamanaka, "Induced Pluripotent Stem Cells: Past, Present, and Future," Cell Stem Cell 10, no. 6 (June 14, 2012): 678–84.

70 Wolinsky, "A Decade of Stem-Cell Research. An Interview with John Gearhart," 14.

Let me conclude with two observations:

The first observation concerns the transformation of science in the American setting and beyond. The *New York Times* in 2014 pointed to the overall shift toward privately funded science. The newspaper had in mind spaceships and telescopes as well as biotech, and suggested that science funding by wealthy donors raised questions about "the social contract that cultivates science for the common good."[71] The stem cell wars indeed suggest that science in the US faces specific difficulties arising from its mobilization for partisan political battles and, more generally, from weakening ties to common national endeavors. This observation seems to integrate well with Daniel Rodgers' assessment of an *Age of Fracture*.

At the same time, however, the stem cell story shows that this fragmentation contains strong elements of continuation. Science funding in the United States historically has been more diverse than in countries such as France or Germany, and the profession has had available to it different potential sources of support. The key difference between the situation in those countries and the US would seem to be that, in the case of stem cell research, other sources can be mobilized because the US has not made a binding legal decision, a situation that frees the research profession to enter new coalitions.

---

71 "Billionaires With Big Ideas Are Privatizing American Science," www.nytimes.com, March 16, 2014, http://www.nytimes.com/2014/03/16/science/billionaires-with-big-ideas-are-privatizing-american-science.html.

The Internationalization of Science, Technology & Innovation (STI): An Emerging Policy Field at the Intersection of Foreign Policy and Science Policy?

Nina Witjes and Lisa Sigl

> You don't solve a problem of nuclear weapons and their relations to the world by saying, "Here is a nuclear core – that's scientific; here is a nuclear weapon – that's military; here is a treaty – that's political." These things all have to live with each other. There are elements that are indeed military or technological or diplomatic, but the process of effective judgement and action comes at a point where you cannot separate them out. It follows that it is also nonsense to talk about the political neutrality of science. (Bundy 1963)

1. The Complex Relationship between International Collaboration in STI and International Relations

Scientific collaboration to improve relations between nation-states and diplomatic efforts to make possible large-scale scientific projects, both have long traditions. After World War II governmental and non-governmental actors each consciously used scientific and technological collaboration as a means of enhancing international relations.[1] With regard to the development, possession, and use of military technology it seemed increasingly necessary to cope with the destructive potential of science and technology through international collaboration. Since science, technology, and innovation (STI) were regarded as having the power to deeply influence international relations, nation-states could no longer retreat to an exclusively national orientation in their STI policies. Building trust between

---

[1] When the US government proposed the 1946 Baruch Plan for the international control of nuclear energy, Russian and American scientists met on a regular basis to assess the dangers of weapons of mass destruction. These meetings were known as the Pugwash conferences.

European nation-states was also one of the core reasons for creating the European Organization for Nuclear Research (CERN), a scientific endeavor that would not have been possible without international cooperation.[2]

Besides a diplomatic motivation for enhancing STI collaboration, an international orientation of scientific efforts was also considered vital for economic reasons. When the Organisation for Economic Co-operation and Development (OECD) started promoting national science policies in the 1960s and the National Innovation System (NIS) became a dominant guiding concept for STI policies, international collaboration was understood as an important part of "national" innovation policies.[3] They were defined as one of five types of relationships that made up an NIS (other than relationships between economic sectors, basic and applied research, different policies and science policy, and policy for economic development).[4] Internationalization thus did not challenge the national framing of STI policies but was seen as compatible with, and in fact essential for, creating a national economic system based on innovation.

In the 1990s, however, international collaboration in STI had become so important that many authors observed a "denationalization" as a prevailing trend in science, particularly through the ongoing internationalization of science funding.[5] The debate raised questions about whether and how far an internationalization of STI would make national policies obsolete and challenge the NIS approach to STI policy making. "[E]mbryonic

---

2 On the development of European scientific institutions, see Arne Pilniok, "The Institutionalization of the European Research Area: The Emergence of Transnational Research Governance and its Consequences," in this volume.

3 While the term was not used before Freeman did so in his study on Japanese innovation in 1987, the nation-centric perspective on STI is implicit in OECD works from the 1960s onwards. See Benoît Godin, "National Innovation System. The System Approach in Historical Perspective," *Science, Technology & Human Values* 34, no. 4 (2009): 468. For Christopher Freeman's study, see his *Technology Policy and Economic Performance* (London: Pinter, 1987).

4 Organisation for Economic Cooperation and Development (OECD), *Science and the Policies of Government* (Paris: OECD, 1963): 26–27; Godin "National Innovation System," 488.

5 Elisabeth Crawford, Terry Shinn, Sverker Sörlin, Nationalization and Denationalization of the Sciences: An introductory Essay," in *Denationalizing Science: The Contexts of International Scientific Practice*, edited by Elisabeth Crawford, Terry Shinn, Sverker Sörlin (Dordrecht: Kluwer, 1993), 1–42.

transnational systems of innovation"[6] seemed to evolve as the European Community indeed took steps towards creating a European system of innovation (through its Single European Act).[7] In the meantime, however, most scholars argued that national policies and institutions continue to play a crucial role in overall STI policies despite higher degrees of internationalization.[8] The two most common explanations for this are that funding frameworks have largely stayed national in scope and that in the process of building a knowledge-based economy, international collaboration has become (and is used as) a strategy to cope with growing global competition.[9]

In this chapter, we will discuss both explanations in view of the European Union's ambition to become "the most dynamic and competitive knowledge-based economy in the world." The EU sees itself as being "confronted with a quantum shift resulting from globalization."[10] Competitiveness on global markets with innovative solutions and commercial products appears as a prime goal in EU-wide strategy papers on the internationalization of STI (most importantly the "International Strategy for Research and Innovation").[11] But other strategic aims have also gained ground, such as international scientific collaboration for dealing with global

6 Bengt-Åke Lundvall, "Introduction," in *National Systems of Innovation: Toward a Theory of Innovation and Interactive Learning*, edited by Lundvall (London: Anthem Press, 1992 [2010], 1–20, see esp. 16.

7 Esben Sloth Anderson and Asger Brændgaard, *Integration, Innovation and Evolution*, in *National Systems of Innovation*, edited by Lundvall, 233–57.

8 Dany Jacobs, "Innovation policies within the framework of internationalization," *Research Policy* 27, no. 7 (1988): 711–24; Ulrich Dolata, "Reflexive Simulation of Disjointed Incrementalism? Readjustments of National Technology and Innovation Policy, Science," *Technology & Innovation Studies* 1, no. 1 (2005): 59–76; Bo Carlsson, "Internationalization of Innovation Systems: A survey of the Literature," *Research Policy* 35, no. 1 (2006): 60.

9 Thomas Mandeville, "Collaboration and the Network Form of Organization in the New Knowledge-Based Economy," in *Handbook on the Knowledge Economy*, edited by David Rooney, Greg Hearn, and Abraham Ninan (Northampton, MA: Edward Elgar, 2005), 165.

10 European Commission, "Extracts from Presidency Conclusions on the Lisbon Strategy by Theme, European Councils: Lisbon (March 2000) to Brussels (June 2004)" (Brussels, 2004), 5.

11 European Commission, "Enhancing and focusing EU international cooperation in research and innovation: A strategic approach Communication from the Commission to the European Parliament, the Council, the European Economic and Social Committee and the Committee of the Regions" (Brussels, Sept. 2012), https://ec.europa.eu/research/iscp/pdf/policy/com_2012_497_communication_from_commission_to_inst_en.pdf.

challenges including climate change, humanitarian crises, forced migration, cyber terrorism, and global diseases.[12] Even if the extent to which internationalization is made possible is often criticized as insufficient, the EU's internationalization objectives are more than a statement of will. While internationalization was primarily addressed at the level of EU member states to support the European Research Area (ERA), most recent research framework programs (FP7, Horizon 2020) reach out to facilitate collaboration with third countries. Internationalization and cooperation with third countries is now a common option in all funding schemes.

For many countries in Europe, EU Framework Programmes have been an important impulse for extending their international orientation. These programmes first opened up cooperation with other EU member states and associated states, and they now aim at facilitating projects that go beyond Europe and help build an international profile. For aligning and coordinating EU efforts in STI internationalization (such as those set out in the 2008 Strategic European Framework for International Science and Technology Cooperation) and member states' activities, a Strategic Forum for International Science and Technology Cooperation (SFIC) was established. It has become an important body for negotiating a common approach (such as common geographic priorities by establishing initiatives for China, India, the US, and Brazil).

The tension between the aim to establish a competitive European Research Area (ERA), on the one hand, and member states' individual aims of staying competitive within and beyond Europe, on the other, make the European context particularly interesting for discussing our second argument: International cooperation is used as a means for dealing with global economic competition.

This chapter helps delineate this tension by exploring how policies of European member states frame international collaboration in STI. We will analyze how member states are building up an infrastructure for STI internationalization, and we will explore entanglements between their STI and foreign policies. In doing so we hypothesize that a new national policy area is developing in these overlapping fields and that this begins to constitute a policy field in its own right (section 3). Our assumption builds on the observation that, since 2000, national governments have been willing to in-

---

12 Inga Ulnicane, "Broadening Aims and Building Support in Science, Technology and Innovation Policy: The Case of the European Research Area," *Journal of Contemporary European Research* 11, no. 1 (2014): 31–49.

vest in new infrastructure for pursuing internationalization parallel to (and distinct from) the EU's transnational effort. Governments are creating platforms for knowledge sharing and coordinating diverse sets of national actors potentially involved in such activities. In reference to EU activities, they aim at maximizing participation in EU funding schemes and they seek to develop a complementary framework for internationalization. A range of European countries is in the process of creating networks of quasi-embassies responsible for international STI relations. These institutions create strategic alliances between policy makers, and they generate and disseminate information and insights for potential cooperation with regions and their institutions. Among them are German Houses for Research and Innovation (*Deutsche Wissenschafts- und Innovationshäuser*, DWIH), the Science and Innovation Network (SIN) in the UK, the swissnex network in Switzerland, and FinNode Innovation Centres in Finland.

It is tempting, therefore, to hypothesize that through STI internationalization, a new policy field is being created that increasingly cultivates institutional and authoritative demarcations of its own. We have to take into account, however, that science, technology, and innovation are governed differently in different national contexts. Historically they build on different political structures, industrial (technological) development opportunities, systems of higher education, and so on.[13] Similarly, foreign policy aims and interests naturally differ between federal states. We have to take into account differences in how STI internationalization is set up and framed in different national contexts. This chapter first contributes to an understanding of how STI internationalization has led to the emergence of a new policy field in the making. Second, we seek to delineate some of the ways in which STI internationalization is pursued politically in different national contexts.

We will make use of a comparative approach by focusing on STI internationalization policies in Germany and the UK to explore two similar but different ways in which EU member states actively build an infrastructure for STI internationalization at home and abroad. We are interested in how STI policies and international relations have come to interrelate in novel ways, since national governments enhanced their STI internationalization efforts, and thus we ask: How are nation-states building up infrastructures for STI internationalization? Has STI internationalization led to the

---

13 Lundvall, "Introduction," 4.

making of a new policy field? How do those active in STI internationalization policy-making assess the interrelation between STI policy and international relations? This leads us to the question of whether and how far the two policy fields have come to mutually shape each other. Finally, we will reflect on implications that a new policy field at the intersection of STI policy and foreign policy could have for further studies on STI internationalization.

Germany and the UK were chosen as case studies because they are among the most active countries in developing explicit internationalization strategies. They have built up physical infrastructures abroad, quasi embassies, for STI collaboration that strongly refer back to the national level, engaging both science policy and foreign policy actors.

2. Theoretical Approach: Bridging Innovation Studies and International Relations

Our analysis is situated at the interface of Innovation Studies and International Relations. Innovation as a field of interest for the social sciences and economics is often dated back to Schumpeter's work in the 1930s, or even to the early 1900s when "anthropologists, sociologists, historians, and economists began theorizing about technological innovation, each from his own respective disciplinary framework."[14] Innovation Studies as a distinct research field, however, only started to develop in the early 1970s, a development fostered by an increasing policy interest (by the OECD and national governments) in how societies can best benefit from innovation.[15] Out of the breadth of approaches to studying innovation (such as change in behavior, inventions, or technological change), this period was characterized by a narrowing down of definitions of innovation and approaches to studying innovation: Innovation was conceived of as "technological innovation defined as commercialized invention,"[16] and the process of innovation became increasingly understood as happening primarily in the

---

14 Benoît Godin, "Innovation Studies: The Invention of a Specialty," *Minerva* 50, no. 4 (2012): 397

15 Jan Fagerberg, Morten Fosaas, Koson Sapprasert, "Innovation: Exploring the knowledge," *Research Policy* 41, no. 7 (2012): 1143.

16 Godin, "Innovation Studies," 397.

framework of National Innovation Systems (NIS).[17] Innovation Studies in the past decades have shown how dominant the NIS approach has become in STI policies today, and how it is challenged by increasing internationalization dynamics.

Since the debate on innovation-related economic and political competition and global innovation leadership is increasingly linked to the transformation of global power dynamics, we enrich these reflections with insights from International Relations. These are helpful for understanding how policies on international cooperation in STI are extended well beyond the traditional realm of STI policies and they have come to touch upon important questions of foreign policy.[18] We conclude that even though it has long been agreed that STI policies are "at the heart of international trade relations, foreign policy, economic strategy, and social interests"[19] and situated within a longstanding debate about the implications of globalizing STI processes for national policies,[20] increasing entanglement between STI and foreign policy remains insufficiently reflected in these disciplines. We aim to overcome this limitation by connecting both fields of study in our analysis.

2.1. The National-Innovation-System Approach: Challenged by STI-Internationalization?

At the end of the past millennium, many authors considered "denationalization" a prevailing trend in science policies. Despite the emergence of new nation-states, "transnational science" seemed to gain the upper hand.[21] It challenged the hegemonic approach of taking the nation-state as a main reference point in analyzing innovation systems and in shaping STI

---

17 Fagerberg, Fosaas, Sapprasert, "Innovation"; Godin, "National Innovation System."

18 Georg Schütte, *Wettlauf ums Wissen: Außenwissenschaftspolitik im Zeitalter der Wissensrevolution* (Berlin: Berlin Univ. Press, 2008); Tim Flink, Ulrich Schreiterer, "Science Diplomacy at the Intersection of S&T Policies and Foreign Affairs. Toward a Typology of National Approaches," *Science & Public Policy* 37, no. 9 (2010): 665.

19 John De la Mothe and Paul R. Dufour, "Techno-Globalism and the Challenges to Science and Technology Policy," *Daedalus* 124, no. 3 (1995): 232.

20 Sylvia Ostry and Richard Nelson, *Techno-Nationalism and Techno-Globalism: Conflict and Cooperation Integrating National Economies* (Washington DC: Brooking Institution Press, 1995).

21 Crawford, Shinn, and Sörlin, "Nationalization and Denationalization;" Carlsson, "Internationalization of Innovation Systems."

policies. As mentioned above, however, internationalization does not con-tradict a basic national approach to understanding STI infrastructure, but rather is an inherent part of it: It has been argued elsewhere that interna-tionalization is often pursued under the banner of strengthening national competitiveness and of increasing national value generation. In the Swe-dish context, for example, key aims were to attract international compe-tence and capital, to use science as a ticket to international cooperation, and to gain international competitive advantage.[22] In this way, the NIS approach has remained the central category, not only for policy makers but for STI-policy analysis as well.

We should mention, however, that a notable shift has taken place in how researchers address science, technology, and innovation. While the research community that studied relationships between knowledge pro-duction and policy issues started out defining itself as a community of science policy researchers, a gradual shift has been taking place towards framing the same research interests as Innovation Studies. This is more than a terminological shift because "innovation" in this context usually infers an integrated study of knowledge-production that reflects the NIS-approach favored by the OECD. Attention has shifted from individual innovation actors (such as firms, universities, and public research labs) to the links and interactions between the various actors making up an inno-vation system. The dominant notion of innovation in this framework (which has been stabilized and co-produced by STI statistics and policy frameworks) is that of innovation as technological innovation for the mar-ket and for national economic growth.[23]

The stronger pace of internationalization in the past decade has pro-voked a new dynamic in the debate. Some caution against "a global retreat into techno-nationalism" that reduces science to its role in national eco-nomic development, and they plead for a new cosmopolitan approach, emphasizing transnational networks.[24] Others still find it legitimate that the Weberian state should remain the defining agent of the national system of

---

22 See Tomas Hellström and Merle Jacob, "Taming Unruly Science and Saving National Competitiveness," *Science, Technology & Human Values* 30, no. 4 (2005): 452.

23 Benoît Godin, "Innovation and Science: When Science Had Nothing to Do with Inno-vation and *Vice-Versa*. Project on the Intellectual History of Innovation" (INRS, Project on the Intellectual History of Innovation, Working Paper No. 16, Montréal, 2014), 38.

24 Charles Leadbeater and James Wilsdon, *The Atlas of Ideas. How Asian Innovation Can Benefit Us All* (London: Demos, 2007), 11

innovation.[25] In defense of the NIS, others argue that innovative activity was never entirely national in scope but that funding and research and development (R&D) activities had such a focus.[26]

Nevertheless, the dispute over the accuracy of the NIS approach has established itself within innovation studies. In a review of empirical studies on the internationalization of innovation systems, Carlsson points to the most common conclusion. The interdependence of national innovation systems grows with the internationalization of corporate R&D, technology transfer, or with the international flow of scientific and technical personnel. National policies and institutions, however, continue to play a role because many institutions relevant to innovation systems are national in scope (such as funding, education, and intellectual property rights regulations). Many suspect that efforts toward European integration will not do away with the predominant national scope of innovation systems in Europe.[27] But international activities on all levels of STI have caused other authors to scrutinize whether we can still accurately speak of NISs. In the context of solving global challenges, it has been argued that we should think more in terms of a Global Systems of Innovation (GSI)[28] or Global Innovation Networks.[29] As one of the key challenges for Innovation Studies, researchers have pointed to the problem of coming to terms with its increasingly international character, observing that key players in innovation (such as multi-national firms) have responded to and advanced economic globalization.

The challenge to IS researchers is to identify, map and analyse these global systems of innovation and their interactions with national and regional systems [...]. This will surely yield important policy implications, just as the development of the NSI concept originally did, not least as we are confronted by ever more urgent global

---

25 Mario Scerri and Helena Lastres, "BRICS Project: Comparative Report on the State and the National Systems of Innovation in BRICS" (Globelics & RedeSist, 2010), 3

26 Ben R. Martin, "Innovation Studies: Challenging the Boundaries" (paper presented at the Lundvall Symposium, Aalborg University, Denmark, Feb. 2012), 9

27 Carlsson, "Internationalization of Innovation Systems," 63.

28 Susan E. Cozzens and Pablo Catalán, "Global Systems of Innovation. Water Supply and Sanitation in Developing Countries" (paper presented at the 4th Globelics International Conference, Mexico City, Sept. 2008).

29 Jason Dedrick, Kenneth L. Kraemer, and Greg Linden, "Capturing Value in a Global Innovation Network: A Comparison of Radical and Incremental Innovation" (unpublished paper, Personal Computing Industry Center, University of California, Irvine, Sept. 2007), http://www.itif.org/files/KraemerValueReport.pdf.

challenges (economic, environmental, demographic, health, security, etc.) and attempt to respond to these.[30]

In the STI policy discourse, however, the systems approach is more successful than ever.[31] The discursive construction of national systems of innovation can even become central for defining and legitimizing the sovereignty of regions or national states. Sharif describes the case of Hong Kong and points out that the NIS approach is used as a rhetorical device for strengthening the Chinese city's claim to political sovereignty.[32]

How dynamics in international relations are affected by STI internationalization, however, is hardly discussed in Innovation Studies beyond the context of closing the innovation gap for staying (or becoming) globally competitive.[33] Many authors have stated that the gap in scientific and technological capabilities is decreasing between the former scientific and technological triad of the EU, the US, and Japan, on the one hand, and newly emerging powers in East Asia and elsewhere, on the other.[34] It seems beyond debate that due to an anxiety over who will take over innovation leadership in the future, new dynamics will arise in international relations. Shifts in the geographies of innovation are often presented rather dramatically. The following quote exemplifies the UK perspective:

The rise of China, India and South Korea will remake the innovation landscape. US and European pre-eminence in science-based innovation cannot be taken for granted. The centre of gravity for innovation is starting to shift from west to east.[35]

The rhetoric of studies oscillates between considering emerging STI powers a threat to the traditional triad of Europe, the US, and Japan, and considering them to provide opportunities for enhanced cooperation with mutual benefits. The rationale behind this is that while international collaboration clearly bears risks that include intellectual property disputes,

---

30 Ben R. Martin, "Innovation Studies: Challenging the Boundaries" (paper presented at the Lundvall Symposium, Aalborg University, Denmark, Feb. 2012), 9.

31 See Lew Perren and Jonathan Sapsed, "Innovation as Politics: The Rise and Reshaping of Innovation in UK Parliamentary Discourse, 1960–2005," *Research Policy* 42, no. 10 (2013): 1815–28; Naubahar Sharif, "Rhetoric of Innovation Policy Making in Hong Kong Using the Innovation Systems Conceptual Approach," *Science, Technology, & Human Values* 35, no. 3 (2010): 408–34.

32 Ibid.

33 See Hellström and Jacob, "Taming Unruly Science."

34 Messner, *Der beschleunigte globale Wandel.*

35 Leadbeater and Wilsdon, *Atlas of Ideas,* 9.

cyber security, or espionage, the biggest risk of all is disengagement. Breaking off relations would be a disaster indeed. More recently, approaches have shifted from earlier optimistic accounts about the benefits of STI internationalization for science towards a more cautious approach of "strategically balancing the benefits that can flow from increased collaboration across the spectrum of research and innovation activity with some of the risks that come with it."[36]

2.2. STI and International Relations within Shifting Geographies of Power

In International Relations, science, technology, and innovation, the three components that make up STI, have almost always been treated as distinct entities. We will first follow this approach to reconstruct how the debate around STI and its role in global affairs has evolved within IR. In this section, we will contribute to a more comprehensive understanding of the reciprocal influence of STI and international politics in order to develop questions for empirical investigation. In considering two major IR theories, realism and liberalism, we see that the overall concern is with technology while science and innovation are considered to be peripheral if they are considered at all. The realist focuses on technology as a fundamental asset in the global distribution of power that defines the material capabilities of nations.[37] The realist also assumes for the state a central role in the governance of technological systems.

Technology came to be entangled with military power in the context of the Cold War, but economic globalization has subsequently led to an understanding of technology as central for mustering economic power in the context of stiff global competition.[38] Fritsch suggests that traditional realist

---

36 Interview with British specialist, no. 2, Sept. 3, 2013. Political support for research on global challenges constitutes an exception as it requires different forms of international collaboration. See Jacob Edler, "International Policy Coordination for Collaboration in S&T" (Manchester Business School Working Paper No. 590, 2010), 5, http://www.mbsportal.bl.uk/secure/subjareas/techinnov/mubs/wp/114382WP590_10.pdf; Patries Boekholdt, Jacob Edler, Paul Cunningham, and Kieron Flanagan, "Drivers of International Collaboration in Research: Final Report" (Technopolis Group, Manchester, 2009), http://ec.europa.eu/research/evaluations/pdf/archive/other_reports _studies_and_documents/drivers_of_international_cooperation_in_research.pdf.

37 Kenneth Waltz, *Theory of International Politics* (New York: Waveland, 1979), 131.

38 Robert Gilpin, *The Political Economy of International Relations* (Princeton: Princeton University Press, 1987).

perspectives on concepts of national interest, power accumulation, and anarchy leave little room for elaborate concepts of technology.[39] Although IR literature brims with examples of how technology such as the nuclear bomb have changed and influenced international relations, its impact is hardly reflected in its theory. This has led authors such as Herrera to refer to technology in international relations literature as "the great residual."[40] For our purposes of charting the relationship between science, technology, and the nation-state, however, we may draw on a realist understanding of the state's role in shaping the governance of national STI systems, and on its understanding of a competition between states as a facilitator of technological innovation. But for the purpose of our study, these realist perspectives should be juxtaposed with liberal approaches towards technology in the international system. Liberal accounts provide important insights into an increased need for cooperation in the context of global interdependence. With respect to the question of whether a new field arises that relates foreign policy and science policy, liberals are aware that state and non-state actors, international and private institutions, create new global structures as well as new forms of global authority and policy.[41]

Innovation is rarely discussed by IR scholars. Taylor observes that "[a]s if in retaliation, most political scientists who discuss technological variables often neglect the enormous body of innovation research that has developed [...] in other social sciences."[42] Taylor is one of few scholars who analyze national innovation rates in the context of domestic tensions and external threats so as to provide an explanation for security-related national differences in innovation rates. He points out that national innovation rates are central to the subfield of security studies within IR. He convincingly relates a nation-state's innovative capacity to its "economic growth, industrial might, and military prowess." He reasons that innovation rates clearly influence the balance of power between states and he tries to grasp (and even calculate) the impact of such rates on war and alliance for-

39 Stefan Fritsch, "Technology and Global Affairs," *International Studies Perspectives* 12, no. 1 (2011): 36.

40 Geoffrey L. Herrera, "Technology and International Systems," *Millennium – Journal of International Studies* 32, no. 3 (2003): 560.

41 Fritsch, "Technology and Global Affairs;" James N. Rosenau and Ernst O. Czempiel, *Governance without Government: Order and Change in World Politics* (Cambridge: Cambridge Univ. Press, 1992).

42 Mark Z. Taylor, "Toward an International Relations Theory of National Innovation Rates," *Security Studies* 21, no. 1 (2012): 115.

mation.[43] Much like in realist accounts of how technology shaped the international system, however, the relevance of innovation for national power remains unexplained and undertheorized in IR.[44] Taylor explains this "black-boxing" of the innovation process by IR scholars' tendency to assume "that the rate and direction of technological innovation are either 1) random, 2) scientifically & technically determined, or 3) structured solely by domestic politics and institutions."[45] The end result of all three approaches is that innovation is treated as exogenous to IR theory. These gaps are certainly enhanced by the neglect of international politics in innovation studies. When approaches emanating from the field conceptualize foreign science policy at all, they consider such policy a means of enhancing international cooperation with the overall aim of succeeding in global competition. But they do not refer to the field of foreign policy. Studies frequently focus on boosters and barriers for international STI collaboration. According to the perspective prominent in IR, the internationalization of science policy aims at attracting researchers from abroad and at integrating international corporations and their innovative research facilities, thereby facilitating access to knowledge, technology, and international networks.[46]

More recent work attempts to bridge this gap between STI and International Relations by drawing on constructivist concepts from science and technology society studies (STS),[47] beginning with the assumption that STI is socially constructed and relates to social processes.[48] Such approaches

---

43 Ibid., 114.

44 Charles Weiss, "Science, Technology and International Relations," *Technology in Society* 27, no. 3 (2005): 295–313.

45 Mark Z. Taylor, "Political Decentralization and Technological Innovation: Testing the Innovative Advantages of Decentralized States," *Review of Policy Research* 24, no. 3 (2007): 231–57.

46 Stefan Kuhlmann, "Forschungs- und Innovationssysteme im internationalen Wettbewerb," in *Der Wettlauf um das Wissen. Zur Außenwissenschaftspolitik*, edited by Georg Schütte (Berlin: Berlin Univ. Press, 2008), 57.

47 Also known as science, technology, and society studies.

48 Karen T. Litfin, "Public Eyes: Satellite Imagery, the Globalization of Transparency, and New Networks of Surveillance," in *Information Technologies and Global Politics: The Changing Scope of Power and Governance*, edited by James N. Rosenau and J. P. Singh (Albany: State Univ. of New York Press, 2002), 65–88; Stefan Fritsch, "Conceptualizing the Ambivalent Role of Technology in International Relations: Between Systemic Change and Continuity," in *The Global Politics of Science and Technology: Concepts from International Relations and Other Disciplines*, edited by Maximilian Mayer, Mariana Carpes, and Ruth Knoblich (Heidelberg: Springer, 2014), 115–38; Ruth Müller and Nina Witjes, "Of Red Threads

aim at a more comprehensive and nuanced understanding of how concepts of science, technology, and innovation relate to systemic changes in international relations.[49] An initial step in this direction would be an investigation of mutual influences between STI and International Relations.

2.3. Soft Power and Science Diplomacy

Soft power, a term coined by Joseph Nye in 1990, provides a conceptual framework for discussing the relationship between international STI cooperation in particular and power dynamics in international relations in general. Soft power differs from traditional conceptions of power in the international system as it emphasizes attraction over coercion or financial interest as instruments for achieving desired outcomes. Attraction is shaped by a country's values, policies, and culture.[50] Rather than getting others to do something they would otherwise not do, soft power stresses a co-opting effect in that it is "getting others to want the outcomes that you want."[51] Within the international system, these "others" increasingly are civil society actors and academic communities rather than nation-states. Regarding the role that science can play as an element of soft power, Nye suggests that in the future, "the factors of technology, education and economic growth are becoming more significant in international power."[52] The term "soft power" recently has been used to address the role of STI in international relations in a broader sense. The Royal Society in the UK and the American Association for the Advancement of Science (AAAS) refer to a soft power of science in that the scientific community "often works beyond national boundaries on problems of common interest" and thus contributes to "emerging forms of diplomacy that require non-traditional

and Green Dragons. Austrian Sociotechnical Imaginaries about STI Cooperation with China," in *The Global Politics of Science and Technology*, edited by Mayer, Carpes, and Knoblich, 47–65; Christian Bueger, "From Expert Communities to Epistemic Arrangements: Situating Expertise in International Relations," in *The Global Politics of Science and Technology*, edited by Mayer, Carpes, and Knoblich, 39–54.

49 Fritsch, "Technology and Global Affairs."

50 Joseph S. Nye, Jr., *Soft Power. The Means to Success in World Politics* (Cambridge, UK: Perseus Book Group, 2004), 5.

51 Ibid.

52 Joseph S. Nye, Jr., "Soft Power," *Foreign Policy* 80 (1990): 154.

alliances of nations, sectors and non-governmental organizations."[53] It seems obvious that science diplomacy should refer to three kinds of activities: *Science for diplomacy*, which means that scientific cooperation can improve international relations; *diplomacy for science*, diplomacy that can facilitate international scientific cooperation; and *science in diplomacy*, science that can provide advice to inform and support foreign policy objectives.

To enhance an understanding of science, technology, and innovation as closely related elements that should not be approached as separated entities in the field of political science, we propose taking science diplomacy as a starting point for an integrated analysis. Nye's quotation provides the clue: Key factors for international power consist of technology, education (including science), and economic growth (including innovative capacities also based on the other two factors). Taking these factors as our guides, we will now apply them to STI diplomacy.

In this section so far, we have pointed out that the two fields of STI policy and foreign policy have come to interact and collaborate in novel ways. Despite a common understanding of science as a transnational endeavor, science policy was for a long time focused on the national level, and foreign policy is still mostly focused on national interests.

In the course of STI internationalization, however, contact points between the policy fields have increased, and the division of labor between them is no longer clear-cut. This is why scientists and policy makers have begun to ask whether the two fields have merely been intertwined through stronger coordination and cooperation, or if new tasks and challenges in the context of globalization have produced a new field altogether. Wagner examined the "foreign policy aspect of science" by investigating the relationship between the scientific and the diplomatic communities, and concludes that the two systems operate in different ways.[54] While the scientific community is working in networks among peers, the diplomatic community works in hierarchies and traditions. While the two communities feel that their agendas increasingly overlap, the question remains whether we are witnessing the emergence of the new field of Foreign Science Policy.

---

53 Royal Society, "New Frontiers in Science Diplomacy: Navigating the Changing Balance of Power" (Royal Society, London, 2010), https://royalsociety.org/~/media/Royal_Society_Content/policy/publications/2010/4294969468.pdf.

54 Caroline Wagner, "Global Science: Dominating the Agendas of National Governments" (paper presented at the annual meeting of the American Association for the Advancement of Science, Boston, 2003).

3. Methodological Approach: Policy Field Analysis

In analyzing STI internationalization strategies by policy makers in Germany and the UK, we follow Mayer et al. who argue that a state-centric approach is the most adequate: First, corporate innovation activities and scientific progress remain entangled with national research policies. Transnational knowledge-production has not yet created a coherent society with a homogenous repertoire of knowledge. Second, in the European Union, the implementation of innovation strategies remains focused on the respective member state's national level. In addition, knowledge-related wealth creation remains closely affiliated with national systems of innovation and their respective priorities. Third, with an eye on the group of emerging economies in Brazil, Russia, India, China, and South Africa (BRICS), Mayer et al. point out that long-term planning and strategic orientation of research and innovation strategies are dominated by national governments through funding and subsidies.[55] Having worked on STI internationalization policies, we found Mayer's arguments to be relevant for this policy area even if internationalization has become a common topic in European STI-policy making. Most STI strategy documents (such as those discussed in section 4 below) assume that efforts and activities will be coordinated on the national level. We have mentioned that institutionalization is usually pursued within the framework of national STI policies. We suppose, therefore, that the strategic objectives of STI internationalization are based on national political interests composed of economic, political, scientific, and other objectives.[56] For an assessment of whether a new policy field is in the making, therefore, a state-centered approach seems most promising.

In the extensive body of literature on public policy and public administration, the term "policy field" (also "policy domain" or "policy area") is widely employed and applied to various contexts. The term, however, lacks a clear definition and differs in its assessment of how new pol-

---

55 Maximilliam Mayer et al., "Sind die BRIC-Staaten aufsteigende Wissensmächte? Herausforderungen für die deutsche Wissenspolitik" (University of Bonn, Center for Global Studies, discussion paper no. 3, Jan. 2011), 8.

56 Birte Fähnrich, *Science Diplomacy. Strategische Kommunikation in der Auswärtigen Wissenschaftspolitik* (Heidelberg: Springer VS, 2013).

icy fields emerge.[57] For operationalizing our own assessment, we adopt the concept of "policy field" as developed by Massey and Huitema in their paper on climate change as an emerging policy field in England.[58] The authors combine Colebatch's concept of "policy" as consisting of three pillars (order, authority, expertise) with an approach in political sociology that also uses a three-pillar construction.[59] According to Massey and Huitema, the key characteristic of a policy field is its substantive authority. It is authorized to "make decisions over an issue or problem so as to produce legitimate policy outputs" including, for example, policy programs and government expenditure on a particular issue.[60] A policy field's second pillar is its institutional order and context that include institutions or organizations such as ministries, bureaus, or government agencies that are devoted to a particular issue so as to "legitimize the products of substantive authority."[61] The authors point to "substantive expertise" as a field's third pillar. Expert knowledge on a particular issue provides the foundation for the other two pillars, the field's substantive authority and its institutional order. Such knowledge may be formal or informal, and it includes studies, task forces, academic papers, and also actors representing nongovernmental organizations (NGOs), think tanks, or policy networks. In our analysis we make use of this definition of a policy field as "a unit of governing within the socio political system of a country where there exist three pillars working in tandem to support each other in the management of a public issue or set of issues."[62]

In the empirical part of this chapter, we will infer these pillars to discuss whether STI internationalization in Germany and the UK meets these criteria for a policy field, how they are similar and how they differ in their

---

57 William N. Dunn, *Public Policy Analysis: An Introduction* (Upper Saddle River, NJ: Pearson Prentice Hall, 2004); Thomas R. Dye, *Understanding Public Policy* (London: Pearson Education, 2004).

58 Eric Massey and Dave Huitema, "The Emergence of Climate Change Adaptation as a Policy Field: The Case of England," *Regional Environment Change* 13, no. 2 (2012): 341–52.

59 Hal K. Colebatch, *Policy*, vol. 3 (Buckingham: Open Univ. Press, 2009); Paul Burstein, "Policy Domains: Organization, Culture, and Policy Outcomes," *Annual Review of Sociology* 17 (1991): 327–50; David Knoke and Edward O. Laumann, "The Social Organization of National Policy Domains: An Exploration of Some Structural Hypotheses," in *Applied Network Analysis: A Methodological Introduction*, edited by Peter V. Marsden and Lin Marsden (Beverly Hills and London: Van Sage Publications, 1983), 255–70.

60 Massey and Huitema, "Emergence of Climate Change," 343.

61 Ibid.

62 Ibid.

approaches to STI internationalization. We do so by drawing on eight expert interviews with specialists in STI internationalization policy for Germany, the UK, or both countries. Interviewees include staff working for STI policy research institutions, in offices of research funding organizations abroad, and for STI internationalization networks. They also include policy makers and experts for building bilateral STI cooperation programs. Many of our interviewees had been or are currently involved as consultants in the design and implementation of STI internationalization policies. Our analysis of expert interviews is supplemented with an extensive analysis of policy documents (such as government strategy papers) and other relevant texts (such as conference proceedings).

4. Case Studies

Let us turn to our two case studies, STI internationalization in Germany and in the UK. We will discuss these cases, first, with regard to whether or not they fit the criteria for a policy field and, second, using a qualitative analysis of how interviewees assessed the character of the respective policies.

4.1. Germany

In 2008 the German Federal Ministry of Education and Research (BMBF) published the federal government's strategy on internationalization, "Strengthening Germany's role in the global knowledge society."[63] The strategy aimed at strengthening research cooperation with global leaders and the exploitation of innovation potentials. It also sought to strengthen long-term cooperation with developing countries in education, research, and development, and to assume international responsibility in dealing with global challenges such as climate change, conflicts over scarce resources, or migration. In order to achieve these goals, the ministry sought to work out

---

63 Bundesministerium für Bildung und Forschung, "Strengthening Germany's role in the global knowledge society: Strategy of the Federal Government for the Internationalization of Science and Research," (BMBF, 2008), http://www.bmbf.de/pubRD/Internationalisierungsstrategie-English.pdf.

bilateral agreements on cooperation in education and research with more than fifty countries. These were to be coordinated by BMBF's International Bureau and supported by its internet-based information service *Kooperation international.* The German strategy is peculiar in that it explicitly states that German research and innovation (with international support through economic and political relations) should help solve global challenges, and that this aim ought to shape science and foreign policy goals.[64]

A year later, in 2009, the German Federal Foreign Office launched its "Research and Academic Relations Initiative." Its aims are to engage STI in order to "work on problems of global reach, capacity building in developing countries and to promote collaborative research to spread civic virtues and cultural dialogue."[65] Such policy is to be pursued through the German Houses for Research and Innovation (set up by the German Foreign Office in New York, Sao Paulo, New Delhi, Tokio, and Moskau as a contribution to the internationalization of STI), through the German Science Centre in Cairo (in cooperation with BMBF), and through close collaboration with the Alliance of Science Organisations in Germany. The Federal Foreign Office also seeks to expand grant programs for international students and researchers. The initiative pays tribute to the initial aims associated with developing an internationalization strategy, of aligning science and foreign policy goals to tackle global challenges, by proposing that knowledge exchange helps to spread democratic values, and by focusing on conflict regions and countries in transition. If we now apply the three pillars that constitute a policy field, three observations begin to emerge.

---

64 Only recently, BMBF published "International Cooperation: Action Plan of the Federal Ministry of Education and Research" (BMBF, 2014), http://www.bmbf.de/pub/ International_Cooperatin_Action_Plan.pdf. The plan refers to the ministry's 2008 strategy for the "Internationalization of Science and Research." The "Action Plan" resulted from an evaluation of the previous strategy's achievements. It also responds to the "complex systemic changes, and dynamic economic, ecological and social upheavals and challenges which we are facing in the 21st century," which "are altering our perspectives regarding the education, research and innovation policies that we need, at the national and international levels" (p. 3). While the "Action Plan" suggests an even stronger political determination to enhance international cooperation, it is too early to tell how it will come to relate to foreign policy.

65 Federal Ministry of Education and Research, "Research and Academic Relations Initiative," http://www.research-in-germany.org/en/research-landscape/r-and-d-policy-framework/research-and-academic-relations-initiative.html.

First, we see that in Germany the field is associated with substantive authority in that there are at least three explicit strategies that aim at the internationalization of national STI. Although it is too early to assess how the "Action Plan" will facilitate cooperation among German federal ministries, we see that both the 2008 and the 2009 strategies highlight the role of STI for achieving or contributing to foreign policy objectives. While developed to match BMBF's strategy for the "Internationalization of Science and Research," the Federal Foreign Office's initiative has been criticized as an attempt to expand its influence and budget.[66] This strongly suggests that as a policy field, STI internationalization comes with substantive authority that prompts actors to compete for influence within it.

Second, we may discern an institutional order. In addition to the Foreign Office and BMBF, the German Federal Ministry for Economic Cooperation and Development is involved as well. Research organizations collaborate with ministries to implement strategies and initiatives. Organization such as the German Academic Exchange Service (DAAD), other research and intermediary organizations, and German universities and their international networks seek to enhance mobility and internationalization. They are increasingly approached by both the Foreign Office and BMBF for participation in German Research and Innovation Houses (DWHIs) and to provide expertise. With respect to the relationship between STI policy, foreign policy, and the emergence of a new policy field, one interviewee told us that "in addition to the traditionally outward-oriented policy fields of economic, cultural, and security policy, a fourth pillar has emerged, or at least should be established, that of foreign science policy: the representation of Germany in the field of science."[67] One member of an intermediary organization also referred to a foreign science policy and saw it as a new yardstick for measuring a country's performance, pointing out that foreign science policy is "a policy field with a future and will increase in significance for the twenty-first century."[68] Regarding their assessment of what this emerging policy field of foreign science policy may have to offer for science, both interviewees considered it essential that an opportunity was created for science to gain access to the sphere of inter-

---

66 Tim Flink and Ulrich Schreiterer, "Science Diplomacy at the Intersection of S&T Policies and Foreign Affairs: Toward a Typology of National Approaches," *Science & Public Policy* 37, no. 9 (2010): 665–77.

67 Interview with German specialist, no. 1, 6 June 2014

68 Interview with German specialist, no. 4, 28 July 2014.

national politics.[69] Such access they considered an opportunity for helping shape funding programs but also for scientific advice in international negotiations. In general, most interviewees referred to foreign science policy as providing a new opportunity for the scientific community to participate in processes of political decision making in line with science *for* diplomacy and science *in* diplomacy (also see section 2.2.).

One interviewee, however, an employee of a large German research funding organization, criticized the heavy political rhetoric used by stakeholders when referring to foreign science policy and the role of such rhetoric in tackling global challenges and in science diplomacy. "I think that one should not be active in countries with a dictatorship," he argued. "But an institution like mine, as a mechanistic system, does not have such a moral approach. For instance, if a Smart City project were planned in Saudi Arabia involving huge amounts of money [...] they would not care if people stoned women there."[70]

German research and funding organizations provide a wide range of expertise relevant to STI internationalization policy. Within BMBF, information structures have been created through programs such as "Research in Germany" and *Kooperation international*. Another important element that informs German STI internationalization policy is the network of science representatives that are based at German embassies abroad and provide country-related expertise to support policy decision-making. Overseas offices by German research and intermediary organizations also provide an important network for sharing information and knowledge relevant in many fields of policy. Among them are the Alexander von Humboldt Foundation, the Fraunhofer-Society, the German Academic Exchange Service (DAAD), the German Council of Science and Humanities (*Wissenschaftsrat*), the German National Academy of Sciences Leopoldina, the German Rectors' Conference (the association of German university presidents, *Hochschulrektorenkonferenz*), the German Research Foundation (DFG), the Helmholtz Association, the Leibniz Association, the Max Planck Society, and the Association of German Chambers of Industry and Commerce (DIHK). Although lacking an independent strategy for the internationalization of STI, the Federal Ministry for Economic Cooperation and Development (BMZ) has available a rich trove of insights about countries and regions in which it is active, knowledge of interest to BMBF and the Fed-

---

69 Interviews with German specialists, no. 1, 6 June 2014, and no. 2, 18 June 2014.
70 Interview with German specialist, no. 5, 15 July 2014.

eral Foreign Office. One interviewee pointed to the significance of well-established contacts between Germany and emerging economies such as China, South Korea, or India. Because the BMZ has been a longstanding partner and technological adviser there (officially denominated as *wissenschaftlich-technische Zusammenarbeit*), it facilitates access to them for other German stakeholders now that these countries are developing rapidly.

4.2. United Kingdom

In the UK, an initial strategy for international engagement in research and development was published two years earlier than in Germany. It followed the setup of a coordination structure labelled "Global Science and Innovation Forum" (GSIF) in 2005. The forum institutionalized cooperation and coordination of governmental actors and other stakeholders (such as the Research Councils and the British Council) in the area of STI internationalization. The founding of the council marks a milestone for a coordinated policy approach to STI internationalization.

Subsequent to the council's creation, a global network of science and innovation officers was established in British embassies, high commissions, and consulates. By 2011, Science and Innovation Network (SIN) consisted of 90 staff in 25 countries and territories. Most staff is located in Europe and in the US but SIN is also well-developed in China, Japan, India, Canada, Southeast Asia, and South Korea.[71] SIN has continued to grow. In 2012 it employed 93 staff in 47 embassies and consulates in 28 countries.[72] SIN, however, is not the only British organization with offices abroad to deal with STI collaboration. The Department for International Development, the British Council, and the Research Councils also have offices abroad, often in close collaboration with SIN. Geographic priority-setting reflects former colonial relations and an alignment with the Commonwealth of Nations. Cooperation with Canada has been a priority and a

---

71 Foreign and Commonwealth Office, Department for Business, Innovation & Skills, "Science and Innovation Network Report, April 2010 to March 2011" (Sept. 30, 2011), 8, https://www.gov.uk/government/uploads/system/uploads/attachment_data/file/32496/11-1014-science-innovation-network-report-2010-2011.pdf.

72 Department for Business, Innovation & Skills and Foreign & Commonwealth Office, "UK Science and Innovation Network Report" (2013), https://www.gov.uk/government/uploads/system/uploads/attachment_data/file/267820/bis-13-1331-uk-science-and-innovation-annual-report.pdf.

special program has been set up with India in 2006, the UK-India Cooperation in Education & Research (UKIERI).

To understand how STI internationalization policy operates in the UK it is necessary to recognize one of the distinctive features of STI policy making. After a government reform in 1995, the STI internationalization agenda was transferred from the cross-governmental Office of Science to a single government department, the Department of Trade and Industry, which in 2007 was replaced by the Department for Business, Innovation and Skills (BIS). That department started to work more closely with the Foreign and Commonwealth Office, particularly by helping fund and run SIN. In the context of STI internationalization, cross-governmental coordination of science policies appeared to have been strengthened. One example of this is the introduction of Chief Scientific Advisors (CSAs) who are assigned to government departments to provide another layer of coordinating efforts by the BIS and the Foreign Office in SIN. For an assessment of whether or not STI internationalization in the UK can be said to have gained substantive authority, the peculiarity of UK policy-making seems crucial. CSAs all work under a Government Advisor who counsels the Prime Minister. The various Government Advisors coordinate their activities in the Chief Scientific Advisors Council.

Through these structures, CSAs have a say in budget negotiations for SIN. One of our interviewees explained that the "key things are finance. There are government reports [...] that determine finances of five year periods."[73] All CSAs were in contact with ministers "who created a good voice for science."[74] British policy-making has created the room for substantive authority to unfold and make a difference, particularly through SIN, through Chief Scientific Advisors, but also in special funding programs such as the UKIERI program mentioned above.

Because of its new structures (that include the GSIF, SIN, and CSAs), the UK has a strong institutional order for policies in the field of STI internationalization, an order that continues to grow and evolve so as to improve coordination among its active players. At the same time, the UK also wields substantive expertise. It is home to well-known science policy research institutes such as the Science Policy Research Institute at the University of Sussex and the Manchester Institute of Innovation Research. Both institutes have joined British think tanks such as Demos and Nesta in making STI internationalization a research focus. The Royal Society has set up a Science Policy Centre to deal with topics of science diplomacy, and it regularly publishes on international science and innovation systems. Complementing the work of these institutions, SIN staff produces strategic knowledge about target countries. One of our interviewees concluded that STI internationalization "has become an academic field of study in its own right."[75]

In our interviews, assessments of the relationship between the goals of science policy and foreign policy oscillated between enthusiastic accounts of the diplomatic power of international scientific collaboration, on the one hand, and sober accounts of balancing risks and benefits, on the other. One interviewee suggested that "scientists are some of the best ambassadors for improving international relations nowadays. [...] That is because science is international."[76] Interviewees frequently referred to the European Organization for Nuclear Research (CERN) because, as one of them put it, the project had made it "inconceivable to go to war again." He concluded that "science has contributed to making the world a safer place."[77] Other interviewees talked about their "growing awareness of darker sides" such as cyber-espionage, making it necessary to balance diplomatic and economic interests.[78]

Compared to other European countries, the UK is characterized by a strong "science as diplomacy" approach to STI internationalization. This is reflected in the terminology our interviewees chose to use in discussing STI internationalization. German interviewees spoke of "foreign science policy" while their British counterparts referred to "science diplomacy" as "certainly a new policy field."[79] Another aspect specific to the UK is the open association of economic interests with STI internationalization. One interviewee subsumed most STI-related activities under his country's economic interests. "All the governments are interested in creating jobs and improving the economy," he said, "and that is almost the number one area of political interest."[80]

---

75 Ibid.
76 Ibid.
77 Interview with British specialist, no. 1, 9 September 2013.
78 Interview with British specialist, no. 2, 3 September 2013.
79 Interview with British specialist, no. 2, 3 September 2013.
80 Interview with British specialist, no. 4, 18 August 2014.

5. Conclusion

This chapter builds on the observation that since STI internationalization became a central policy goal of the European Union's Lisbon Strategy in 2000, national STI policies are increasingly designed to shape the EU's external relations and those of its member states. We have shown that while the EU sets agendas and funding frameworks for STI internationalization, Germany and the UK are one step ahead of the EU in building physical and virtual infrastructures for supporting the internationalization of their respective national STI portfolios. STI policies thus can be regarded as an arena in which foreign affairs are shaped on different policy levels (EU and national) and at a different pace and intensity in different member states. Going one step further we conclude that for the cases of Germany and the UK, a new policy field has emerged within the last decade in which STI policies and foreign policies are increasingly intertwined. We suggest that STI internationalization can be understood as a policy field that features three necessary preconditions: substantive authority, institutional order and substantive expertise.[81] The two countries, however, apply different terms when referring to their STI internationalization policies and activities. In Germany, activities are labelled as "foreign science policy" while in the UK the term "science diplomacy" is used. Although some studies treat both terms as equivalent[82] we suggest that each indicates a different approach towards international relations and how this approach impinges on the design of STI internationalization policies.

The ways in which this policy field is structured in Germany and the UK reflects different approaches. When it comes to substantive authority, referring to the existence of specific policy programmes and government expenditure that are related to STI internationalization, we see a transverse policy field where issues from other policy fields, such as foreign policy and economic policy, are brought together. Simultaneously, this fusion of issue areas has in both cases led to new forms of inter-ministerial cooperation and government expenditures for internationalization activities.

One central finding is that the political nodes from which STI internationalization activities emanate are different in both countries due to their distinct techno-political and socio-economic histories that are reflected in

---

81 See Massey and Huitema, "The Emergence of Climate Change Adaptation," and our discussion above.
82 Flink and Schreiterer, "Science Diplomacy."

the respective institutional order for STI internationalization. In Germany, BMBF was the frontrunner in drawing up an internationalization strategy for research and innovation. The Federal Foreign Office has become a second node in managing the Foreign Science Policy Initiative and cooperates with BMBF and the international network of German Houses of Research and Innovation.

The central function of the BMZ is that of a facilitator due to the fact that it has the longest-standing institutionalized contacts with former developing countries that have now became global STI powers (such as China and India). While there is overlap between ministries' agendas, it seems that STI internationalization has prompted new forms of collaboration while leading to increased competition for funding and influence.

In the UK, the institutional order of this transverse policy field is strongly shaped by the central role SIN plays in STI internationalization policies. As a global network much larger than the German Houses of Research and Innovation, SIN seems to have become a much stronger reference point for stakeholders concerned with internationalization. We also see cooperation between different stakeholders facilitated by the Global Science and Innovation Forum (GSIF) that coordinates governmental and non-governmental actors in the field. Personal and professional links between the Chief Scientific Advisors (CSAs) allow for science policy makers to act across the government with other departments. Another remarkable difference between Germany and the UK is that the involvement of economics in international STI policies is much stronger in the UK than in Germany. The British SIN is operated by the Department of Business, Innovation and Skills (BIS) and the Foreign and Commonwealth Office (FCO). This difference can be attributed to the overall structure of science policy-making in the UK where the Minister of Science and Universities reports to BIS, and an economic logic cross-cuts strategy papers and the overall design of STI internationalization policy-making. Thus, although both labels ("foreign science policy" and "science diplomacy") carry the term "science", STI internationalization is approached in different ways. The UK follows a more integrated approach that is led by an understanding of "innovation" as technological innovation for the market (integrating science, technology, and innovation into one overall agenda). The German approach to STI seems to be less uniform in that it derives from different nodes and has a much stronger focus on the international-relations dimension of science policies, particularly with respect to developing countries and countries in transition.

Regarding substantive expertise, it can be stated that while stakeholders in the UK can resort to some of the oldest institutions concerned with international cooperation in STI, such as the Science Policy Research Unit at the University of Sussex or the Manchester Institute of Innovation Research, no such support is available in Germany. Here the structure of institutions and organizations that inform the knowledge base for STI internationalization reflects the three central institutions involved, with each of them having its own research foundation, organization or in-house expertise at its disposal. However, in both cases a knowledge base that provides stakeholders with the necessary expertise on STI-internationalization is present.

Our conclusion that STI internationalization is a policy field in its own right and that it evolves in different ways in different countries, leads to three important implications for future research. First, this relatively new policy field involves an overlap between two policy fields, and these two fields have had different outlooks. While science policy traditionally has been national, foreign policy has always involved international relations. Future studies should expand their frame of analysis to provide a comprehensive picture by integrating approaches from International Relations (IR), Science, Technology, and Society Studies (STS), and Innovation Studies.

Second, future research would have to adopt a more refined understanding of the range of STI policy fields that set science policies, technology policies, and innovation policies in relation to each other. This would allow for a more precise working out of the interrelations between STI policies and internationalization policies. It will make a difference for priorities in foreign policies whether STI internationalization is merely understood in terms of fostering cooperation between excellent scientists, in terms of cooperation for economic development, or in terms of creating new markets for European innovations. It seems clear that international, STI-related collaboration has consequences for global power relations. Within the changing dynamics of such relations, STI provides another arena for nation-states to deal with issues of competition and cooperation.

Third, such an approach should take into consideration that foreign policy goals and strategies, as well as economic policies, have implications for how international scientific collaborations are forged and what knowledge will be produced. It should aim to understand, furthermore, how STI internationalization as a policy field might become co-productive of other transnational innovation systems. This would call for case studies that investigate collaborations as they are facilitated by current policy frameworks. The central question would be whether frameworks continue to refer back to national foreign policy aims or call for more transnational policy frameworks in order to meet innovation needs to face current global challenges. The latter would challenge the dominant national innovation-system approach in innovation studies.

Section IV:
Global Science

The Institutionalization of the European Research Area: The Emergence of Transnational Research Governance and its Consequences

Arne Pilniok

1. Introduction

For a long time in the twentieth century, the relationship between science and the nation-state could be described in terms of a dichotomy: The research system increasingly served as a major example of communications on a global scale, while public governance of research and its institutions remained bound to nation-states—even after the establishment of the supranational European Community.[1] Nation-states provided the complete legal and financial framework for public research organizations. But this has altered considerably within the last decades. This chapter thus argues that a profound structural change in the relationship between member states, national research systems, and the European Union has taken place. Its major argument is the emergence of a new transnational research policy in Europe. This hybrid between the levels of governance was created in a second phase of European research policy, which transformed the national embeddedness of this policy area mainly through European institution-building. Just as the history of European integration at large can be written as a history of institution building, this is also true for the European research policy and its consequences.[2]

---

1 See e.g. Rudolf Stichweh, "Science in the System of World Society," http://www.fiw.uni-bonn.de/demokratieforschung/personen/stichweh/pdfs/5_43stwscienceworldsoc.pdf.

2 Johann Olsen has impressively demonstrated this point. Cf. Johann Olsen, *Governing through Institution Building: Institutional Theory and Recent European Experiments in Democratic Organization* (Oxford: Oxford Univ. Press, 2010); see also Åse Gornitzka and Julia Metz, "European institution building under inhospitable conditions – the unlikely establishment of the European Institute of Technology," in *Building the Knowledge Economy in*

This chapter will first draw a general picture of how the European Community, in a first phase of European research policy, changed the relationship between European and national levels in research (sect. 2.). In the initial decades of European integration the European Community installed a Joint (Nuclear) Research Centre and built up administrative capacities with a Directorate-General for Research as one of the Commission services in the 1970s. Some years later, the European Community installed direct research funding under the wing of its own administration. For decades the institutionalization of these "Framework Programmes" provided the reference for a path-dependent development. The European Research Area defines the focus for the second phase of integration as, from 2000 onwards, the European Union created the European Research Area as an all-embracing *Leitbild* (model) of research policy, thereby building on ideas from the 1970s. The European Research Area sought to attain a whole bundle of objectives that centered around the core idea of a structural integration of research, research funding, infrastructure, and policy. The leading paradigm was the creation of an "internal market for research,"[3] that exhibited the strong connection between research and economic goals in European policy, and is reminiscent of another central European success story, the creation of the internal market in 1992. Second, this chapter will turn to the different forms of institution-building in more detail, focusing on those which have taken place within this framework in the last two decades (sect. 3.). This includes novel forms of European research funding, hybrid research funding jointly administered at the European and the national level, and new institutions for coordinating and integrating research policy. Consequently, national matters, such as research funding, were increasingly intertwined with and embedded in supranational policies. This development led to inevitable adjustments by national systems of research regarding standards, structures, and procedures. At the same time, new forms of supranational and hybrid research funding were implemented "in-between" the EU and the member states. To a considerable extent, however, they were dominated by European narratives and European administrative culture. This led to conflicts between EU administra-

---

Europe, edited by Meng-Hsuan Chou and Åse Gornitzka (Cheltenham: Elgar, 2014), 111: "Supranational institution building is at the heart of European integration."

3 Cf. Meng-Hsuan Chou, "Constructing an Internal Market for Research through Sectoral and Lateral Strategies: Layering, the European Commission and the Fifth Freedom," *Journal of European Public Policy* 19 (2012): 1052–70.

tions and member states, which considered research policy a sensitive area of integration.[4] The final section of this chapter discusses some consequences arising from these institutions, including effects on researchers, research organizations, organizations that fund research, and on nation-states (sect. 4.).

2. The First Phase of European Research Policy: A Complex Story Cut Short

Since the central aim of this paper is to attain a deeper understanding of the second phase of this development, a closer look at the first phase is necessary. This section of the chapter briefly analyzes the complex history of European research policy as the changing relationship between the European Union and nation-states can only be fully understood in contrast to the initial model for European integration. Three aspects of this phase are relevant for the argument of the chapter: While research policy was an issue from day one of the European Communities, for a long time the EC's approach resulted in a lot of talk but little action.[5] This changed with the emergence of a paradigmatic model of European research funding in the 1980s (2.1.). The institutionalization of the traditional form of European research funding via Framework Programmes and their specific design created a path dependency for subsequent developments (2.2.). The political concept of the European Research Area marked a new approach, and this development requires a closer look (2.3.), before its institutional details are analyzed (in section 3.).

---

4 Cf. Åse Gornitzka, "Networking Administration in Areas of National Sensitivity: The Commission and European Higher Education," in *European Integration and the Governance of Higher Education and Research*, edited by Alberto Amaral et al. (Dordrecht: Springer, 2009), 109.

5 "European Communities:" European Economic Community (EEC), European Atomic Energy Community (EAEC), and European Coal and Steel Community (ECSC).

2.1. Much Talk, Little Action: The Emergence of the EC's Research Policy

At the beginning of the EC, research policy was organized in sectors according to the different steps of integration.[6] The nucleus was clearly provided by the European Atomic Energy Community, although the two other founding treaties also provided research clauses regarding coal and steel as well as agriculture. Compared to their extensive need for resources, the possibilities in this sector of research were evidently limited for the six founding member states, notwithstanding national sensitivities related to nuclear policy. Cooperation was expected to provide added value. As a result, legal provisions, designed inter alia, provided a coordinating mechanism to prevent parallel and thus redundant nuclear research. For decades this approach drove European research policy, although the structure of state-driven nuclear research was better suited for such coordination than more common areas of research on a smaller scale. The creation of the Joint Nuclear Research Centre provided the first step towards an institutional development which would eventually evolve into a Commission-oriented research service for all of the Union's policy areas.[7]

Based on initial approaches at the beginning of the 1970s by Commissioner Altiero Spinelli, the first forays into a non-sectoral European research policy were made in 1974.[8] Ralf Dahrendorf, a German sociologist who was Commissioner at that time, presented the idea of a "European Scientific Area."[9] The central motive for coordinating national research policies was once again to avoid unnecessary research so as to efficiently use national public research funds, a rationale that would help justify the creation of the European Research Area 25 years later. Four non-binding resolutions called for coordination with the European Science Foundation

---

6 For an overview, see Hans-Heinrich Trute and Arne Pilniok in EUV/AEUV, edited by Rudolf Streinz (Munich: Beck, 2012), Art. 179 Rn. 3–8.

7 On the Joint Research Centre, see Arne Pilniok, "Zwischen Wissenschaft, Politik und Verwaltung: Die Gemeinsame Forschungsstelle der Europäischen Kommission als Element der europäischen Wissensinfrastruktur," *Die Öffentliche Verwaltung* 65 (2012): 662–71.

8 See Meng-Hsuan Chou, "The Evolution of the European Research Area as an Idea in European Integration," in *Building the Knowledge Economy in Europe*, edited by Meng-Hsuan Chou and Åse Gornitzka (Cheltenham: Elgar, 2014), 37.

9 Ralf Dahrendorf, *Towards a European Science Policy* (Southampton: Univ. of Southampton, 1973).

and created initial funding for a science-based research policy.[10] However, the Committee for Research into European Science and Technology (CREST), also launched by resolution in 1974 and comprised of representatives of national research ministries, soon became aware of problems connected with the coordination of national research policies.[11] Difficulties arose from the fact that the initiative for research topics remained with research groups and organizations and could hardly be influenced by government funding. Decisions about research topics, and regarding what solutions were promising, were matters that remained (and continue to be) confined to the scientific community, which were spread out over the various member states, even if this resulted in an "inefficient" doubling of efforts for research groups and organizations.

This approach to European research policy was characterized by its intergovernmental style. Member states demonstrated their pre-dominance through the creation of CREST, and through the committee's composition and responsibilities. In this sensitive policy area, the European Commission and the Parliament were relegated to an inferior role. At this time, many institution-building initiatives at the European level were deliberately delegated to bodies outside the Community framework. Research organizations such as the European Molecular Biology Laboratory were explicitly founded as intergovernmental institutions.[12] The same is true for the European Science Foundation, which has had a tense relationship with the European Commission since its founding.[13]

10 Álvaro De Elera, "The European Research Area: On the Way Towards a European Scientific Community?," *European Law Journal* 12 (2006): 559–74; 561; Chou, *Building the Knowledge Economy in Europe*, 38.

11 For an analysis of the CREST Committee, see Arne Pilniok, *Governance im europäischen Forschungsförderverbund* (Tübingen: Mohr Siebeck, 2011), 173–76.

12 Cf. Luca Guzzetti, "The 'European Research Area' Idea in the History of Community Policy-Making," in *European Science and Technology Policy: Towards Integration or Fragmentation*, edited by Henri Delanghe, Ugur Muldur et al. (Cheltenham: Elgar, 2009), 71. Luca Guzzetti stated in his *A Brief History of European Union Research Policy* (Luxembourg: Office for Official Publ. of the European Communities, 1995), 115: "The battle between Community and intergovernmental research [...] has never ended." For an extensive historical analysis, see the contributions in *History of European Scientific and Technological Cooperation*, edited by John Krige and Luca Guzzetti (Luxembourg: Office for Official Publ. of the European Communities, 1997).

13 Thomas Banchoff, "Institutions, Inertia and European Union Research Policy," *Journal of Common Market Studies* 40 (2002): 1–21, 7; Sandra Lavenex, "Switzerland in the European Research Area: Integration without Legislation," *Swiss Political Science Review* 15 (2009): 629–51, 638.

2.2. The Appearance of the Traditional Community Approach

The start of direct research funding by the EC in the 1980s marked a turning point from talk to action. Building on a pilot programme that had started some years earlier, the first Framework Programme began in 1984. Its content and structure was deeply intertwined with the economic rationale for European integration. The programme was industry-oriented and a consequence of the prevailing narrative of the economic competition between the Community, the United States, and Japan. Some of its sub-programmes (such as the ESPRIT programme on information technology) gained considerable attention as examples of changes in funding structures between the Community and the member states.[14] In subsequent years, framework programmes gradually evolved as additional financial resources were allocated to research budgets.[15] This also included a broadened thematic approach and expanded funding for the social sciences and humanities.

All of this evolved into the paradigmatic model that would define the first phase of supranational European research policy:[16] direct funding of research projects by the Community administered by the European Commission. The member states had (and retain) institutional forms to co-direct research funding. For a long time, the Council was in charge of adopting the relevant legal acts. The content and the conditions for awarding Community funds thus became part of institutional negotiation processes. The implementation of the different lines of funding within the framework programmes was accompanied by "comitology" committees staffed by national bureaucracies.[17] The committees' longstanding legal regime provided it with considerable veto power over the Commission, a model that had significant consequences for Commission services: Several Directorates-General (most notably of course the Directorate-General for

---

14 Edgar Grande, "Innovationspolitik im europäischen Mehrebenensystem: Zur neuen Architektur des Staatlichen," in *Innovationspolitik in globalisierten Arenen*, edited by Klaus Grimmer, Stefan Kuhlmann et al. (Wiesbaden: VS Verlag für Sozialwissenschaften, 1999), 93; Id., "Von der Technologie- zur Innovationspolitik – Europäische Forschungs- und Technologiepolitik im Zeitalter der Globalisierung," in *Politik und Technik*, edited by Georg Simonis et al. (Wiesbaden: Westdeutscher Verlag, 2001), 375.

15 For an overview: Chou, *Building the Knowledge Economy in Europe*, 40.

16 See especially Banchoff, *Institutions, Inertia and European Union Research Policy*, passim.

17 For details on the role of comitology committees in research policy, see Pilniok, *Governance im europäischen Forschungsförderverbund*, 118–44.

Research) were in charge of preparation and evaluation of research foci and also of day-to-day management of the individual research projects it sponsored. Over the years, Framework Programme budgets grew and the Commission's administrative capacities expanded accordingly. But resources for making informed decisions about strategy and policy were increasingly limited by the burden of managing proposals.[18] Due to budgetary constraints and limitations within the rationale of framework programmes, further expansion of the Commission's staff was unlikely. This situation ultimately prompted institutional reform in the European governance of research funding, a reform that came to shape the second phase of European research policy.

2.3. When the Time is Ready: The Advent of the European Research Area

While Commissioner Dahrendorf had proposed the idea of a "European Scientific Area" to integrate the research policy domain in the early 1970s, his ideas had fallen on deaf ears.[19] In 2000, however, Commissioner Philippe Busquin drew on Dahrendorf's ideas to introduce the overall concept of the European Research Area (ERA).[20] Busquin proposed this new strategic approach to strengthen the Union's role in research policy and it was soon supported by all institutions, especially the Council. The European Commission, Council working groups, and expert groups developed conceptual ideas to flesh out the new policy framework. Policy development was prompted by comparison with the United States and Japan, and by the perceived significance of science for politics and economic prosperity. The Commission also pointed to a lack of coordination between member states and the supranational level, and to insulated national research systems and heterogeneous legal and administrative rules that prevented integration and cooperation. This was vividly expressed by the formula of the "15+1" research systems to be replaced by the European Research Area.[21] The Commission now used a very broad definition of the ERA as a

---

18 Pilniok, *Governance im europäischen Forschungsförderverbund*, 89–97.

19 For a theoretical account see Chou, *Building the Knowledge Economy in Europe*, 27–50.

20 Communication from the Commission, "Towards a European Research Area", *COM* (2000) 6 final.

21 Ibid. (with reference to the number of member states in 2000).

unified research area open to the world based on the internal market, in which researchers, scientific knowledge and technology circulate freely and through which the Union and its Member States strengthen their scientific and technological bases, their competitiveness and their capacity to collectively address grand challenges.[22]

Within this framework, the Union proposed, inter alia, increased integration and coordination of all national activities. Member states were to integrate with a network of actors at all levels of research governance. Even if the main content of the ERA has changed over the years, the framework has paved the way for a new era of institution building.

European research policy has always been ahead of competences granted by treaties, from coordinating measures in the 1970s to the introduction of the European Research Area. The lack of synchronicity of the political and legal development is a central feature of European research policy in general and of the European Research Area in particular: Research policy (including the Framework Programmes) was part of common strategic objectives for a long time, but the ERA did not become the Union's legal objective before 2009 when it was part of the Treaty of Lisbon. Article 179 of the Treaty on the Functioning of the European Union calls for the completion of this goal and empowers the Union to do so within certain limits. The Union (particularly the Commission) used the Treaty to establish a range of initiatives to re-design existing instruments, most notably the Framework Programme, and to devise new institutional structures. This marked the transition to the second phase of EU research policy, which will be analyzed in more detail in the following section.

3. The Second Phase: The Emergence of New Supranational Governance Structures

The second phase was induced by the advent of the European Research Area in 2000 and, in contrast to the first phase, it considerably changed the relationship between the levels of research governance within the European Union. Subsequent relations between the Union and the member

---

22 Communication from the Commission, "A Reinforced European Research Area Partnership for Excellence and Growth", *COM* (2012) 392, 3.

states are underlined in this section. First, institutions for research funding evolved at the Union level, including the European Research Council's aim to support basic research (sect. 3.1.). Second, different forms of joint research funding emerged that were shouldered by the member states' research funding organizations together. These were initiated, structured and controlled by the Commission on behalf of the European Union. Among them are ERA-Networks and Article 185 initiatives (3.2.). Joint technology initiatives and joint programming initiatives, furthermore, created new institutional forms at the intersection of science, member states, and EU actors (3.3.). Initiatives include national research organizations, but they are now embedded in new policy environments, and this has consequences discussed in section 4.

3.1. ERC as a Supranational Research Funding Institution

In the first phase of EC research policy, the relationship between the European Community and the member states could at least partly be characterized by a division of labor in funding matters. EC funding was aimed at applied, industry-oriented research projects, even if the social sciences were gradually included to pay tribute to social problems. This approach was not only a result of the emergence of EC research policy, but also due to the limited competences provided by the Single European Act.[23] At the same time, member states retained responsibility for funding basic research, mostly through autonomous research councils of various organizational forms.[24] Only with the Treaty of Amsterdam in 1999 was the Union's competence for funding broadened from technology-related research to basic research.

But since the mid-1990s, research organizations of the member states had called for an autonomous European research funding agency.[25] This

23 Trute and Pilniok, *EUV/AEUV*, Art. 187 Rn. 5.

24 Hans-Heinrich Trute and Thomas Groß provide a comparative analysis in "Rechtsvergleichende Grundlagen der europäischen Forschungspolitik," *Wissenschaftsrecht* 27 (1994): 203–48. See also Dietmar Braun, "The Role of Funding Agencies in the Cognitive Development of Science," *Research Policy* 27 (1998): 807–21; Chris Caswill, "Old Games, Old Players – New Rules, New Results," in *Changing Governance of Research and Technology Policy: The European Research Area*, edited by Jakob Edler et al. (Cheltenham: Elgar, 2003), 64–80.

25 Trute and Groß, *Rechtsvergleichende Grundlagen der europäischen Forschungspolitik*, 234.

demand was certainly a response to the bureaucratic model of research funding established by the Framework Programme, which relates to the first phase of European research policy. The introduction of the concept of the European Research Area accelerated discussions about this new form of research funding and created new contexts,[26] but its institutionalization within the European Research Council did not take place until 2007.[27] Most member states initially opposed the idea because they wished to retain competence for the funding of basic research instead of shifting it to the European level.[28] This new institution represented the first European funding programme that defined scientific excellence as the single criterion for funding individuals based on peer-review. This created parallel funding with that of member states and had consequences for research institutions and researchers.[29] The approach "embodies a break with established principles and rules of resource distribution in the EU's research policy domain."[30]

The organizational structure of the ERC reflects specific European legal requirements as well as the Union's administrative culture.[31] In contrast to national research councils, the ERC was not established as a permanent institution. Following a Commission proposal, it was instead incorporated in the multi-annual Framework Programme and funded via the specific programmes "Ideas" (Framework Programme 7) and "Excellent Science" (Framework Programme 8) respectively.[32] Therefore its existence is limited

---

26 For a detailed account of the institutional creation see Åse Gornitzka and Julia Metz, "Dynamics of Institution Building in the Europe of Knowledge: The Birth of the European Research Council," in *Building the Knowledge Economy in Europe*, edited by Meng-Hsuan Chou and Åse Gornitzka (Cheltenham: Elgar, 2014), 81–110.

27 For a discussion of the prehistory, see for example Maria Nedeva, "Between the Global and the National: Organising European Science," *Research Policy* 42 (2013): 220–30.

28 Gornitzka and Metz, *Dynamics of institution building in the Europe of Knowledge*, 90.

29 See the discussion in section 4 below.

30 Gornitzka and Metz, *Dynamics of Institution Building in the Europe of Knowledge*, 81.

31 For a detailed account see Thomas Groß et al., *Regelungsstrukturen der Forschungsförderung: Staatliche Projektfinanzierung mittels Peer-Review in Deutschland, Frankreich und der EU* (Baden-Baden: Nomos, 2010), 95–120; Thomas Groß and Remzi N. Karaalp, "The European Research Council: A Legal Evaluation of Research Funding Structures," in *The Changing Governance of Higher Education and Research*, edited by Dorothea Jansen and Insa Pruisken (Cheltenham: Springer, 2015), 179–87.

32 See Council Decision of 3 December 2013 establishing the specific programme implementing Horizon 2020 – the Framework Programme for Research and Innovation (2014–2020) and repealing Decisions 2006/971/EC, 2006/972/EC, 2006/973/EC, 2006/974/EC and 2006/975/EC, OJ L 347/965.

for the seven-year life span of the Framework Programmes, and has to be renewed by the European institutions accordingly.[33] The decision to embed the ERC in the Framework Programme, which was sketched by the Treaties of the European Union (TFEU), created path dependencies in the concrete institutional design.[34] Consequently, the funding is provided from supranational sources. However, although the ERC is financially detached from the member states, it is closely attached to the Commission. New dependencies have thus been created. This is not only true in financial matters but also in matters of organization.[35]

The ERC consists of the Scientific Council, on the one hand, and the European Research Council Executive Agency, on the other. The Scientific Council is composed of 22 scientists, who are responsible for the annual work programmes and for the scientific parts of the selection procedure.[36] The executive agency is entrusted with the administrative implementation and the execution of this part of the Framework Programme. As in all executive agencies, management is selected by the Commission, and there is a detailed micro-management of the agencies' work by the Commission services. Member State representatives do not play an active role in ERC work. One of the Framework Programme's comitology committees (whose members are delegates sent by member states) is indeed devoted to the ERC. But both legally and in practice the committee retains a rather symbolic role.[37] National research councils, furthermore, are not explicitly involved in the operation of the ERC.

The idea of the "research council," in other words, in Europe has been translated into a unique structure, which is determined by the administrative framework of the Commission. Especially the detailed financial regulation governs all relevant activities of the Union, including the research

---

33 Commission decision of 12 December 2013 establishing the European Research Council, OJ C 373/23; Commission implementing decision of 17 December 2013 establishing the European Research Council Executive Agency and repealing Decision 2008/37/EC, OJ L 346/58.

34 Gornitzka and Metz, *Dynamics of Institution Building in the Europe of Knowledge*, 93.

35 Thomas König's analysis in his "Funding Frontier Research: Mission Accomplished?," *Journal of Contemporary European Research* 11 (2015): 124–35.

36 Commission decision of 12 December 2013 establishing the European Research Council, OJ C 373/23. This can also be seen as symbolic rejection of the European standard that all member states are represented, cf. Ernst-Ludwig Winnacker, *Europas Forschung im Aufbruch: Abenteuer in der Brüsseler Bürokratie* (Berlin: Berlin Univ. Press, 2012), 126.

37 For a brief account of the actual work, see Gornitzka and Metz, *Dynamics of Institution Building in the Europe of Knowledge*, 93.

sector, and provides for a variety of internal control mechanisms. While regulation constitutionalizes this area of European administrative law,[38] in research funding it expresses a general mistrust towards funding recipients. Controversial discussion of the absence of institutional autonomy for the ERC is ongoing.[39] The TFEU offer alternative legal options for independent organizational structures, which are used in other contexts of research policy.[40] Out of a perceived need for power and control within the administrative framework, therefore, the Commission created a path dependency that might be hard to overcome in the future. Ernst Winnacker, the first secretary-general of the ERC, in his autobiography on "adventures in Brussels bureaucracy" charts some of the policy's rather absurd consequences.[41]

Given these difficulties, it comes as a surprise that the ERC presents the new funding lines as success stories. Indicators of this perceived success are the high number of applications, which results in a success rate lower than that of most national research councils, and the reputation gained in the scientific community. But this is only half the truth. At the same time, the institutionalization of Europe-wide funding for basic research creates funding parallel to that provided by research councils in member states, and it has consequences for research policies, research funding institutions, research organizations, and researchers in member states.[42] Certain effects are hard to ignore, especially with a view to funding criteria and standards. Given the broad legal framework on standards and procedures for the selection of proposals, the ERC's scientific council created a "best of" list of national rules.[43] This generated a new European point of reference. One example is the common language. As most of the traditional and new instruments of funding, the ERC's funding lines presuppose, for practical reasons, that English be the working language for

---

38 See Paul Craig, "The Constitutionalisation of Community Administration," in *European Law Review* 28 (2003): 840–69.

39 See e. g. Armin von Bogdandy and Dietrich Westphal, "Der rechtliche Rahmen eines autonomen Europäischen Forschungsrates," *Wissenschaftsrecht* 37 (2004): 224–38; Ralph Alexander Lorz and Mehrdad Payandeh, *Die Institutionalisierung des Europäischen Forschungsraums* (Tübingen: Mohr Siebeck, 2012), 15.

40 See Art. 187 TFEU on which the "Joint Technology Initiatives" are based.

41 Winnacker, *Europas Forschung im Aufbruch*.

42 For a detailed discussion, see section 4.

43 See also Gornitzka and Metz, *Dynamics of Institution Building in the Europe of Knowledge*, 98–99.

proposals. Another telling example is the decision to require open access publications by researchers who receive ERC funding. In line with the generally strong EU impetus to establish open access publications as the norm, these stipulations are used to exert leverage on scientific publishers.[44] It is also interesting that embedding new institutions at the European level apparently involves them in powerful European narratives. One of the old themes deeply rooted in the history of European research policy is the use of research to create innovation and a competitive European economy, mostly using an oversimplified conception of the relationship between research and innovation. The ERC, although committed to the idea of a "frontier" of basic research, introduced a specific funding line to foster the economic exploitation of research results.

3.2. Transnational Research Funding in Institutional Varieties[45]

The coordination of national research funding programmes lay at the center of the proposal the Commission made for the creation of a European Research Area.[46] From the EU's perspective, a major objective was to align research funding of member states with joint goals, standards and procedures. Unlike the first phase of European research policy, the different types of research funding organizations were addressed for the first time in the European Research Area. Such an alignment again requires institution-building, and this section analyzes two of the most important institutional forms, the so-called ERA-Nets (a) and the initiatives based on Article 185 of the Treaty on the Functioning of the European Union (b). Despite their institutional variety they share basic principles: Through funding and on the basis of competition, the EU stimulates the creation of networks on different research topics by national research funding organizations. EU funding ensures that they are closely attached to the Commission in ful-

---

44 See the Commission Recommendation of 17 July 2012 on access to and preservation of scientific information and the Communication from the Commission, "European Research Area Progress Report", *COM* (2013) 637, 8

45 This section adapts parts of Arne Pilniok, "Changing European Governance of Research, A Public Law Perspective," in *The Changing Governance of Higher Education and Research*, edited by Dorothea Jansen and Insa Pruisken (Dordrecht: Springer 2015), 222–27.

46 Cf. Communication from the Commission, "*Towards a European Research Area*", *COM* (2000) 6 final; European Commission, "Greenbook European Research Area: New Perspectives," *COM* (2007) 161 final.

filling their purpose, which is the development and implementation of transnational research funding programmes in all scientific areas.

a) ERA-Net

The Sixth Framework Programme introduced the ERA-Net scheme, addressing the national research funding organizations for the first time in the history of European research funding. With several alterations this scheme continues in the Seventh and Eighth Framework Programme. Since 2002 more than 170 networks of research funding organizations have been founded by the European Union, each of them devoted to a specific scientific discipline or scientific topic.[47] These measures reflect a change in relationship between the European and national level of research funding, as well as new forms of governance introduced by the European Union.

The consequence of the integration of the ERA-Net scheme into the Framework Programme was to start a competition between national research funding organizations due to the fact that the Financial Regulation of the Union[48] requires a competitive procedure for European grants. Since there are budgetary alternatives, one can assume that a competitive governance structure was explicitly chosen by the Union—and especially by the Commission—because of its advantages in a multi-level system. The rules of the competition for European grants are structured by the Framework Programme, the respective Specific Programme and—with regard to the most important criteria—the work programme issued by the Commission. According to the work programme, the Commission issues calls for proposals. While these calls originally were open to any scientific subject, disciplines and topics were more specifically addressed in the Seventh Framework Programme. Applications require the participation of at least three research funding organizations from different member states. The selection of applications is based, inter alia, on the creation of adequate internal governance structures by the funding organizations as well as on a long-term commitment to cooperate. This governance arrangement forces

---

47 European Commission, *The ERA-Net scheme from FP6 to Horizon 2020* (Brussels 2014).

48 Regulation (EU, Euratom) No 966/2012 of the European Parliament and of the Council of 25 October 2012 on the financial rules applicable to the general budget of the Union and repealing Council Regulation (EC, Euratom) No 1605/2002, OJ 2012 L 298/1.

the funding organizations into intensive cooperation even at the preparation phase of an application. The self-coordination process requires the funding organizations to know about the funding structures in the other member states. The institutionalization of competition mobilizes the strategic interests and the decentralized knowledge of the funding organizations.[49] The competition between the consortia of the member states' research funding organizations is based on incentives, not coercion. It thus allows a cooperative initiative to be designed that adequately respects the heterogeneity of funding structures in the member states as well as to the different approaches to scientific disciplines and topics. In this respect, the competition fosters innovative solutions while simultaneously allowing for differentiated integration.

The vertical governance structure between the Commission and the consortium of the participating research funding organizations is determined by an agreement whose conclusion is required by the Financial Regulation. For all grants within the Framework Programme the Commission uses a model agreement, with different variants for specific funding lines. The agreement fixes the working programme and the deliverables, on the one hand, and the Union's financial contribution, on the other. The most important role in the agreement is the coordinator of the consortium, who is responsible for obligations towards the Commission. These obligations include reporting obligations that serve the Commission not only for control purposes but also for generating knowledge about the research funding landscape within the EU. Additionally, extensive financial control mechanisms are established by the model agreement, as well as unilateral rights of the Commission.

As recommended, but not required, by the Commission, the horizontal governance structure between the participating research funding organizations is usually governed by a consortium agreement in which the internal decision-making structure and the distribution of work packages within the consortium are laid out. Consequently, the consortia exhibit characteristics of an organization combined with features of a rather loosely coupled network. Based on the equal participation of all funding organizations involved, internal governance is dominated by negotiations within the institutional structure laid out by the consortium agreement. In sum, this

---

49 On the different strategies in diverse organizational settings of integration see Benedetto Lepori et al., "Logics of Integration and Actor's Strategies in European Joint Programs," *Research Policy* 43 (2014): 391–402.

creates a complex network of contracts administered by the Commission, integrating a huge number of research funding organizations across Europe. As a consequence of this diversity, the Commission installed mechanisms for monitoring the networks and their mutual exchanges.[50]

The Commission and the consortia of research funding organizations typically agree on procedural integration measures that follow a step-by-step approach. Initially, the cooperation focuses on stocktaking exercises that evaluate the heterogeneity of research funding, its organization and its procedures in the specific scientific discipline being addressed. Their goal is to get to know the situation in the other member states and to create mutual trust. This is supposed to be complemented by the exchange of best practices in research funding and to lead to extensive self-evaluations by all disciplines and member states, which are of high value to the Commission as well as to the scientific community at large. Finally, all networks of research funding organizations are urged to implement a joint research programme, which has become a funding requirement since then.[51] Therefore, a range of models has evolved whose main differences consist in how far financial aspects are handed over to the consortium. Through the co-funding of these joint programmes the Commission sets additional incentives for establishing joint research funding programmes that include a "common pot". This "common pot" is used for funding, irrespective of the nationality of the successful applicants and thus creates a denationalization of research funding and Europe-wide competition. Overall, the Commission states that in 2014 around three billion euros of joint research funding have been allocated in the past decade via the ERA-Networks.[52]

b) Initiatives Based on Article 185 TFEU

Another branch of institution building includes so-called Article-185-Initiatives. Article 185 TFEU requires that the Union participate in research and development programmes undertaken by several member

---

50 See http://netwatch.jrc.ec.europa.eu/home.

51 Article 26 of the Regulation No. 1291/2013 of the European Parliament and of the Council of 11 December 2013 establishing Horizon 2020 – the Framework Programme for Research and Innovation and repealing Decision No. 1982/2006/EC, OJ 2013 L 347/104.

52 European Commission, The ERA-Net Scheme from FP6 to Horizon 2020 (Brussels 2014).

states. Although introduced in the Single European Act in 1987, it has not been referred to after 2002. Given the objectives of the European Research Area in general, and the alignment of public research funding in the member states in particular, this competence gained significance for the Commission. In the context of the Sixth Framework Programme, a first experimental measure on the basis of Article 185 TFEU was taken, which revealed the presuppositions and the problems of the integration of national and European research funding. The Horizon 2020 Framework Programme is used as a legal basis for the institutionalizing of five initiatives.[53] They address a broad range of problems to be tackled by interdisciplinary research, such as oceanography or metrology, the science of measurement.

Article 185 TFEU allows for the participation of the EU in the joint action of several, but not necessarily all, member states. The norm aims at integrating the research funding programmes of several member states. Thus, the scope is limited to existing programmes. Research funding programmes are understood by the Commission as "clearly defined activities or measures (whether or not formally called "programmes") on a specific theme or in a specific area, with an earmarked budget and implemented over a set period following clear procedures."[54] Integration in this area can be intensified without requiring the consent of all member states and therefore constitutes an instrument of variable geometry. At the same time,

53 Decision No. 862/2010/EU of the European Parliament and of the Council of 22 September 2010 on the participation of the Union in a Joint Baltic Sea Research Programme undertaken by several member states, OJ 2010 L 256/1; Decision No. 553/2014/EU of the European Parliament and the Council of 15 May 2014 on the participation of the Union in a Research and Development Programme jointly undertaken by several member states aimed at supporting research and development performing small and medium-sized enterprises, OJ 2008 L 201/25; Decision No. 554/2014/EU of the European Parliament and of the Council of 15 May 2014 on the participation of the Union in the Active and Assisted Living Research and Development Programme jointly undertaken by several member states, OJ 2014 L 169/14; Decision No. 555/2014/EU of the European Parliament and of the Council of 15 May 2014 on the participation of the Union in a European Metrology Programme for Research and Innovation undertaken by several member states, OJ 2014 L 169/27; Decision No. 556/2014/EU of the European Parliament and of the Council of 15 May 2014 on the participation of the Union in a second European and Developing Countries Clinical Trials Partnership Programme (EDCTP2) undertaken by several member states, OJ 2014 L 169/38.
54 Communication from the Commission to the Council and the Parliament, "The Framework Programme and the European Research Area: Application of Article 169 and the Networking of National Programmes," COM (2001) 282 final, 5.

it is in the interest of the Union to incorporate a significant number of member states since they are necessary to attain a substantial funding volume. Research funding programmes have to be publicly financed, either by the member states or publicly financed research funding agencies in the member states. The participation of the Union is mainly left open by Article 185 TFEU; nevertheless, financial participation is the key driver for the participation of the member states and their research funding organizations. The article also allows the Union and participating member states to create common organizational structures. Decisions on the basis of Article 185 TFEU are taken through the ordinary legislative procedure according to the Articles 188 par. 2 and 289 TFEU. This constitutes a strong position for the European Commission given its monopoly on initiatives, which is limited by both the necessary consent of the participating member states and the requirement of being laid out in the Framework Programme. Unlike the ERA-Net scheme, competitive elements can be found only in the Commission's exertion of its right of initiative, which is based on the previous experiences of member states cooperating in a specific field of research funding. The ordinary legislative procedure requires time-consuming political bargaining processes and provides less flexibility in the course of implementation, as compared to the ERA-Net scheme. Substantially higher transaction costs limit this formalized course of integration to only a small number of scientific fields.

The regulatory structure of these institutions is complex: Within this normative scheme, the Framework Programmes contain a general decision of the Union to participate in joint research funding programmes with several member states. The Framework Programme outlines common criteria for participation, while the Specific Programmes for the implementation of the Framework Programme substantiates the particular scientific fields that are envisaged for a joint programme of several member states. The decisions taken so far contain similar elements and follow a common structure: The decision itself lays down the conditions for the financial participation of the Union concerning the joint research funding programme. Annex I of each decision describes this research funding programme in general terms and prescribes some basic rules for the procedures. Annex II of each decision adopts the internal decision-making structure of the "dedicated implementation structures," which have to be established by participating member states in order to administer the funds. The decisions are supplemented with general and annual agreements between the Commission and dedicated implementation structures. Thus, normatively complex governance structures emerge along both the vertical and horizontal dimension.

From a vertical perspective, the relationship between the Commission and the dedicated implementation structure, which is established by the participating member states according to private law rules, is mainly focused on financial issues. The decisions define several conditions for the financial contribution of the Union. These are, inter alia, the establishment of a joint research funding programme as laid out in the decisions, the establishment of an "appropriate and efficient governance model,"[55] and formal commitments by the member states to contribute in sum the same amount as the Union. The strong financial focus of the Council decisions is also underlined by the extensive rules—stemming from financial regulation—to protect the financial interests of the Union. These "safeguards" for Union funding are a response to the problems revealed by the first initiative based on Article 185 TFEU, especially the reluctance of participating member states to follow-through on their planned contributions.[56]

The dedicated implementation structures have to be organized according to available corporate law. Consequently, they are governed by both public and private law. In the emerging field of European corporate law, the European Economic Interest Grouping (EEIG) is one legal structure that was introduced to facilitate the cooperation of business actors in the single market. For its use by the member states and governmental actors, EEIGs have the disadvantage of unlimited liability for all participating parties, which in German public law, for example, is forbidden for governmental actors. Nevertheless, one can find EEIGs among the dedicated implementation structures in several forms of non-profit organizations, according to the laws of different member states. This individualized construction allows for a description of internal decision-making structures on only a very general level: Usually, all implementation structures prescribe a general assembly that consists of representatives from all participating research funding organizations. The main implementation tasks are entrusted to an executive board, which is supported by a secretariat. Finally, all structures contain advisory bodies to consult with the relevant scientific community from fields that are funded.

---

55 See Article 2c of the Decision 742/2008/EC.

56 For details, see Commission Staff Working Document, "Progress Report on the European and Developing Countries Clinical Trials Partnership", *SEC* (2008) 2723.

Through their joint research programmes, the Union and participating member states together distribute several billion euros to these new actors. The central element of the joint research funding is the annual working programme that defines planned calls for proposals and sets a timetable for funding activities.[57] This reflects the formative influence of the Union's Framework Programme on these initiatives driven by the member states. The basic elements of the funding programmes are codified in the Annex to the Council's decision. The development of, and the decision about, the working programme follows the respective decision-making structure of the dedicated implementation structure. Generally speaking, it is the right of the general assemblies, representing all participating member states, to decide on the working programme, as this is one of the most important decisions when creating a new funding programme. The dedicated implementation structure has to submit the annual working programme to the Commission as a precondition for concluding the annual financial agreement. The Commission thus has a strong role with hierarchical rights.

The common denominator is a centralization of the selection procedures by the dedicated implementation structure, while the administration of the selected research projects is left to participating national research organizations. The basic principles of the selection procedures are fixed in the respective Council's decision, but leeway is given to dedicated implementation structures for creating specific procedures, guided by the principles of equal treatment and transparency. The criteria for awarding grants are prescribed by the Council's decisions as well, resembling the respective criteria in the Framework Programme and the Specific Programmes. Scientific excellence is proclaimed to be the primary criterion. Proposals for research projects have to be reviewed by independent experts. Based on these reviews, the dedicated implementation structure—as a rule, their general assemblies—creates a ranked list of proposals, which is binding for the allocation of funding both from the Union's contribution and from national budgets earmarked for the respective joint programme. Difficulties arise regarding the question of whether there is judicial supervision of these funding decisions. Since Article 263 par. 4 TFEU comprises only organs foreseen by the TFEU, no legal control by the European Court of First Instance is provided for. Judicial supervision is restricted to measures

---

57 Pilniok, *Governance im europäischen Forschungsförderverbund*, 334.

available in Member State hosting dedicated implementation structures according to their respective civil procedure law.

3.3. Transnational Planning of Infrastructure and Funding

New forms of governance structures, which change the relationship between the Union and the member states, have also emerged regarding the joint planning of research infrastructures and research funding. The institutions sketched above are complemented by the approaches for developing a coordinated, yet more loosely-coupled approach to aligning funding programmes of member states. In the framework of the European Research Area the Commission launched its Joint Programming Initiative in 2008.[58] A key actor beside the Commission is a sub-committee of the European Research Area Council (formerly CREST) that represents national research ministries. The joint programming initiative's characteristic feature is its voluntary nature for the member states. In certain areas, which are defined by societal needs instead of scientific disciplines, the member states are supposed to develop a so-called Strategic Research Agenda. This approach therefore represents a top-down approach to research funding, fostering certain projects to "steer" research towards socio-economic solutions. So far, ten Joint Programming Initiatives have been implemented.[59] Not surprisingly, these initiatives lead to initial forms of institution building. All initiatives have developed specific governance structures by creating executive boards, management positions as well as advisory boards; secretariats strengthen the institutionalization as well. As a common denominator, standards were developed, e.g. for peer review processes.[60] Despite the non-binding character compared to the ERA-Net and Article 185 initiatives described above, the Joint Programming Initiatives were successful in aligning a considerable amount of research funding. Joint calls for research funding are published on a regular basis by the transnational network formed by the member states involved.

---

58 Communication from the Commission, "Towards Joint Programming in Research: Working together to tackle common challenges more effectively," COM (2008) 468 final.

59 http://ec.europa.eu/research/era/joint-programming-initiatives_en.html.

60 See, for example, http://ec.europa.eu/research/era/docs/en/voluntary_guidelines.pdf.

In the process of defining the concept of the European Research Area, the European Commission soon attended to the research infrastructures.[61] Large-scale research infrastructures in particular are an obvious example of the advantages of a division of labor within the European Research Area as a means to maximize the efficient use of public funds. To facilitate cooperation in this area, the Commission again invented a new institution and set up the European Strategy Forum for Research Infrastructures (ESFRI) in 2002, formally as an expert group of the Commission.[62] The members of the ESFRI are representatives ("senior policy officials") of the member states' research ministries and of DG Research of the Commission. The ESFRI has developed a complex organizational structure, with a number of working groups under its umbrella.[63] The activities of the ESFRI are connected with the funding conducted via the Framework Programme's funding line for research infrastructures: The competition for European funding from the Framework Programme is predominantly restricted to those research infrastructure projects chosen by the ESFRI.[64] These activities also led to new legislation on research infrastructures, which created a special framework for legal entities established by several member states for the joint operation of research infrastructures.[65] In sum, these examples confirm the analysis that traditional tasks of the nation-state are relocated and embedded in a transnational European sphere. This has consequences at all levels of research governance.

---

61 Commission Staff Working Paper, "A European Research Area for Infrastructures," *SEC* (2001) 356.

62 Register of Commission expert groups available at http://ec.europa.eu/transparency /regexpert/index.cfm?Lang=EN.

63 Procedural guidelines of ESFRI, http://ec.europa.eu/research/infrastructures/pdf/esfri /how_esfri_works/esfri_procedural_guidelines.pdf.

64 ESFRI, Strategy Report on Research Infrastructures – Roadmap 2010 (Luxembourg: Publications Office of the European Union, 2011); on the nexus between acceptance in the roadmap and funding the Part I Nr. 4 of the Council Decision of 3 December 2013 establishing the specific programme implementing Horizon 2020—the Framework Programme for Research and Innovation (2014–2020) and repealing Decisions 2006/971/EC, 2006/972/EC, 2006/973/EC, 2006/974/EC and 2006/975/EC, OJ 2013 L 347/965.

65 Council Regulation (EC) No 723/2009 of 25 June 2009 on the Community legal framework for a European Research Infrastructure Consortium, OJ L 206/1.

4. Consequences: Denationalization of National Science Systems

As analyzed above, the recent developments in the European governance of research can be characterized as institutional differentiation in research funding. It is no longer the "standard procedure" of the EU's Framework Programme that is representative of European research policy. Just like the European Research Council, the networks that connect European and national research funding are institutional novelties. The same is true for the new forms of joint planning for research funding and infrastructure between the member states. These developments have consequences on different levels: The new European initiatives lead to a higher diversity of available funds. Hence, from the perspective of the researchers and the research organizations, the changing landscape of research funding is relevant due to shifting relations between institutional and competitive funding (4.1.). The new forms of research funding and research policy obviously also have consequences for member states and their research funding organizations (4.2.). National research funding organizations are involved in complex and overlapping networks with their European partner organizations. Finally, given the vitality of the European research policy's development in recent decades, this raises the question of where this development is headed. Of course, one can only speculate. Current discussions and political demands, however, point to a next phase of Europeanization (4.3.). In contrast to previous decades, the EU now strives for a comprehensive use of legislative competence in research policy. This includes legislative acts that affect research organizations and funding procedures at the national level.

4.1. Consequences for Researchers and Research Organizations

At the level of research institutions, the second phase of European research policy brings about different consequences. As a general European trend in the aftermath of the introduction of New Public Management agendas in member states, the funding of nearly all research organizations is based on performance indicators.[66] This development is actively supported by the European Union. As part of its agenda of creating "effective national research systems," the European Commission in its recent "ERA Progress Report 2013" has called for competitive funding and performance-based institutional assessment.[67] The complete or partial association of institutional funding (including external funding) with performance puts pressure on competitive strategies within organizations. The changing European environment of research funding leads to an increased heterogeneity of funding opportunities and calls for adjustments. Organizations need to develop expertise in different fields of science and their specific structures of transnational funding organizations as well as their programmes and criteria. From the perspective of research institutions, furthermore, inter-organizational competition within the European Research Area and its new institutional settings is stiffening. Different lines of funding offered by the European Research Council allow successful applicants to pick for their project a research institution irrespective of nationality. This creates competition between research organizations for the best working conditions. In keeping with predictions by member states opposing this institutional design, distribution of ERC projects among research organizations has been uneven. When compared to the number of their researchers and institutions of higher education, member states in the south and the east of the European Union host considerably fewer ERC projects than other states. These developments suggest a certain path dependency that restricts competition: Despite efforts to balance such factors, reputations that research institutions have earned over decades or centuries have come to play a key role.

Research organizations and (more importantly) researchers consequently see themselves confronted with an increased variety of actors who decide about funding and who fix the criteria for funding decisions. Paradoxically a greater heterogeneity and the emergence of a more uniform model of scientific standards may occur at the same time. On the one hand, all transnational research organizations such as the ERA-Nets have

---

66 Cf. Jürgen Enders et al., "Turning Universities into Actors on Quasi-markets: How New Public Management Affect Academic Research," in *The Changing Governance of Higher Education and Research*, edited by Dorothea Jansen and Insa Pruisken (Dordrecht: Springer, 2015), 89–103.

67 Communication from the Commission, "European Research Area Progress Report," COM (2013) 637, 3–4.

to create their own standards. They are usually based on a fusion of national standards that apply in those member states that are involved. This is one of the reasons why usually the first step in all of these initiatives is a comparative assessment of procedural and substantial standards in the relevant field of science.[68] The EU approach of theme- or field-specific funding initiatives generally allows for the adaptation of differences between disciplines. This leads to a certain homogeneity, however, partly as a result of the European Commission's requirement to provide financial support. Out of practical necessity, for example, English has been the dominant language, which does not always fit well with the needs of all disciplines, e.g. in the humanities.[69] The distribution of national funds through these networks follows European standards and procedures, and researchers as well as research organizations have to adjust.

Within research systems, adaptations to European research policy may also be found with regard to the development and strengthening of transnational fields of science.[70] As pointed out above, scientific communication is global, but this does not always include transnational institution-building in the research system, such as the emergence of European academic associations, journals, congresses, networks and the like. The development of new forms of transnational research funding within the framework of the European Union certainly is not the only factor relevant to these developments. But these new forms positively reinforce and accelerate them. Funding success in new areas of research is increasingly dependent on the relevant fields' integration and on their achievements in Europe-wide networks.

Consequences are also apparent for science-driven research councils.[71] In a simplified view, these research funding organizations have in common that they were traditionally funded by the nation-state and oriented towards their "clients" according to the specific mission of the organization. The various initiatives within the European Research Area framework

---

68 Ibid.

69 Cf. Roland Broemel, Arne Pilniok et al., "Disciplinary differences from a legal perspective," in *Governance and Performance in the German Public Research Sector: Disciplinary Differences*, edited by Dorothea Jansen (Dordrecht: Springer, 2010), 19–41.

70 For an enlightening analysis see Johan Heilbron, "European Social Science as a Transnational Field of Research," in *Routledge Handbook of European Sociology*, edited by Sokratis Koniodos and Alexandros Kyrtsis (London: Routledge, 2014), 67.

71 For a typology of research funding organizations see Trute and Groß, *Rechtsvergleichende Grundlagen der europäischen Forschungspolitik*, 212–34.

sketched above foster horizontal cooperation with other research councils. Research councils such as the German DFG, of course, have always been in touch with similar institutions in Europe and elsewhere. In order to tap funding lines such as ERA-Net, however, the creation of stable networks and their juridification by contracts becomes inevitable. This calls for an organizational strategy as well as for additional organizational resources. Research funding organizations need to integrate their staff dealing with European and national research funding, and these measures are likely to impact the organization's view of how research should be done. Key feature of these networks, developing best practice guidelines in research funding and mutual agreements on common procedures, are likely to impact "purely" national and regional research funding programmes. This is all the more likely because integration among the staff of these organizations will further "Europeanize" cognitive patterns towards research and its funding.

4.2. Consequences for National Research Policies

Adaption within changing European governance of research is necessary at the national level of research bureaucracy as well. Nation-states and their bureaucracies have to take into account a highly complex set of actors. Much like during the first phase of European research policy, they need to follow and staff the European Commission with its numerous comitology committees and expert groups as well as many other organizations. ERA-Networks, Joint Programming Initiatives, Initiatives based on Article 185, and Joint Undertakings actively involve representatives of member states. In Germany, this usually involves the Federal Ministry of Research and Education. Participation at the European level demands internal re-organization and adequate staffing as well as setting up an internal flow of information to adapt to and serve in an altered institutional environment. All of this opens up opportunities for political steering as well, both for nation-states as actors in the context of European initiatives and for the nation-states themselves. Influence within boards and committees, however, is usually limited. Setting up a transnational research funding institution presupposes compromises and new common rules. Participation requires significant funds, which might reduce resources available for "autonomous" national research projects. European research funding no longer complements national funding but meshes with it.

A telling example of challenges posed by the European Research Area is the German federal government's development of a national ERA strategy in July 2014.[72] Although the process of integrating European research policy is only partly based on binding European law, EU policy changes call for an active national response to it. Sometimes it is only a re-labeling of well-known policy, such as changes to the German constitution allowing for permanent cooperation between the federal level and the various German states in financing institutions of higher education. But other changes include the alignment of federal research funding with European programmes, inter alia within governance forms analyzed above, or the commitment to open access strategies, which is induced by European initiatives.

Changes analyzed in this chapter also have consequences within the federal structures of nation-states. European research policy and legislation is embedded in a multi-level system of governance.[73] Each territorial level has its own, autonomous decision-making power with respect to research-related legislation and funding. In federally-organized member states, furthermore, competence for research is frequently allocated to individual states or the federal level shares it with them. Difficulties in allocating authority to these levels are reflected in how competences are coordinated. Although *Länder* (state) authorities are partly involved in European affairs, the German federal structure is not very well suited for dealing with them. The Europeanization of research policy and funding might accelerate the loss of authority by German states to govern higher education, even if the German constitution grants this authority. At the federal level, European integration establishes new interdependencies that limit options for a self-sufficient research policy.

---

72 Strategy of the Federal Government on the European Research Area (ERA), available at http://www.bmbf.de/pubRD/Strategy_of_the_Federal_Government_on_the_Europea n_Research_(ERA.pdf.

73 For a detailed analysis, see Pilniok, *Governance im europäischen Forschungsförderverbund*, 13–114.

4.3. The Future: Towards a Third Phase of European Research Policy?[74]

In the near future, the frontier between European and national research policies could shift even further. European institutions currently are testing the limits through a more extensive use of EU competences in research policy. According to the model implemented during the first phase, direct funding via the Framework Programme allocated to the EU level the authority for additional funding and for coordinating research policies. Since 2000, the beginning of the second phase of European research policy left the constitutional text unaltered and complicated governance structures. Governance structures were characterized by a combination of legal and non-legal elements. The ERA-Net Programme sketched above, for example, has been governed by different legal acts (including the Framework Programme and Specific Programmes), by the EU's financial regulation, and by horizontal agreements between research funding organizations as well as vertical agreements with the European Commission. Allowing for differences in the details, the same is true for new forms of institutionalizing and funding research policy across borders. According to Article 182 of TFEU, the multi-annual Framework Programme needs to define all EU activities. Starting with their sixth edition, Framework Programmes contributed substantially to the implementation of the ERA. Therefore, all measures for realizing the ERA, including financial support for beneficiaries at the European or national level, are governed by EU legislation in the shape of the Framework Programmes and their acts implementing them, as well as the Financial Regulation. The ERA, it should be noted, has instigated a substantial body of legislation.

When considering ERA-related legislation, it is important to distinguish between two different types of legislation. The most prominent involves measures to implement research funding by the European Union (Framework Programme, Specific Programmes, implementing measures). Legally binding non-funding measures are exceptions.[75] Regulations based on Article 185 TFEU (interlacing European and national research funding through specific organizations) constituted hybrids between the two. From

---

74 This section draws on Arne Pilniok, "'The measures necessary for the implementation of the European Research Area' - what is the future role of legislation in EU research policy," in *The Future of Research and Innovation*, edited by Rene von Schomberg (Luxembourg: Publications Office of the European Union, forthcoming).

75 Namely the Scientific Visa Directive and the ERIC regulations.

the perspective of member states, these initiatives as well as those based on the ERIC are characterized by their variable geometry and their voluntary nature. European law merely offers governance structures, structures that participating member states or national organizations may or may not fill.[76] With the exception of the visa directive, most norms relevant for all member states or research-performing organizations such as the European Charter and Code for Researchers, are voluntary and enforced through non-legal mechanisms such as a competition for reputation.[77] In discussing ERA-related legislating, therefore, there is a need to distinguish between future legal measures not related to funding that are binding for member states (directives) or for research organizations (regulations).

The Treaty of Lisbon, however, changed the EU research policy's legal framework by allowing the enactment, inter alia, of legislative measures necessary for the implementation of the European Research Area (Article 182 par. 5 TFEU). By the time of its introduction in 2000, the Commission pointed out that the "full panoply of instruments available to the Union should be brought into play," including regulations and directives.[78] The Commission and the European Parliament semantically inferred the common market in calling for a "completion" of the European Research Area. As a result, discussion about the use of Union legislation to strengthen the ERA has intensified. Under the heading "A Maastricht for Research," members of the European Parliament have called for an ERA framework directive.[79] Akin to the promotion of research policy in member states, EU research policy is largely driven by the executive.[80] Indeed, given its co-legislative role under current TEU provisions, an expansion of legislative measures would enhance the European Parliament's role. No wonder that Member State governments rigidly oppose such demands and instead prefer instruments that remain legally non-binding. In keeping with this view, the ERAC states that "[t]he use of legislation to address obstacles is not

---

76 Pilniok, *Governance im europäischen Forschungsförderverbund*, 393.

77 Ibid., 239.

78 Communication from the Commission, *Towards a European Research Area*, 22.

79 Amalia Sartori and Luigi Berlinguer, "Towards a Maastricht for Research," http://www.eurekanetwork.org/c/document_library/get_file?uuid=d2bd3b88-69cb-491e-8756-539b44e82ab7&groupId=10137.

80 Åse Gornitzka, "Executive governance of EU research policy. WZB Discussion Paper," http://bibliothek.wzb.eu/pdf/2012/iv12-502.pdf.

widely supported by MS and should be used only where clear and significant need is agreed, i.e. only as last resort."[81]

If the European Union enacts legislation detached from research
funding and its organizational structures, a third phase of European research policy might emerge. For the first time, this would establish for the
EU direct legislative access to research organizations and research personnel. European legislation aiming at the "completion of the European Research Area" would drastically enhance the legal pluralism at the national
and organizational level. EU legislation, however, is subject to the same
general restrictions as public research governance in member states: As a
functional system in a differentiated society, the research system is governed mainly by the intrinsic rationality of researchers (e.g. new knowledge,
truth, reputation, etc.). The research system's self-governance is essential
and direct intervention remains impossible. Also, knowledge asymmetries
between research systems and policy makers continue to persist. Public
research governance is mainly limited to shaping framework conditions
that may (or may not) structure and influence behavior within it. Governance usually focuses on shaping organizations (of research and of research
funding) through funding such organizations, through creating funding
programmes, and by establishing frameworks with conditions, rules, and
procedures for recruiting and employing researchers. The European Research Area consists of a hodgepodge of research-performing organizations, among them associations representing scientific disciplines, each
with its own internal standards and norms.[82] The landscape of organizations that provide funding for research is heterogeneous as well.[83] Such
organizations are intermediaries decoupled from the political process. Consequently, they enjoy some autonomy from government, which complicates matters for the latter when that government seeks, for example, to
implement an ERA framework directive. Given this array of actors, there
can be no single solution for any given problem that can be enshrined in
EU law. In order to ensure the implementation of legal rules, governments
will have to sustain some flexibility.

---

81 ERAC Opinion on the development of an ERA Framework, http://register.consilium
   .europa.eu/doc/srv?l=EN&f=ST%201215%202011%20INIT.
82 On the consequences for the law, see Broemel, Pilniok et al., *Governance and Performance in
   the German Public Research Sector*, 19–41.
83 Trute and Groß, *Rechtsvergleichende Grundlagen der europäischen Forschungspolitik*, 222–32.

If the EU moves forward with legislative acts to further implement a European Research Area, this would alter the relationship between science and the nation-state even more. During its first phase, European research policy "added on" to research policies in nation-states that preserved authority over key aspects of research governance. As elaborated throughout this chapter, this connection was eroded during the second phase. While formally leaving their authority intact, both research organizations and funding organizations in the various national states were forced to adapt to the emergence of transnational research funding, to structural changes in the research system itself, and to university reforms based on New Public Management. Further EU legislation transforming national laws governing research and its funding will hasten the departure of traditional relationships between science and the nation-state.

5. Conclusion

Other contributions to this volume point out that science and the state have been intertwined from the institutionalization of modern science in seventeenth-century England.[84] During the late twentieth century, European integration has changed the structure of nation-states by embedding in a supranational legal order policy areas including science. The same holds true for other areas, but due to the characteristics of the research system in a functionally differentiated society, we may observe a number of features that distinguish it from others. With respect to the public governance of research, the European integration process may be conceived of as a process of institution building. As argued throughout this chapter, we can see a second phase of institutionalization in EU research policy. In the first phase, the level of institution building within the Community framework was comparatively low. As a paradigmatic model, direct European support of research added another layer of funding. During its first phase, EC research policy was limited to directing certain fields of research to stipulate economic integration and progress. During the second phase of EU policy, the division of labor changed. Within its framework of the European Research Area, the EU now envisioned deep integration and complex hori-

---

84 Andreas Franzmann, Axel Jansen, Peter Münte, "Legitimizing Science: Introductory Essay," in this volume.

zontal and vertical governance structures. Structural changes have had consequences for research teams, research organizations, funding organizations, and research policies developed by member states. At this time, we may perhaps anticipate a third phase of EU research policy. In the near future, that phase may be driven by the ambition to legislate into existence an integrated ERA, not only of funding for research, but also of research organization and of personnel.

Universalized Third Parties: "Scientized" Observers and the Construction of Global Competition between Nation-States[1]

Tobias Werron

1. Introduction

The title of this volume, *Legitimizing Science: National and Global Publics, 1800–2010* evokes both active and passive connotations. The more intuitive ones, perhaps, are passive: How has modern science *been* legitimized in the past two hundred years? In contrast, an active understanding of "legitimizing science" implies the question: How has science contributed to the formation and legitimization of other societal structures? This is the question at the center of the present chapter. Highlighting the active rather than the passive understanding also tends to shift the attention from national science-state relationships to the global impact of modern science. While much of the literature focuses on the role and legitimacy of science in the nation-state, therefore, my article will focus on the role of science in the making of globalization.

In recent years, neo-institutionalist world polity scholars have analyzed important aspects of this process, dubbing it "scientization" and arguing that since the mid-to-late nineteenth-century modern science has contributed to the rise of a global style of reasoning that has created a common global environment, a "world polity," for all nation-states.[2] However, they and other students of globalization have so far largely neglected a significant dimension to this story that I wish to highlight here: the role of science and scientists in the formation of new forms of global competition

---

1 This chapter builds and partly draws on Tobias Werron, "What Do Nation-States Compete for? A World-Societal Perspective on Competition for 'Soft' Global Goods," in *From Globalization to World Society. Neo-institutional and Systems-theoretical Perspectives*, edited by Boris Holzer, Fatima Kastner, and Tobias Werron (London: Routledge, 2014), 85–106.

2 Gili S. Drori, John W. Meyer, Francisco O. Ramirez, and Evan Schofer, eds., *Science in the Modern World Polity: Institutionalization and Globalization* (Stanford: Stanford Univ. Press, 2003).

between nation-states since the mid-to-late nineteenth century. To make sense of these forms, I combine globalization research with recent developments in the sociology of competition. At the core of my argument is the rise of what I call universalized third parties: "scientized" observers who present themselves as disinterested third parties external to the system of nation-states and create competition by constantly comparing and evaluating nation-states according to universalistic criteria such as economic growth, human rights protection, or scientific, athletic and artistic excellence. I argue that the forms of competition resulting from these long-term processes, and the roles science and scientists play in the construction of these forms, should be studied in order to understand how today's national-science-nation-state relationships are embedded in globalization processes and how world-societal institutions are affected by their national enactment or rejection.

To develop this perspective, I connect insights from world society research with ideas from my own previous work on a historical sociology of competition. I start with a short introduction to, and interpretation of, the scientization thesis by the neo-institutionalist approach to world society led by John W. Meyer. This approach is particularly relevant here because it sees science not only as a self-generating global system but also as a force that structures world societal processes. The second section introduces a sociological concept of global competition. Based on Georg Simmel's concept of "pure" competition, it explains the aforementioned notion of the universalized third parties and their historical role in the social construction of global forms of competition between nation-states since the mid-to-late nineteenth century, pointing, among other things, to the impact of international organizations, telecommunication technologies and imaginations of global publics. The third section uses these conceptual insights to identify and highlight three historical trends towards new forms of nation-state competition since the mid-to-late nineteenth century: First, competition for "modernity prestige," second, competition for "specific cultural achievement prestige," and third, competition for attention and legitimacy. The final section summarizes the argument and concludes with some remarks on the position of this approach within the current literature on the science/nation-state relationship.

2. "Scientization" and "Rationalized Others:" A World-Societal View of the Nation-State/Science Relationship

In a globalization context, and somewhat analogous to the distinction between a passive and active understanding of "legitimizing science," two instructive connotations of scientization can be distinguished: First, the term can be used to frame the analysis of the formation of modern science as an autonomous global field, in line with what system theorists are interested in when investigating the emergence, internal differentiation and globalization of modern science.[3] Second, it can draw attention to the impact of science on its environment, particularly on other global fields or global society at large. In the latter case, it is close to what historian Lutz Raphael has called "the scientization of the social."[4] The present article will focus on the second meaning of the term.

Within the topic of the societal impact of science, I suggest distinguishing four sub-themes: First, academic socialization of experts and professions (scientization by education); second, tangible forms of influence of scientists/scientific professions on other actors or fields, particularly by means of expertise, such as in political counseling, applied research, or popularization of scientific knowledge in the mass media (scientization by expertise); third, the societal authority, or legitimacy, of science (scientization by reputation); and fourth, the capacity of science to structure expectations in global society at large (scientization as global structuration). To a degree, the last two reflect the other dimensions of scientization, as they will always rest on more tangible forms of academic-scientific socialization and inter-field roles and expertise. However, the authority and global impact of science also have their own causes and effects, and it is these effects that are put into focus by the neo-institutionalist approach to world society.

The neo-institutional scientization-thesis argues that the global authority of science links with the formation of global institutions by "scientized" observers to create a world culture (also referred to as "world polity" or

---

3 Rudolf Stichweh, "The Sociology of Scientific Disciplines: On the Genesis and Stability of the Disciplinary Structure of Modern Science," *Science in Context* 5, no. 1 (1992): 3–15; "Genese des globalen Wissenschaftssystems," *Soziale Systeme* 9, no. 1 (2003): 3–26.

4 Lutz Raphael, "Die Verwissenschaftlichung des Sozialen als methodische und konzeptionelle Herausforderung für eine Sozialgeschichte des 20. Jahrhunderts," *Geschichte und Gesellschaft* 22, no. 2 (1996): 165–93.

"world society"[5]) that serves as a common environment for all nation-states.[6] Neo-institutionalists argue that in a stateless world society, scientific knowledge and academic experts have unrivalled legitimacy in establishing universalistic norms and models of rational actorhood, as well as considerable leeway in imposing those criteria and models on governments and other actors around the world. John W. Meyer has called these experts "rationalized others," which is supposed to highlight that their influence is based not so much on responsible action, but on observing and criticizing, and not on acting themselves, but on the theorization and modeling of the actions of others.[7]

The abstract models created by those rationalized others imply high expectations of what a modern state is and how a modern state should act, including, for instance, the expectation to adopt state-of-the-art economic policies, to guarantee the protection of human rights, or, in the field of science policy, that each nation-state should have a national science foundation, universities, and other organizations specialized on the production of knowledge and tertiary education.[8] Rationalized others and rationalized models of actorhood appear in various social roles and organizational forms such as international governmental organizations (IGOs), international nongovernmental organizations (INGOs), applied scientists of all kinds, journalists with a universalistic perspective, and other experts and counselors with an academic-scientific background. Scientization in these terms, then, is not only about activities of scientists and academics, it is also about the long-term institutionalization of a "scientized" style of reasoning that can influence local, national and regional structures around the world. The main effects usually highlighted by world polity scholars is a surprising degree of isomorphism between nation-states and other actors on an "official" global level, combined with regular decoupling between

5 Neo-institutionalists tend to use these three terms mostly interchangeably; for possible nuances cf. George Thomas, "World Polity, World Culture, World Society," *International Political Sociology* 3, no. 1 (2009): 115–19.

6 Evan Schofer and Elizabeth H. McEneaney, "World Society and the Authority and Empowerment of Science," in *Science in the Modern World Polity: Institutionalization and Globalization*, edited by Gili S. Drori et al. (Stanford: Stanford Univ. Press, 2003), 23–42.

7 John W. Meyer, "Rationalized Environments," in *Institutional Environments and Organizations: Structural Complexity and Individualism*, edited by John W. Meyer and W. Richard Scott (Thousand Oaks: Sage, 1994), 28–54.

8 Martha Finnemore, "International Organizations as Teachers of Norms: The United Nations Educational, Scientific, and Cultural Organization and Science Policy," *International Organization* 47, no. 4 (1993), 565–97.

expectations on the global level and actual local practices.[9] There are, however, further consequences that feature less prominent in world polity research but are equally important for an understanding of scientization, one of which is the rise of global forms of competition since the mid-to-late nineteenth century.[10]

3. Universalized Third Parties: Social Construction of "Soft" Forms of Global Competition between Nation-States

How is the neo-institutionalist view of scientization connected to the historical rise of global forms of competition? I suggest combining the scientization thesis with insights from the sociology of competition and historical globalization research. This leads to a number of conceptual remarks.

(1) Competition as a triadic social form. First, it is useful to distinguish competition more clearly than usual from other forms of struggle. Following Georg Simmel, the "pure" form of competition can be conceived as a triadic social form, where at least two parties strive for the favor of a third party.[11] This understanding draws attention to the active role of third parties in creating and sustaining competitive social relationships, and thus helps distinguish competition from more "direct" (dyadic) forms of struggle, such as conflicts or "hard" forms of competition for territories, natural resources, that do not necessarily require the participation of third parties. On this basis, Simmel was able to stress an often-overlooked characteristic

---

9 John W. Meyer, "The World Polity and the Authority of the Nation-State," in *Studies of the Modern World-System*, edited by Albert Bergesen (New York: Academic Press, 1980), 109 37; "World Society, Institutional Theories, and the Actor," *Annual Review of Sociology* 36, (2010): 1–20. For a detailed recent discussion of the regular decoupling between isomorphic global structures on the one hand and actual local practices on the other, see Boris Holzer, "The Two Faces of World Society: Formal Structures and Institutionalized Informality," in *From Globalization to World Society. Neo-Institutional and Systems-Theoretical Perspectives*, edited by Boris Holzer, Fatima Kastner, and Tobias Werron (London: Routledge, 2014), 37–60.

10 The neglect of these forms of competition is explained in Werron, "What Do Nation-States Compete for?"

11 Georg Simmel, "Soziologie der Konkurrenz," *Neue Deutsche Rundschau (Freie Bühne)* 14, no. 10 (1903): 1009–23; for an English translation, see Georg Simmel, *Conflict: The Web of Group-Affiliation*, trans. Kurt H. Wolff and Reinhard Bendix (New York: Free Press, 1955), 57–85.

of such forms, namely that they also depend upon the third party's favor being perceived as scarce. He illustrated this point with a counter-example: Believers who try to win God's favor by surpassing others' "good deeds" may be said to participate in some sort of contest or rivalry (*Wettstreit*). However, according to Simmel they don't actually compete because, from a Christian-theological standpoint, "there is room for everybody in God's mansion." In other words: God's, the third party's, favor is not perceived as scarce and thus not able to produce competition; conversely, God's favor could be scarce if it were perceived as such.[12]

(2) *Public forms of competition.* Simmel developed the concept of competition as a triadic social form more than a hundred years ago but since then, surprisingly little work has been done on such forms in social theory or historical sociology. A certain pre-theoretical confidence in the discernibility of contested goods seems to have prevented even constructivist approaches from undertaking serious research on these indirect forms.[13] Simmel's model is a good starting point for such an endeavor as it draws attention to third parties that produce and distribute their limited amounts of favor between the competitors. It was with regard to the third party that Simmel could attribute to competition a "synthetic power," that is, the ability to create a "connex of minds" between the competitors, on the one hand, and a large number of potential third parties such as consumers, voters, and newspaper readers, on the other.[14]

Simmel failed to spell out what difference it makes for the sociological form of competition when competitors and third parties do not or cannot always interact face-to-face. Market sociologist Ezra Zuckerman raised a similar question a few years ago when he criticized the state of research in New Economic Sociology for ignoring two important aspects of the social construction of markets implicit: "the presence of an audience confronting focal actors and the competition among such actors for the favor of this audience. Without an audience," Zuckerman pointed out, "legitimacy loses its value and, indeed, its meaning."[15] To fill this void, he analyzed the categories used by market analysts and showed that the selection of catego-

---

12 Georg Simmel, *Soziologie: Untersuchungen über die Formen der Vergesellschaftung* (Frankfurt: Suhrkamp, 1992), 334.

13 For details, see Tobias Werron, "On Public Forms of Competition," *Cultural Studies <=> Critical Methodologies* 14, no. 1 (2014): 62–76.

14 Simmel, *Soziologie*, 332.

15 Ezra W. Zuckerman, "The Categorical Imperative: Securities Analysts and the Illegitimacy Discount," *American Journal of Sociology* 104, no. 5 (1999): 1398–438.

ries such as associating individual companies with "industries" may significantly influence the reaction of the audience and the development of stock market prices ("categorial imperative").

In developing a general sociological model of public forms of competition, three insights may and should be added to Zuckerman's argument. First, explicating audience expectations is useful not only with regard to markets but in any social situation where two or more providers of some product or offer (such as stocks, political decisions, newspapers, art, or athletic performance) try to win the favor of an indefinite, and thus basically unknowable, audience. Second, as competition for an audience cannot be based on face-to-face contact alone, it calls attention to media technologies and public communication processes that constantly address and create the audience as a "hidden" third party.[16] Third, the construction of competition for an audience's favor is also a product of imaginations of the audience that project the audience as a "public" of attentive and critical individuals rather than an undifferentiated "mass."[17] Precisely because it cannot be directly observed, the very fiction (imagination, assumption) of the audience enables public communication processes to construct such forms of competition.[18]

(3) Global forms of competition. Finally, these insights allow a connection to be made between the sociological model of competition and the neo-institutionalist scientization-thesis: The neo-institutionalist concept of the "rationalized others," when integrated into this model of public forms of competition, can be re-conceptualized as universalized third parties that

16 Tobias Werron, "Zur sozialen Konstruktion moderner Konkurrenzen: Das Publikum in der 'Soziologie der Konkurrenz,'" in *Georg Simmels große "Soziologie"*, edited by Hartmann Tyrell, Otthein Rammstedt, and Ingo Meyer Meyer (Bielefeld: Transcript, 2011). For a reading of the historical discourse about competition in the light of this model, see Werron, "Why Do We Believe in Competition? A Historical-Sociological View of Competition as an Institutionalized Modern Imaginary," in "Competition," edited by Eva Hartmann and Poul F. Kjaer, special issue, *Distinktion: Scandinavian Journal of Social Theory* 16 (forthcoming 2015).

17 Gabriel Tarde, "The Public and the Crowd," in *Gabriel Tarde: On Communication and Social Influence: Selected Papers*, edited by Terry Clark (Chicago: Univ. of Chicago Press, 1969), 277–96.

18 Particularly interesting and influential examples of imaginations of the audience are the diverse forms of *audience research* that emerged and spread in the twentieth century, such as opinion polls or market research: "Market shares," "viewer ratings," or "approval ratings" imagine the audience's favor as a "good" that can be divided into fragments or quantities, pushing firms and other actors to compete for these statistically imagined audiences.

participate in the production of global forms of competition by observing, comparing, and evaluating performance according to universal standards while addressing and imagining global audiences. Historically, this argument fits with empirical insights from globalization scholars, particularly by the Canadian media historians Dwayne Winseck and Robert Pike on what they call the "global media system."[19] The term includes, first of all, the invention and diffusion of a number of new communication technologies, beginning with the telegraph system and the rotation press in the mid-nineteenth century, but it also points to sociocultural prerequisites and consequences, such as the rise of news agencies and the proliferation of daily newspapers and specialized journals. Most notably, in our context, it was this media-technological infrastructure that first enabled the separation of the speed of communication from the speed of transport of goods and people. This "dematerialization of communication"[20] also enabled the imagination of universal and global audiences such as a political "world public opinion" or global "consumer publics" that refer to humanity as a whole rather than local or national audiences.[21]

Combining these insights leads to a model of competition visualized in the chart below (Figure 6).[22] The model further differentiates Simmel's triadic model of competition—at least two competitors compete for the favor of third parties—into four elements, thus effectively transforming it into a quadripartite model that consists of at least two competitors (nation-states in this case), mediating universalized third parties, and a global audience imagined and addressed by these third parties. Analytically, this model describes the minimal prerequisites of global competition; empirically, it allows an unlimited number of nation-states and third parties to participate in one of the aforementioned positions.

19 Dwayne R. Winseck and Robert M. Pike, *Communication and Empire: Media, Markets, and Globalization, 1860–1930* (Durham: Duke Univ. Press, 2007).

20 Roland Wenzlhuemer, "Editorial: Telecommunication and Globalization in the Nineteenth Century," in *Global Communication*, edited by Roland Wenzlhuemer (Cologne: Center for Historical Social Research, 2010), 7–18.

21 The rise of such imaginations is nicely captured in Ian Clark, *International Legitimacy and World Society* (Oxford and New York: Oxford Univ. Press, 2007).

22 For details, see Tobias Werron, "On Public Forms of Competition," *Cultural Studies <=> Critical Methodologies* 14, no. 1 (2014): 62-76.

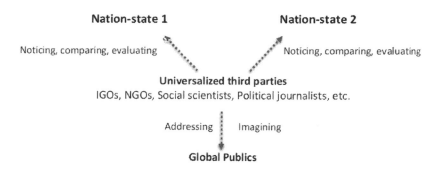

Fig. 6. Model of global competition between nation-states.

4. Historical Trends: Competition between Nation-States for "Soft" Global Goods since 1900

Let me now illustrate the advantages of this model by identifying and outlining three long-term historical trends since the mid-to-late nineteenth century. First, a trend towards competition for what I call modernity prestige, second, a trend towards competition for what I call cultural achievement prestige, and third, competition for attention and legitimacy.

4.1. Competition for Modernity Prestige: "Managerial States" and the Quantification of State Performance

The first trend reflects the influence of universalized third parties that deal with nation-states as quasi-individual units. As part of this trend, the modern "marriage" between state and nation (to quote Ernest Gellner's description of the modern ideal of the nation-state[23]) is taken as a given, which allows comparing national "societies" as a whole according to universalistic criteria such as economic growth, the protection of human rights, the level of corruption, and the protection of the environment.

As mentioned above, world polity scholars have studied such observers, calling them "rationalized others" and holding them responsible for

---

23 Ernest Gellner, *Nations and Nationalism* (Oxford: Basil Blackwell, 1983).

the "scientization" of world society.[24] Our emphasis on the role of third parties in competition, however, draws attention to a consequence that figures much less prominently in world polity research, namely, that these observers also "scarcify" their critical sympathy for all nation-states by comparing and ranking them while addressing global audiences. For instance, when Transparency International ranks all states on a Global Corruption Perceptions Index, which is then picked up by newspapers around the world, or, when UN experts measure states according to indicators such as a Human Development Index (HDI), then states are not only confronted with global norms and universal expectations ("good government," "human development," "efficient social infrastructure"), they are also subjected to continuous comparisons in which they find themselves set to compete for favorable attention from a global audience. Rationalized others not only produce modernity prestige, in other words, they also construct it as a scarce resource, which implies that the reputational gain of one country may be perceived as a reputational loss of another. Such perceptions rest on socially constructed scarcities and I suggest calling such constellations "artificial zero-sum games."

Let me illustrate this point by explaining how such scarcities are socially constructed in rankings and other statistical representations. The above chart shows the detail of a chart representing the Human Development Index (HDI)—next to the GDP probably the most prominent of global indices invented by international organizations. The HDI ranks nation-states according to a mix of economic, educational, and health criteria that it uses to generate an overall performance value. Looking at the chart's two right-hand columns you will notice that for the countries ranked 1 to 10, the alleged difference in performance and ranking is due to differences smaller than one tenth of a full index point. If we focused on the first decimal place only, we would be unable to see a difference in performance, and all countries would reach the same rank. It is only through the overly exact quantification of state performance in the generation and assessment

---

24 John W. Meyer, "Rationalized Environments," in *Institutional Environments and Organizations: Structural Complexity and Individualism*, edited by John W. Meyer and W. Richard Scott (Thousand Oaks: Sage, 1994), 28–54.

Rank			HDI	
New 2014 estimates for 2013 [10]	Change in rank between 2014 report to 2013 report[10]	Country	New 2014 estimates for 2013 [10]	Change compared between 2014 report and 2013 report [10]
1	—	🇳🇴 Norway	0.944	▼ 0.011
2	—	🇦🇺 Australia	0.933	▲ 0.002
3	—	🇨🇭 Switzerland	0.917	▲ 0.001
4	—	🇳🇱 Netherlands	0.915	—
5	—	🇺🇸 United States	0.914	▲ 0.002
6	—	🇩🇪 Germany	0.911	—
7	—	🇳🇿 New Zealand	0.910	▲ 0.002
8	—	🇨🇦 Canada	0.902	▲ 0.001
9	▲ (3)	🇸🇬 Singapore	0.901	▲ 0.002
10	—	🇩🇰 Denmark	0.900	—
11	▼ (3)	🇮🇪 Ireland	0.899	▼ 0.017

Fig. 7. Human Development Index (HDI), 2014, cropped to show positions 1 to 11,
https://en.wikipedia.org/wiki/Human_Development_Index
(accessed September 25, 2015).

of such rankings that we come to distinguish between levels of performance that we would otherwise ignore. By combining scientific methods, statistical thinking, and massive data collection, rankings represent the logic of the "scientific" construction of competition for modernity prestige in an ideal-typical fashion.

While ranking may appear as ideal-typical representations, the historical rise of competition for modernity prestige goes far beyond the history of rankings. This can be nicely illustrated using the example of competition between "national economies," and particularly with regard to the role that economists have played in changing or shaping our understanding economic competition between nation-states.

Around 1900, observers of the state-system tended to assume that states, especially large and powerful ones, were involved in a continual zero-sum-game of power that included competition for economic advantage. Max Weber, for example, held that "the economic community is just another form of the struggle between nations," expressing the "perpetual fight for the advancement of their national race."[25] Such opinions are almost indistinguishable from the mercantilist understanding of state-competition that dominated political thinking about competition in the early modern state system.[26] Against this backdrop, it becomes all the more apparent that the current mainstream view of economic competition between nation-states is dominated by the views of universalized third parties such as neo-classical economists, political economists, and other economic experts and professions that have proliferated since the late nineteenth century.[27] Through elaborated statistical concepts and models such as the GDP,[28] these observers not only account for the very emergence of more or less self-contained "national economies" in the early to mid-twentieth century,[29] they also create new forms of competition among these national economies.

How this has affected our understanding of nation-state competition can be demonstrated using the example of a debate among economists in the 1990s about the (new) concept of "national competitiveness." In 1994, Nobel Prize winning economist and *New York Times* columnist Paul Krugman in his *Foreign Affairs* article "Competitiveness: A Dangerous Obsession" argued that nation-states do not compete. National competitiveness is a misleading concept, he suggested, since we cannot consider states

---

25 Max Weber, "Der Nationalstaat und die Volkswirtschaftspolitik (1895)," in *Gesammelte politische Schriften* (Tübingen: Mohr Siebeck, 1971), 14.

26 Jean-Baptiste Colbert, Minister of Finances under Louis XIV, argued that trade was a continual fight, both violent and peaceful, between states for a larger share of the cake; cf. Edmund Silberner, *La guerre dans la pensée économique du 16. au 18. siècle* (Paris: Sirey, 1939), 35.

27 Marion Fourcade, "The Construction of a Global Profession: The Transnationalization of Economics," *American Journal of Sociology* 112, no. 1 (2006): 145–95.

28 Daniel Speich, "The use of global abstractions: National income accounting in the period of imperial decline," *Journal of Global History* 6, no. 1 (2011): 7–28.

29 Timothy Mitchell, "Society, Economy, and the State Effect," in *State/Culture: State-Formation After the Cultural Turn*, edited by George Steinmetz (Ithaca, NY: Cornell Univ. Press, 1999), 76–97.

to be competitors in a zero-sum game for limited market shares.[30] Much like Ricardo had observed in the early nineteenth century, states could profit from free trade and could all grow together rather than grow at the expense of each other. Setting aside attempts to justify protectionist politics, Krugman asked why they should see each other as competitors. In response, other economists acknowledged that states do not compete in a straightforward economic sense. They also pointed out, however, that the status of states depends on their economies, because they constantly compete for the prestige that comes with a thriving national economy characterized by constant growth, low debt, low unemployment rates, and adequate currency rates.[31] Economist Robert Gilpin nicely summarized this interpretation's core assumption: "Although nations may not compete with one another in a narrow economic sense, nations can be said to compete in a broader sense, that is, in their ability to manage their economic affairs effectively."[32] According to this view, nation-states compete economically because they are constantly observed and compared by economists and other universalized third parties, creating competition between "managerial states"[33] for the prestige of having a well-organized and "competitive" national economy. In other words: They compete because economists see it that way.

We may apply this reasoning, not only to national economies, but also to environmental policies, education systems, and systems of science, anti-corruption politics, social security policies, human rights practices, and other fields continuously monitored by universalized third parties. Such developments have a discernible historical trajectory. Historian Thomas Bender has noted that around 1900, "there was constant comparison among nations, especially among bureaucrats in the relevant ministries eager to keep up with the newest social legislation [...] competition among home ministries on social policy was not unlike that waged among war

---

30 Paul Krugman, "Competitiveness: A Dangerous Obsession," *Foreign Affairs* 73, no. 2 (1994): 28–44.

31 Malcom H. Dunn, "Do Nations Compete Economically? A Critical Comment on Prof. Krugman's Essay 'Competitiveness: A Dangerous Obsession,'" *Intereconomics* 29, no. 6 (1994): 303–08.

32 Robert Gilpin, *Global Political Economy* (Princeton: Princeton Univ. Press, 2001), 182. For a systematic exploration of this idea, see Hans-Werner Sinn, *The New Systems Competition* (Oxford: Basil Blackwell, 2003).

33 Walter C. Opello and Stephen J. Rosow, *The Nation-State and Global Order: A Historical Introduction to Contemporary Politics* (Boulder and London: Lynne, 2004), 139.

ministries."[34] In the twentieth century, these forms have multiplied and globalized, creating ever-new forms of competition for modernity prestige. And although we may doubt the significance of its consequences, we can hardly ignore that "scientized" competition has gained stability and visibility at the level of world society.

4.2. Competition for Cultural-Achievement Prestige: The Global Banalization of National Differences

The first trend towards competition between nation-states for modernity prestige built on an understanding of the modern nation-state as a quasi-individual unit, or "national society," where state and culture are supposed to build a more or less harmonious marriage. The second trend does not depend on such a rather specific understanding of the nation-state. Instead, it draws attention to non-state actors that use (or non-state performances that involve) national symbols and identities such as artists, television series, touristic attractions, and sports clubs, all of which may be observed and compared by universalized third parties. These phenomena point to an autonomous use of national symbols and differences in non-political functional systems such as the arts, mass media, tourism, and sports, and they call for an integration of insights on functional differentiation with research on a "banal" and everyday dimension of nationalism.

"Banal nationalism" was first identified and analyzed by Michael Billig and has since become an important topic of nationalism research.[35] Billig's original aim was to draw attention to the reproduction of national identities in allegedly de-nationalized western countries by means of inconspicuous markers such as "we" or "them" in the daily press. This everyday "banalization" of national symbols and identities, however, is also at work in global fields that build national symbols and identities into universalistic scopes of comparison. In a world-societal perspective, one may speak of a global banalization of the national, which manifests itself in fields such as sports (more on this example below), "nation branding" in tourism or

---

34 Thomas Bender, *A Nation Among Nations* (New York: Hill & Wang, 2006), 290.

35 Michael Billig, *Banal Nationalism* (London: Sage, 1995). Other references to the subject include John Hutchinson, *Nations as Zones of Conflict* (London: Sage, 2004), 147–149, and Rogers Brubaker, "Ethnicity, Race, and Nationalism," *Annual Review of Sociology* 35 (2009): 28.

product markets,[36] categories of literary critics, national framing of pop-rock music,[37] and the "methodological nationalism" of social scientists.[38] Competition for national achievement prestige is a regular characteristic of these processes but has not yet been investigated. But its logic and general historical trajectory can be illustrated using the example of modern sports.

Modern sports first emerged in the late nineteenth century in national or regional frameworks. For instance, football became a "national game" in Great Britain, baseball a "national pastime" in the US. In these localized competitive cultures, however, universalistic events (world events) and universalistic indicators of achievement (such as records and statistics) evolved that established an interpretative framework for the globalization of these sports in the twentieth century. Sports journalists and publicists played the role of universalized third parties, creating specialized journals and yearbooks, collecting data, reporting on contests, and devising categories and criteria (such as statistics, records, and legends) for a universalistic comparison and evaluation of performance.[39] When these sports globalized in the twentieth century, they began to make use of national identities not only to frame competition (in league systems and cup competitions) but also to compare the performance of representatives of national associations within a global competitive framework, imagining them as sources of identification for national audiences,[40] while global associations started

---

36 Göran Bolin and Per Ståhlberg, "Between Community and Commodity: Nationalism and Nation Branding," in *Communicating the Nation*, edited by Anna Roosvall and Inka Salovaara-Moring (Göteborg: Nordicom, 2010), 79–101.

37 Motti Regev, "Pop-Rock Music as Expressive Isomorphism: Blurring the National, the Exotic, and the Cosmopolitan in Popular Music," *American Behavioral Scientist* 55, No. 5 (2011), 558–73.

38 Andreas Wimmer and Nina Glick Schiller, "Methodological Nationalism, the Social Sciences, and the Study of Migration: An Essay in Historical Epistemology," *International Migration Review* 37 (2003): 576–610.

39 For details of this argument, see Tobias Werron, "World Sport and Its Public: On Historical Relations of Modern Sport and the Media," in *Observing Sport*, edited by Ulrik Wagner and Rasmus Storm (Schorndorf: Hofmann, 2010), 33–59; Werron, *Der Weltsport und sein Publikum: Zur Autonomie und Entstehung des modernen Sports* (Weilerswist: Velbrück, 2010); Werron, "How are Football Games Remembered? Idioms of Memory in Modern Football," in *European Football and Collective Memory*, edited by Wolfgang Pyta and Nils Havemann (Basingstoke: Palgrave, 2015), 18–39.

40 Barbara J. Keys, *Globalizing Sport: National Rivalry and International Community in the 1930s* (Cambridge: Harvard Univ. Press, 2006); Mark Dyreson, "Globalizing the Nation-Making Process: Modern Sport in World History," *The International Journal of the History of Sport* 20 (2003), 91–106.

to take responsibility for the organization and standardization of competition.[41] By 1900, publicist William Thomas Stead could observe that "sports which twenty years ago were almost exclusively national have now become international, and every year increases the number of events in which the primary interest of sport is reinforced by national rivalry."[42] Within a few decades "world sports" had evolved to provide the basis for a broad spectrum of collective identities as villages, cities, regions, nations, and even continents adopted its universalistic rationale of comparison. Accordingly, it has become a typical characteristic of global sports events that the audience is imagined and simultaneously addressed in two different but complementary roles: as a universalistic "expert" or "referee" imagined to be interested in excellent performance, and as a particularistic "partisan" imagined to be identifying with the representatives of "his" or "her" nation.[43]

Although such forms are not directly concerned with competition between states, this historical outline illustrates that they are relevant for a historical-sociological perspective on competition among nation-states. By making national symbols regular features of global fields, they have helped institutionalize the idea that the world consists, or should consist, of "nations." At the same time, it is important to note that national identities here are part of specific, autonomous global domains where they are one among many possible sources of identification. Historical research on global competition for prestige through cultural achievement, prestige attributed by universalized third parties, may considerably refine our understanding of a paradoxical effect noted by globalization theorists. As both Roland Robertson and Saskia Sassen have observed, globalization concurrently strengthens and weakens nation-states and their national identities.[44]

---

41 Such as the International Association of Federation Football (FIFA). See Christiane Eisenberg, "Der Weltfußballverband Fifa im 20. Jahrhundert: Metamorphosen eines 'Prinzipienreiters'," *Vierteljahrshefte für Zeitgeschichte* 54 (2006), 209–30.

42 William Thomas Stead, *The Americanization of the World* (New York and London: Horace Markley, 1901), 334.

43 Daniel Dayan and Elihu Katz, *Media Events: The Live Broadcasting of History* (Cambridge: Harvard Univ. Press, 1992), 41.

44 Roland Robertson, "Globalization Theory 2000+: Major Problematics," in *Handbook of Social Theory*, edited by George Ritzer and Barry Smart (London: Sage, 2001), 458–71; Saskia Sassen, *Territory, Authority, Rights: From Medieval to Global Assemblages* (Princeton: Princeton Univ. Press, 2005).

4.3. Competition for Attention and Legitimacy: Staging Conflicts for the Favor of a "World Public Opinion"

My remarks on global sources of national differences inferred a trend that I would like to discuss next: the competition for attention and legitimacy. To grasp the significance of this trend, consider the probability of the survival of existing states in the current state system. Given a "realist" understanding, threats to the survival of states should be expected to come exclusively from powerful states with the military capacity to conquer weaker ones, and there are, of course, still more than enough examples for this kind of power competition in the international system. Both the stability of existing states and (secessionist) ambitions of national movements, however, today depend less on military strength than on the ability to win the favor of universalized third parties and world public opinion.

This trend evolved in the late nineteenth century when the idea of a world public opinion and of a global audience (addressed in everyday political communication) inspired the formation of "internationalist" political movements.[45] From then on, legitimacy within the international system was no longer solely a matter of "international relations" between "great powers" but was increasingly influenced by external observers. Long-term effects of this trend are reflected in the staging of media scandals in the press, cooperation of national movements with non-governmental organizations (NGOs) and human rights activists, membership in the Unrepresented Nations and Peoples Organization (UNPO), and, more recently, social scientific research on "public accountability" that elaborates general concepts and models to legitimize and de-legitimize state behavior.[46] All of these observers have established themselves as universalized third parties, drawing attention to norm violations and attempting to de-legitimize behavior not in accordance with human rights and, in particular, the norm of national self-determination.

Literature on globalization has mainly focused on the increasing normative pressure exerted by these observers, often displaying a rather optimistic belief in the civilizing effects of this new form of "global govern-

---

45 Cf. Ian Clark, *International Legitimacy and World Society*, (Oxford: Oxford Univ. Press, 2007).

46 David Held, "Democratic Accountability and Political Effectiveness from a Cosmopolitan Perspective," in *Global Governance and Public Accountability*, edited by David Held and Mathias Koenig-Archibugi (Malden, MA, and Oxford: Blackwell, 2005), 364-391.

ance."[47] In contrast to this optimistic view, the model of global competition presented here also points to competitive effects and stresses that the attention and legitimacy produced by universalized third parties are often perceived as *scarce* goods. This can produce unintended consequences. How such competition may affect the political struggle of independence movements is impressively shown in Clifford Bob's *The Marketing of Rebellion*,[48] which argues that a movement's chances to be accepted internationally as a conflict party depends on its ability to gain the attention of a global audience—attention mediated by NGOs and universalistic, western-style mass media. In a competitive environment, not all worthy cases are likely to be heard. Not every movement has a Dalai Lama or an equally charismatic leader for the reliable production of attractive television interviews, nor can every movement resort to non-violent strategies to win the sympathy of human rights activists.

Competition for scarce attention and legitimacy might also help to explain why models of the nation-state today seem to converge in a "modest nation-state," as sociologist Michael Mann has put it,[49] or even a "nice" one,[50] a model in which nation-states are expected to live in peace with each other and to accept the existence of smaller peoples or nations. This model has come to shape the institutionalization of the "right of nations to self-determination" in international law (in the 1945 UN Charter, the 1960 Declaration of the Granting of Independence to Colonial Countries and Peoples, and in the 1966 International Covenant on Civil and Political Rights). Since there is no accepted formula for defining a nation's entitlement to this right, however, the actual success of national movements has remained a political rather than legal matter, making political competition for attention and legitimacy through universalized third parties a decisive factor in such conflicts.[51]

This trend has evolved from the late nineteenth and early twentieth centuries. Spectacular examples from the period of the competition's in-

---

47 Richard Price, "Transnational Civil Society and Advocacy in World Politics: Review Article," *World Politics* 55, no. 4 (2003): 579–606.

48 Clifford Bob, *The Marketing of Rebellion* (Cambridge: Cambridge Univ. Press, 2005).

49 Michael Mann, "Has globalization ended the rise and rise of the nation-state?," *Review of International Political Economy* 4, no. 3 (1997): 476.

50 John W. Meyer et al., "World Society and the Nation-State," *American Journal of Sociology* 103, no. 1 (1997): 154.

51 Jörg Fisch, *Das Selbstbestimmungsrecht der Völker* (Munich: C. H. Beck, 2010); Mikulas Fabry, *Recognizing States* (Oxford: Oxford Univ. Press, 2010).

ception date to World War I, when US President Woodrow Wilson and socialist leader Vladimir Lenin sought to rally the support of world public opinion by proclaiming their respective versions of the right to national self-determination.[52] Competition for favorable attention of universalized third parties also is part of a perceptive analysis by historical sociologist Julian Go, who argues that after the diffusion of post-colonial nationalism, "official" imperial strategies were associated with a pronounced risk of losing global political legitimacy, leading US governments after 1945 to accept national self-determination in their colonies or spheres of influence (such as in the Philippines or in Iran). The US switched to more subtle forms of power politics.[53] Both examples indicate that competition for attention and legitimacy through universalized third parties may considerably influence politics even of "great powers," relegating them to a somewhat more modest status in the global political field.

5. Conclusion: Studying the Global Impact of Science from a Historical-Sociological Perspective

Let me conclude with a short summary and outlook. From a globalization perspective, this chapter has argued that we should study in more detail what neo-institutionalist world-polity scholars call scientization, that is, the global authority of science and its capacity to structure world-societal processes. This view urges us to consider in more detail that much of the societal impact of science has to do with the fact that scientists and scientized observers dominate a largely state-less world society on a global discursive level. Combining this view with insights from the sociology of competition, I have argued for a model of global competition that draws attention to the constitutive role of scientized third parties, or universalized third parties, in the construction of global forms of competition. This model helped me identify long-term historical trends towards competition between nation-states for "soft" global goods since the mid-to-late nine-

---

52 Erez Manela, *The Wilsonian Moment: Self-Determination and the International Origins of Anticolonial Nationalism* (Oxford: Oxford Univ. Press, 2007).

53 Julian Go, "Global Fields and Imperial Forms: Field Theory and the British and American Empires," *Sociological Theory* 26, no. 3 (2008): 201–29.

teenth century: competition for modernity prestige, for cultural achievement prestige, and for attention and legitimacy.

A general insight to be drawn from this analysis is that science can impact political processes not only by producing new knowledge or technology, but also by infiltrating societal structures with a universalistic style of reasoning that is hard to resist, given modern science's global authority and legitimacy. To arrive at a more sophisticated understanding of the societal impact of these forms of competition, however, we would have to investigate in more detail the various forms of public comparison that have become a regular feature of global processes, from media scandals and organizational reports to international rankings, study their historical trajectories in different societal fields, and analyze how they interact with each other and how they are interpreted (or ignored) in different national contexts.

Against this background, it seems fair to say that there are three equally important and complementary outlooks on the nation-state/science-relationship: first, case studies and comparative research on national science systems; second, research on science as an autonomous, self-generating global system; and third, the globalization-by-scientization perspective highlighted here, which draws attention to the global authority and global societal impact of modern science. I have used this approach to draw attention to the rise of global forms of competition between nation-states since the mid-to-late nineteenth century. Each perspective captures important facets of the nation-state/science-relationship, but only a combination of all three of them can hope to do justice to its actual complexities. To date, they have operated more or less separately from each other. Bringing them more closely together and integrating them into a common research perspective is a worthwhile challenge for future research.

Figures

Contributors

Andreas Franzmann is affiliated with the Department of Sociology at Frankfurt University as a Privatdozent. His most recent book explores the professionalization of scientists through their occupational habitus (*Die Disziplin der Neugierde*, Bielefeld: Transcript, 2012). In an earlier monograph, he investigated the role of intellectuals during the Dreyfus affair in France between 1894 and 1906 (*Der Intellektuelle als Protagonist der Öffentlichkeit. Krise und Räsonnement in der Affäre Dreyfus*, Frankfurt: Humanities Online, 2004). He has taught at the University of Frankfurt, at the University of Bielefeld, and in Tübingen. In 2012/13 he was a Visiting Scholar at the University of California, Los Angeles (UCLA). He currently cooperates with Axel Jansen in a research project on the "Public Context of Science since 1970" (funded by the Volkswagen Foundation).

Axel Jansen is a *Privatdozent* in American Studies at Frankfurt University. He has published a monograph on American volunteers in the European war zone between 1914 and 1917 (*Individuelle Bewährung im Krieg: Amerikaner in Europa, 1914–1917*, Frankfurt and New York: Campus, 2003). His second book is an analytical biography of Alexander Dallas Bache (*Alexander Dallas Bache: Building the American Nation through Science and Education in the Nineteenth Century*, Frankfurt and New York: Campus, 2011). He has taught North American history at universities in Europe and in the US. In 2013/14, he was a visiting fellow at Wolfson College, Cambridge (UK). At Tübingen University and at the University of California, Los Angeles (UCLA), he currently works with Andreas Franzmann on a research project on the "Public Context of Science since 1970," which is sponsored by the Volkswagen Foundation.

Dieter Langewiesche is professor emeritus in the Department of History at Tübingen University. He has published widely on the history of European nationalism and liberalism. His key publications include *Liberalism in Germany* (Princeton: Princeton Univ. Press, 2000), *Reich, Nation, Förderation. Deutschland und Europa* (Munich: Beck, 2008), and *Die Monarchie im Jahrhundert Europas. Selbstbehauptung durch Wandel im 19. Jahrhundert* (Heidelberg: Winter, 2013). Dieter Langewiesche has received several distinguished prizes. In 1996, the German Research Foundation (DFG) awarded him the *Gottfried Wilhelm Leibniz-Preis*, the most prestigious German award in the social sciences and humanities. From 1997 to 2000 he was released from his duties in Tübingen to help build up the University of Erfurt as that university's *Prorektor*.

Fabian Link is assistant professor (*wiss. Assistent*) in the history of science and the humanities at Frankfurt University. In his research he focuses on the history of archaeology and medieval studies during the Nazi regime and on the history of the social sciences during the early Cold War. He has published a monograph on the history of castle research in Germany during the Nazi regime (*Burgen und Burgenforschung im Nationalsozialismus. Wissenschaft und Weltanschauung 1933–1945*, Cologne: Böhlau, 2014). He currently works on a monograph on the history of the "Frankfurt School" during the 1950s.

Peter Münte studied philosophy, linguistics, and sociology at the University of Frankfurt. In his published dissertation, he focused on the founding of the Royal Society. For several years, he worked at the Institute of Science and Technology Studies, University of Bielefeld. His recent publications include "Das Mediationsverfahren als sozialtechnologische Form herrschaftstechnischer Versachlichung und inszenierter Herrschaftsfreiheit. Eine Analyse eines Entwurfs der Vereinbarung über eine Mediation zum Ausbau des Flughafens Wien," in *Mikrostrukturen der Governance: Beiträge zur materialen Rekonstruktion von Erscheinungsformen neuer Staatlichkeit*, edited by Alfons Bora and Peter Münte (Baden-Baden: Nomos, 2012), and "Die Autonomie der Wissenschaft im Ordnungsdiskurs der Moderne: Ein Versuch über den Formenwandel der modernen Wissenschaft," in *Autonomie revisited: Beiträge zu einem umstrittenen Grundbegriff in Wissenschaft, Kunst und Politik*, edited by Martina Franzen et al., (Weinheim: Beltz Juventa, 2014), 143–65.

Arne Pilniok is assistant professor (*Juniorprofessor*) at the Universität Hamburg's Law School, where he teaches German and European constitutional and administrative law. He was a member of the German Research Foundation's research group 517 (Governance of research) from 2005 to 2010. He has published a book on the legal structure of European research policy (*Governance im europäischen Forschungsförderverbund*, Tübingen: Mohr Siebeck, 2011). In 2015/16, he is a Visiting Scholar at the University of California, Berkeley, funded by the German Academic Exchange Service. His research currently focuses on the comparative law of democracy.

Lisa Sigl is research assistant and lecturer in the fields of Science and Technology Studies and Higher Education Studies at the Technical University of Dortmund and at the University of Vienna. In her dissertation she explored the implications of precarious working conditions of early-stage researchers for work cultures in the life sciences. From 2012 to 2014, she led a research group on the "Internationalization of Science, Technology and Innovation" at the Austrian Institute for International Affairs. She now works with Liudvika Leisyte (TU Dortmund) on the commercialization of research in Germany, and she teaches on the politics of innovation. Recent publications include "On the Tacit Governance of Research by Uncertainty: How Early Stage Researchers Contribute to the Governance of Life Science Research," *Science, Technology & Human Values* (2015).

Rudolf Stichweh is Dahrendorf Professor for the Theory of Modern Society and the Director of the *Forum für Internationale Wissenschaft* at the University of Bonn. He is a key proponent of the systems-theory approach in sociology and he contributed significantly to the sociology of science and to the sociology of the professions. He came to Bonn via the Max-Planck-Institut for the Study of Societies in Colgne (1985–89), the *Maison des Sciences de l'Homme in Paris* (1987), the Max-Planck-Institute for the History of European Law in Frankfurt (1989–94), a professorship at the Faculty for Sociology at the University of Bielefeld (1994–2003), and a professorship at the University of Lucerne (2003–12). His major publications include *Zur Entstehung des modernen Systems wissenschaftlicher Disziplinen* (Frankfurt: Suhrkamp, 1984); *Der frühmoderne Staat und die europäische Universität* (Frankfurt: Suhrkamp, 1991); *Wissenschaft, Universität, Professionen* (Frankfurt: Suhrkamp, 1994); *Die Weltgesellschaft* (Frankfurt: Suhrkamp, 2000); and *Der Fremde* (Frankfurt: Suhrkamp, 2010).

Shiju Sam Varughese is assistant professor at the Centre for Studies in Science, Technology and Innovation Policy (CSSTIP) in the School of Social Sciences of the Central University of Gujarat, Gandhinagar, India. His areas of expertise include public engagement with science, social history of knowledge, and cultural studies of science and technology. He has recently edited a volume (with Satheese Chandra Bose) that theorizes regional modernities in South Asia (*Kerala Modernity: Ideas, Spaces and Practices in Transition*, Hyderabad: Orient Blackswan, 2015). Currently he is working on a monograph on science and media for Oxford University Press, New Delhi.

Tobias Werron in March 2015 was appointed professor for Science Studies and Politics at the *Forum Internationale Wissenschaft*, University of Bonn. Previously, he had been an assistant professor at universities in Bielefeld and Lucerne and a visiting scholar at Boston University. He completed his doctorate in 2008 with a dissertation on the autonomy and emergence of modern sport, and his habilitation in 2014 with a book on the sociology of global competition. His research interests include globalization/world society studies, the sociology of competition, the sociology of the media, and the interface between science and politics. He is working on a book on the sociology of competition and preparing a research project on international rankings with a particular focus on the role of science and scientists in the production and dissemination of rankings.

Nina Witjes is a researcher at the Austrian Institute for International Affairs in Vienna and head of a research group on "Internationalization of Science, Technology and Innovation." She is completing her PhD thesis on "The Co-Production of Science, Technology and International Politics: Exploring Emergent Fields of Knowledge and Policy" at the Munich Centre for Technology in Society (TU Munich). As a Research Fellow at the University of Freiburg in 2012, she worked on the "Universality and Potential of Acceptance of Social Science Knowledge: On the Circulation of Knowledge between Europe and the Global South" (sponsored by the German Ministry for Education and Science). Recent publications include "Sociotechnical Imaginaries of Big Data: Commercial Satellite Imagery and its Promise of Speed and Transparency" (with Philipp Olbrich, in *Big Data: Innovation, Ethics and Transformation of the Digital Revolution*, edited by Michael Mulqueen and Andrej Zwitter, London: Palgrave Macmillan, forthcoming).

Michaela M. Hampf,
Simone Müller-Pohl (eds.)
Global Communication Electric
Business, News and Politics
in the World of Telegraphy
2013. 386 pages. ISBN 978-3-593-39953-9

Willibald Steinmetz,
Ingrid Gilcher-Holtey,
Heinz-Gerhard Haupt (eds.)
Writing Political History Today
2013. 413 pages. ISBN 978-3-593-39806-8

Sebastian Jobs
Welcome Home, Boys!
Military Victory Parades in
New York City 1899–1946
2012. 276 pages. ISBN 978-3-593-39745-0

Stefan B. Kirmse (ed.)
One Law for All?
Western models and local practices
in (post-)imperial contexts
2012. 297 pages. ISBN 978-3-593-39493-0

Isabel Heinemann (ed.)
Inventing the Modern American Family
Family Values and Social Change in
20th Century United States
2012. 335 pages. ISBN 978-3-593-39640-8

Jörg Feuchter, Friedhelm Hoffmann,
Bee Yun (eds.)
Cultural Transfers in Dispute
Representations in Asia, Europe and
the Arab World since the Middle Ages
2011. 335 pages. ISBN 978-3-593-39404-6

Ralph Jessen, Hedwig Richter (eds.)
Voting for Hitler and Stalin
Elections Under 20th Century Dictatorships
2011. 349 pages. ISBN 978-3-593-39489-3

Gábor Klaniczay, Michael Werner,
Otto Gécser (eds.)
Multiple Antiquities –
Multiple Modernities
Ancient Histories in Nineteenth
Century European Cultures
2011. 611 pages. ISBN 978-3-593-39101-4

Axel Jansen
Alexander Dallas Bache
Building the American Nation through Science
and Education in the Nineteenth Century
2011. 353 pages. ISBN 978-3-593-39355-1

Sebastian Jobs, Alf Lüdtke (eds.)
Unsettling History
Archiving and Narrating in Historiography
2010. 253 pages. ISBN 978-3-593-38818-2